LECTURES ON QUANTUM GRAVITY

Series of the Centro de Estudios Científicos

Series Editor: Claudio Teitelboim
Centro de Estudios Científicos
Valdivia, Chile

IONIC CHANNELS IN CELLS AND MODEL SYSTEMS, 1986
Edited by Ramón Latorre

PHYSICAL PROPERTIES OF BIOLOGICAL MEMBRANES AND
THEIR FUNCTIONAL IMPLICATIONS, 1988
Edited by Cecilia Hidalgo

PRINCIPLES OF STRING THEORY, 1988
Lars Brink and Marc Henneaux

QUANTUM MECHANICS OF FUNDAMENTAL SYSTEMS 1, 1988
Edited by Claudio Teitelboim

QUANTUM MECHANICS OF FUNDAMENTAL SYSTEMS 2, 1989
Edited by Claudio Teitelboim and Jorge Zanelli

TRANSDUCTION IN BIOLOGICAL SYSTEMS, 1990
Edited by Cecilia Hidalgo, Juan Bacigalupo, Enrique Jaimovich,
and Julio Vergara

JOAQUIN LUCO: Dos Historias de Una Vida, 1991
Ana Tironi and Pedro Labarca, Ediciones Pedagógicas Chilenas S.A.

QUANTIZATION OF GAUGE SYSTEMS, 1992
Marc Henneaux and Claudio Teitelboim, Princeton University Press

QUANTUM MECHANICS OF FUNDAMENTAL SYSTEMS 3, 1992
Edited by Claudio Teitelboim and Jorge Zanelli

FROM ION CHANNELS TO CELL-TO-CELL CONVERSATIONS, 1997
Edited by Ramón Latorre

THE BLACK HOLE 25 YEARS AFTER, 1998
Edited by Claudio Teitelboim and Jorge Zanelli, World Scientific

BLACK HOLES AND THE STRUCTURE OF THE UNIVERSE, 2000
Edited by Claudio Teitelboim and Jorge Zanelli, World Scientific

THE PATAGONIAN ICEFIELDS: A Unique Natural Laboratory for
Environmental and Climate Change Studies, 2002
Edited by Gino Casassa, Francisco Sepúlveda and Rolf M. Sinclair

LECTURES ON QUANTUM GRAVITY, 2004
Edited by Andrés Gomberoff and Donald Marolf

PUMPS, TRANSPORTERS AND ION CHANNELS
Studies on their Structure, Function and Cell Biology, 2004
Edited by Francisco V. Sepúlveda and Francisco Bezanilla

LECTURES ON QUANTUM GRAVITY

Edited by
ANDRÉS GOMBEROFF
Centro de Estudios Científicos (CECS)
DONALD MAROLF
University of California, Santa Barbara (UCSB)

 Springer

Andres Gomberoff
Centro de Estudios Cientificos
Avenida Arturo Prat 514
Valdivia, Chile

Donald Marolf
Physics Department
University of California
Santa Barbara, CA 93106

Series Editor:
Claudio Teitelboim
Centro de Estudios Cient'íficos
Valdivia, Chile

Any opinions, findings, and conclusions or recommendations expressed in this material are those of the authors and do not necessarily reflect the views of the National Science Foundation.

Cover image by Andres Gomberoff.

Library of Congress Cataloging-in-Publication Data

Lectures on quantum gravity / edited by Andrés Gomberoff, Donald Marolf.
        p. cm. – (Series of the Centro de Estudios Científicos)
    Includes bibliographical references and index.
    ISBN 0-387-23995-2
    1. Quantum gravity.   I. Gomberoff, Andrés.   II. Marolf, Donald.   III. Series.

    QC178.L38   2005
    530.14'3–dc22

                                                                        2004065095

ISBN-10: 0-387-23995-2   e-ISBN 0-387-24992-3   Printed on acid-free paper.
ISBN-13: 978-0387-23995-2   ISSN: 1571-571X

Printed in the United States of America.   (SBA)

9 8 7 6 5 4 3 2 1   SPIN 11329367

springeronline.com

Foreword

The Centro de Estudios Cientificos (CECS) began a new phase of its existence at the end of 1999, when it moved to the city of Valdivia, 800 kilometers South of the capital of Chile, Santiago. The letter "S", which stood for Santiago in the original acronym has been maintained to provide a sense of historical continuity, and it is now - when necessary - explained as arising from the plural in the word Científicos.

Valdivia used to be part of the "frontier" in the early days of the country and one still breathes frontier air in it. This frontier air has inspired the Center to undertake new and bolder challenges in science and exploration, such as an unprecedented airborne exploration of the Amundsen Sea in West Antarctica and the development of a state of the art Transgenic Facility.

However, in the midst of all this excitement and frenetic activity, we were distinctly reminded by the physicists who came to Valdivia from many countries to take part in the School of Quantum Gravity that, as Richard Feynman used to say: "there is nothing better in life than eating cookies and talking about Physics".

Claudio Teitelboim
Director, Centro de Estudios Científicos
Valdivia, April 2004.

Contents

Contributing Authors

Thibault Damour Institut des Hautes Études Scientifiques (IHÉS).

Andrés Gomberoff Centro de Estudios Científicos (CECS).

Marc Henneaux Université Libre de Bruxelles.

Ted Jacobson University of Maryland.

Juan Maldacena Institute for Advanced Study.

Robert C. Myers Perimeter Institute for Theoretical Physics, University of Waterloo and McGill University.

Hermann Nicolai Max-Planck-Institut für Gravitationsphysik, Albert-Einstein-Institut.

Rafael D. Sorkin Syracuse University.

Washington Taylor Center for Theoretical Physics, Massachusetts Institute of Technology (MIT).

Claudio Teitelboim Centro de Estudios Científicos (CECS).

Robert M. Wald University of Chicago.

Frank Wilczek Center for Theoretical Physics, Massachusetts Institute of Technology (MIT).

Preface

The 2002 Pan-American Advanced Studies Institute *School on Quantum Gravity* was held at the Centro de Estudios Cientificos (CECS), Valdivia, Chile, January 4-14, 2002. The school featured lectures by ten speakers, and was attended by nearly 70 students from over 14 countries. A primary goal was to foster interaction and communication between participants from different cultures, both in the layman's sense of the term and in terms of approaches to quantum gravity. We hope that the links formed by students and the school will persist throughout their professional lives, continuing to promote interaction and the essential exchange of ideas that drives research forward.

This volume contains improved and updated versions of the lectures given at the School. It has been prepared both as a reminder for the participants, and so that these pedagogical introductions can be made available to others who were unable to attend. We expect them to serve students of all ages well.

ANDRES GOMBEROFF AND DONALD MAROLF

Acknowledgments

We are greatly indebted to all of the many people and organizations who made the school such a success. Key support provided by the US National Science Foundation and the US Department of Energy via a grant to Syracuse University through the Pan-American Advanced Studies Institutes program is gratefully acknowledged. We also thank the Millennium Science Initiative (Chile), Fundación Andes, the Tinker Foundation, and Empresas CMPC for their support through CECS. We hope that our school has lived up to the standards of all of these fine programs and that this proceedings will help to encourage the future growth of both PASI and CECS.

We thank the administrative staff and colleagues at CECS for their hard work and for providing a marvelous environment for the school.

Last but not least, we would like to express our thanks to Syracuse University and its excellent staff for their support and assistance in all phases of the project. We thank Penny Davis in particular for her many hours of effort. Although it was often behind the scenes, coordination with participants and North American funding agencies by Penny and other staff members was crucial for making this school a reality.

THE THERMODYNAMICS OF BLACK HOLES

Robert M. Wald

Enrico Fermi Institute and Department of Physics
University of Chicago
5640 S. Ellis Avenue
Chicago, Illinois 60637-1433
rmwa@midway.uchicago.edu

Abstract

We review the present status of black hole thermodynamics. Our review includes discussion of classical black hole thermodynamics, Hawking radiation from black holes, the generalized second law, and the issue of entropy bounds. A brief survey also is given of approaches to the calculation of black hole entropy. We conclude with a discussion of some unresolved open issues.

This article is based upon an article of the same title published in Living Reviews in Relativity, http://www.livingreviews.org.

1

1.   Introduction

During the past 30 years, research in the theory of black holes in general
relativity has brought to light strong hints of a very deep and fundamental
relationship between gravitation, thermodynamics, and quantum theory. The
cornerstone of this relationship is black hole thermodynamics, where it appears
that certain laws of black hole mechanics are, in fact, simply the ordinary laws
of thermodynamics applied to a system containing a black hole. Indeed, the
discovery of the thermodynamic behavior of black holes – achieved primarily
by classical and semiclassical analyses – has given rise to most of our present
physical insights into the nature of quantum phenomena occurring in strong
gravitational fields.

The purpose of this article is to provide a review of the following aspects of
black hole thermodynamics:

- At the purely classical level, black holes in general relativity (as well as
  in other diffeomorphism covariant theories of gravity) obey certain laws
  which bear a remarkable mathematical resemblance to the ordinary laws
  of thermodynamics. The derivation of these laws of classical black hole
  mechanics is reviewed in section 2.

- Classically, black holes are perfect absorbers but do not emit anything;
  their physical temperature is absolute zero. However, in quantum theory
  black holes emit Hawking radiation with a perfect thermal spectrum. This
  allows a consistent interpretation of the laws of black hole mechanics as
  physically corresponding to the ordinary laws of thermodynamics. The
  status of the derivation of Hawking radiation is reviewed in section 3.

- The *generalized second law* (GSL) directly links the laws of black hole
  mechanics to the ordinary laws of thermodynamics. The arguments in
  favor of the GSL are reviewed in section 4. A discussion of entropy
  bounds is also included in this section.

- The classical laws of black hole mechanics together with the formula
  for the temperature of Hawking radiation allow one to identify a quantity
  associated with black holes – namely $A/4$ in general relativity – as playing
  the mathematical role of entropy. The apparent validity of the GSL
  provides strong evidence that this quantity truly is the physical entropy
  of a black hole. A major goal of research in quantum gravity is to provide
  an explanation for – and direct derivation of – the formula for the entropy
  of a black hole. A brief survey of work along these lines is provided in
  section 5.

- Although much progress has been made in our understanding of black
  hole thermodynamics, many important issues remain unresolved. Pri-

mary among these are the "black hole information paradox" and issues related to the degrees of freedom responsible for the entropy of a black hole. These unresolved issues are briefly discussed in section 6.

Throughout this article, we shall set $G = \hbar = c = k = 1$, and we shall follow the sign and notational conventions of [1]. Although I have attempted to make this review be reasonably comprehensive and balanced, it should be understood that my choices of topics and emphasis naturally reflect my own personal viewpoints, expertise, and biases.

2.   Classical Black Hole Thermodynamics

In this section, I will give a brief review of the laws of classical black hole mechanics.

In physical terms, a black hole is a region where gravity is so strong that nothing can escape. In order to make this notion precise, one must have in mind a region of spacetime to which one can contemplate escaping. For an asymptotically flat spacetime (M, g_{ab}) (representing an isolated system), the asymptotic portion of the spacetime "near infinity" is such a region. The *black hole* region, $\mathcal{B}$, of an asymptotically flat spacetime, (M, g_{ab}), is defined as

$$\mathcal{B} \equiv M - I^-(\mathcal{I}^+), \tag{1}$$

where $\mathcal{I}^+$ denotes future null infinity and I^- denotes the chronological past. Similar definitions of a black hole can be given in other contexts (such as asymptotically anti-deSitter spacetimes) where there is a well defined asymptotic region.

The *event horizon*, $\mathcal{H}$, of a black hole is defined to be the boundary of $\mathcal{B}$. Thus, $\mathcal{H}$ is the boundary of the past of $\mathcal{I}^+$. Consequently, $\mathcal{H}$ automatically satisfies all of the properties possessed by past boundaries (see, e.g., [2] or [1] for further discussion). In particular, $\mathcal{H}$ is a null hypersurface which is composed of future inextendible null geodesics without caustics, i.e., the expansion, θ, of the null geodesics comprising the horizon cannot become negatively infinite. Note that the entire future history of the spacetime must be known before the location of $\mathcal{H}$ can be determined, i.e., $\mathcal{H}$ possesses no distinguished local significance.

If Einstein's equation holds with matter satisfying the null energy condition (i.e., if $T_{ab}k^a k^b \geq 0$ for all null k^a), then it follows immediately from the Raychauduri equation (see, e.g., [1]) that if the expansion, θ, of any null geodesic congruence ever became negative, then θ would become infinite within a finite affine parameter, provided, of course, that the geodesic can be extended that far. If the black hole is *strongly asymptotically predictable* – i.e., if there is a globally hyperbolic region containing $I^-(\mathcal{I}^+) \cup \mathcal{H}$ – it can be shown that this implies that $\theta \geq 0$ everywhere on $\mathcal{H}$ (see, e.g., [2, 1]). It then follows that the surface area, A, of the event horizon of a black hole can never decrease with time, as discovered by Hawking [4].

It is worth remarking that since $\mathcal{H}$ is a past boundary, it automatically must be a C^0 embedded submanifold (see, e.g., [1]), but it need not be C^1. However, essentially all discussions and analyses of black hole event horizons implicitly assume C^1 or higher order differentiability of $\mathcal{H}$. Recently, this higher order differentiability assumption has been eliminated for the proof of the area theorem [3].

The area increase law bears a resemblance to the second law of thermodynamics in that both laws assert that a certain quantity has the property of never decreasing with time. It might seem that this resemblance is a very superficial one, since the area law is a theorem in differential geometry whereas the second law of thermodynamics is understood to have a statistical origin. Nevertheless, this resemblance together with the idea that information is irretrievably lost when a body falls into a black hole led Bekenstein to propose [5, 6] that a suitable multiple of the area of the event horizon of a black hole should be interpreted as its entropy, and that a *generalized second law* (GSL) should hold: The sum of the ordinary entropy of matter outside of a black hole plus a suitable multiple of the area of a black hole never decreases. We will discuss this law in detail in section 4.

The remaining laws of thermodynamics deal with equilibrium and quasi-equilibrium processes. At nearly the same time as Bekenstein proposed a relationship between the area theorem and the second law of thermodynamics, Bardeen, Carter, and Hawking [7] provided a general proof of certain laws of "black hole mechanics" which are direct mathematical analogs of the zeroth and first laws of thermodynamics. These laws of black hole mechanics apply to stationary black holes (although a formulation of these laws in terms of isolated horizons will be briefly described at the end of this section).

In order to discuss the zeroth and first laws of black hole mechanics, we must introduce the notions of stationary, static, and axisymmetric black holes as well as the notion of a Killing horizon. If an asymptotically flat spacetime (M, g_{ab}) contains a black hole, $\mathcal{B}$, then $\mathcal{B}$ is said to be *stationary* if there exists a one-parameter group of isometries on (M, g_{ab}) generated by a Killing field t^a which is unit timelike at infinity. The black hole is said to be *static* if it is stationary and if, in addition, t^a is hypersurface orthogonal. The black hole is said to be *axisymmetric* if there exists a one parameter group of isometries which correspond to rotations at infinity. A stationary, axisymmetric black hole is said to possess the "t–ϕ orthogonality property" if the 2-planes spanned by t^a and the rotational Killing field ϕ^a are orthogonal to a family of 2-dimensional surfaces. The t–ϕ orthogonality property holds for all stationary-axisymmetric black hole solutions to the vacuum Einstein or Einstein-Maxwell equations (see, e.g., [8]).

A null surface, $\mathcal{K}$, whose null generators coincide with the orbits of a one-parameter group of isometries (so that there is a Killing field ξ^a normal to $\mathcal{K}$) is called a *Killing horizon*. There are two independent results (usually referred to as "rigidity theorems") that show that in a wide variety of cases of interest, the event horizon, $\mathcal{H}$, of a stationary black hole must be a Killing horizon. The first, due to Carter [9], states that for a static black hole, the static Killing field t^a must be normal to the horizon, whereas for a stationary-axisymmetric black hole with the t–ϕ orthogonality property there exists a Killing field ξ^a of the form

$$\xi^a = t^a + \Omega \phi^a \qquad (2)$$

which is normal to the event horizon. The constant Ω defined by Eq. (2) is called the *angular velocity of the horizon*. Carter's result does not rely on any field equations, but leaves open the possibility that there could exist stationary black holes without the above symmetries whose event horizons are not Killing horizons. The second result, due to Hawking [2] (see also [10]), directly proves that in vacuum or electrovac general relativity, the event horizon of any station-ary black hole must be a Killing horizon. Consequently, if t^a fails to be normal to the horizon, then there must exist an additional Killing field, ξ^a, which is normal to the horizon, i.e., a stationary black hole must be nonrotating (from which staticity follows [11, 12, 13]) or axisymmetric (though not necessarily with the t–ϕ orthogonality property). Note that Hawking's theorem makes no assumptions of symmetries beyond stationarity, but it does rely on the properties of the field equations of general relativity.

Now, let $\mathcal{K}$ be any Killing horizon (not necessarily required to be the event horizon, $\mathcal{H}$, of a black hole), with normal Killing field ξ^a. Since $\nabla^a(\xi^b\xi_b)$ also is normal to $\mathcal{K}$, these vectors must be proportional at every point on $\mathcal{K}$. Hence, there exists a function, κ, on $\mathcal{K}$, known as the *surface gravity* of $\mathcal{K}$, which is defined by the equation

$$\nabla^a(\xi^b\xi_b) = -2\kappa\xi^a. \qquad (3)$$

It follows immediately that κ must be constant along each null geodesic gen-erator of $\mathcal{K}$, but, in general, κ can vary from generator to generator. It is not difficult to show (see, e.g., [1]) that

$$\kappa = \lim(Va), \qquad (4)$$

where a is the magnitude of the acceleration of the orbits of ξ^a in the region off of $\mathcal{K}$ where they are timelike, $V \equiv (-\xi^a\xi_a)^{1/2}$ is the "redshift factor" of ξ^a, and the limit as one approaches $\mathcal{K}$ is taken. Equation (4) motivates the terminology "surface gravity". Note that the surface gravity of a black hole is defined only when it is "in equilibrium", i.e., stationary, so that its event horizon is a Killing horizon. There is no notion of the surface gravity of a general, non-stationary black hole, although the definition of surface gravity can be extended to isolated horizons (see below).

In parallel with the two independent "rigidity theorems" mentioned above, there are two independent versions of the zeroth law of black hole mechanics. The first, due to Carter [9] (see also [14]), states that for any black hole which is static or is stationary-axisymmetric with the t–ϕ orthogonality property, the surface gravity κ, must be constant over its event horizon $\mathcal{H}$. This result is purely geometrical, i.e., it involves no use of any field equations. The second, due to Bardeen, Carter, and Hawking [7] states that if Einstein's equation holds with the matter stress-energy tensor satisfying the dominant energy condition, then κ must be constant on any Killing horizon. Thus, in the second version of the zeroth law, the hypothesis that the t–ϕ orthogonality property holds is eliminated, but use is made of the field equations of general relativity.

A *bifurcate Killing horizon* is a pair of null surfaces, $\mathcal{K}_A$ and $\mathcal{K}_B$, which intersect on a spacelike 2-surface, $\mathcal{C}$ (called the "bifurcation surface"), such that $\mathcal{K}_A$ and $\mathcal{K}_B$ are each Killing horizons with respect to the same Killing field ξ^a. It follows that ξ^a must vanish on $\mathcal{C}$; conversely, if a Killing field, ξ^a, vanishes on a two-dimensional spacelike surface, $\mathcal{C}$, then $\mathcal{C}$ will be the bifurcation surface of a bifurcate Killing horizon associated with ξ^a (see [15] for further discussion). An important consequence of the zeroth law is that if $\kappa \neq 0$, then in the "maximally extended" spacetime representing a stationary black hole, the event horizon, $\mathcal{H}$, comprises a branch of a bifurcate Killing horizon [14]. This result is purely geometrical – involving no use of any field equations. As a consequence, the study of stationary black holes which satisfy the zeroth law divides into two cases: "extremal" black holes (for which, by definition, $\kappa = 0$), and black holes with bifurcate horizons.

The first law of black hole mechanics is simply an identity relating the changes in mass, M, angular momentum, J, and horizon area, A, of a stationary black hole when it is perturbed. To first order, the variations of these quantities in the vacuum case always satisfy

$$\delta M = \frac{1}{8\pi}\kappa\delta A + \Omega\delta J. \tag{5}$$

In the original derivation of this law [7], it was required that the perturbation be stationary. Furthermore, the original derivation made use of the detailed form of Einstein's equation. Subsequently, the derivation has been generalized to hold for non-stationary perturbations [11, 16], provided that the change in area is evaluated at the bifurcation surface, $\mathcal{C}$, of the unperturbed black hole (see, however, [17] for a derivation of the first law for non-stationary perturbations that does not require evaluation at the bifurcation surface). More significantly, it has been shown [16] that the validity of this law depends only on very general properties of the field equations. Specifically, a version of this law holds for any field equations derived from a diffeomorphism covariant Lagrangian, L. Such a Lagrangian can always be written in the form

$$L = L\left(g_{ab}; R_{abcd}, \nabla_a R_{bcde}, ...; \psi, \nabla_a \psi, ...\right), \tag{6}$$

where ∇_a denotes the derivative operator associated with g_{ab}, R_{abcd} denotes the Riemann curvature tensor of g_{ab}, and ψ denotes the collection of all matter fields of the theory (with indices suppressed). An arbitrary (but finite) number of derivatives of R_{abcd} and ψ are permitted to appear in L. In this more general context, the first law of black hole mechanics is seen to be a direct consequence of an identity holding for the variation of the Noether current. The general form of the first law takes the form

$$\delta M = \frac{\kappa}{2\pi}\delta S_{\text{bh}} + \Omega \delta J + ...,\qquad(7)$$

where the "..." denote possible additional contributions from long range matter fields, and where

$$S_{\text{bh}} \equiv -2\pi \int_{\mathcal{C}} \frac{\delta L}{\delta R_{abcd}} n_{ab} n_{cd}.\qquad(8)$$

Here n_{ab} is the binormal to the bifurcation surface $\mathcal{C}$ (normalized so that $n_{ab}n^{ab} = -2$), and the functional derivative is taken by formally viewing the Riemann tensor as a field which is independent of the metric in Eq. (6). For the case of vacuum general relativity, where $L = R\sqrt{-g}$, a simple calculation yields

$$S_{\text{bh}} = A/4,\qquad(9)$$

and Eq. (7) reduces to Eq. (5).

The close mathematical analogy of the zeroth, first, and second laws of thermodynamics to corresponding laws of classical black hole mechanics is broken by the Planck-Nernst form of the third law of thermodynamics, which states that $S \to 0$ (or a "universal constant") as $T \to 0$. The analog of this law fails in black hole mechanics – although analogs of alternative formulations of the third law do appear to hold for black holes [18] – since there exist extremal black holes (i.e., black holes with $\kappa = 0$) with finite A. However, there is good reason to believe that the "Planck-Nernst theorem" should not be viewed as a fundamental law of thermodynamics [19] but rather as a property of the density of states near the ground state in the thermodynamic limit, which happens to be valid for commonly studied materials. Indeed, examples can be given of ordinary quantum systems that violate the Planck-Nernst form of the third law in a manner very similar to the violations of the analog of this law that occur for black holes [20].

As discussed above, the zeroth and first laws of black hole mechanics have been formulated in the mathematical setting of stationary black holes whose event horizons are Killing horizons. The requirement of stationarity applies to the entire spacetime and, indeed, for the first law, stationarity of the entire spacetime is essential in order to relate variations of quantities defined at the horizon (like A) to variations of quantities defined at infinity (like M and J). However, it would seem reasonable to expect that the equilibrium thermodynamic behavior of a black hole would require only a form of local stationarity at the event horizon. For the formulation of the first law of black hole mechanics, one would also then need local definitions of quantities like M and J at the horizon. Such an approach toward the formulation of the laws of black hole mechanics has recently been taken via the notion of an *isolated horizon*, defined as a null hypersurface with vanishing shear and expansion satisfying the additional properties stated in [21]. (This definition supersedes the more restrictive definitions given, e.g., in [22, 23, 24].) The presence of an isolated horizon does not require the entire spacetime to be stationary [25]. A direct analog of the zeroth law for stationary event horizons can be shown to hold for isolated horizons [26]. In the Einstein-Maxwell case, one can demand (via a choice of scaling of the normal to the isolated horizon as well as a choice of gauge for the Maxwell field) that the surface gravity and electrostatic potential of the isolated horizon be functions of only its area and charge. The requirement that time evolution be symplectic then leads to a version of the first law of black hole mechanics as well as a (in general, non-unique) local notion of the energy of the isolated horizon [26]. These results also have been generalized to allow dilaton couplings [24] and Yang-Mills fields [27, 26].

In comparing the laws of black hole mechanics in classical general relativity with the laws of thermodynamics, it should first be noted that the black hole uniqueness theorems (see, e.g., [8]) establish that stationary black holes – i.e., black holes "in equilibrium" – are characterized by a small number of parameters, analogous to the "state parameters" of ordinary thermodynamics. In the corresponding laws, the role of energy, E, is played by the mass, M, of the black hole; the role of temperature, T, is played by a constant times the surface gravity, κ, of the black hole; and the role of entropy, S, is played by a constant times the area, A, of the black hole. The fact that E and M represent the same physical quantity provides a strong hint that the mathematical analogy between the laws of black hole mechanics and the laws of thermodynamics might be of physical significance. However, as argued in [7], this cannot be the case in classical general relativity. The physical temperature of a black hole is absolute zero (see subsection 4.1 below), so there can be no physical relationship between T and κ. Consequently, it also would be inconsistent to assume a physical relationship between S and A. As we shall now see, this situation changes dramatically when quantum effects are taken into account.

3.   Hawking Radiation

In 1974, Hawking [28] made the startling discovery that the physical temperature of a black hole is not absolute zero: As a result of quantum particle
creation effects, a black hole radiates to infinity all species of particles with a
perfect black body spectrum, at temperature (in units with $G = c = \hbar = k = 1$)

$$T = \frac{\kappa}{2\pi}. \tag{10}$$

Thus, $\kappa/2\pi$ truly is the *physical* temperature of a black hole, not merely a
quantity playing a role mathematically analogous to temperature in the laws of
black hole mechanics. In this section, we review the status of the derivation of
the Hawking effect and also discuss the closely related Unruh effect.

The original derivation of the Hawking effect [28] made direct use of the
formalism for calculating particle creation in a curved spacetime that had been
developed by Parker [29] and others. Hawking considered a classical spacetime
(M, g_{ab}) describing gravitational collapse to a Schwarzschild black hole. He
then considered a free (i.e., linear) quantum field propagating in this background
spacetime, which is initially in its vacuum state prior to the collapse, and he
computed the particle content of the field at infinity at late times. This calculation involves taking the positive frequency mode function corresponding to a
particle state at late times, propagating it backwards in time, and determining
its positive and negative frequency parts in the asymptotic past. His calculation
revealed that at late times, the expected number of particles at infinity corresponds to emission from a perfect black body (of finite size) at the Hawking
temperature (Eq. (10)). It should be noted that this result relies only on the
analysis of quantum fields in the region exterior to the black hole, and it does
not make use of any gravitational field equations.

The original Hawking calculation can be straightforwardly generalized and
extended in the following ways. First, one may consider a spacetime representing an arbitrary gravitational collapse to a black hole such that the black hole
"settles down" to a stationary final state satisfying the zeroth law of black hole
mechanics (so that the surface gravity, κ, of the black hole final state is constant
over its event horizon). The initial state of the quantum field may be taken to
be any nonsingular state (i.e., any Hadamard state – see, e.g., [15]) rather than
the initial vacuum state. Finally, it can be shown [30] that all aspects of the
final state at late times (i.e., not merely the expected number of particles in each
mode) correspond to black body[1] thermal radiation emanating from the black
hole at temperature (Eq. (10)).

It should be noted that no infinities arise in the calculation of the Hawking
effect for a free field, so the results are mathematically well defined, without any
need for regularization or renormalization. The original derivations [28, 30]
made use of notions of "particles propagating into the black hole", but the

results for what an observer sees at infinity were shown to be independent of the ambiguities inherent in such notions and, indeed, a derivation of the Hawking effect has been given [31] which entirely avoids the introduction of any notion of "particles". However, there remains one significant difficultly with the Hawking derivation: In the calculation of the backward-in-time propagation of a mode, it is found that the mode undergoes a large blueshift as it propagates near the event horizon, but there is no correspondingly large redshift as the mode propagates back through the collapsing matter into the asymptotic past. Indeed, the net blueshift factor of the mode is proportional to $\exp(\kappa t)$, where t is the time that the mode would reach an observer at infinity. Thus, within a time of order $1/\kappa$ of the formation of a black hole (i.e., $\sim 10^{-5}$ seconds for a one solar mass Schwarzschild black hole), the Hawking derivation involves (in its intermediate steps) the propagation of modes of frequency much higher than the Planck frequency. In this regime, it is difficult to believe in the accuracy of free field theory – or any other theory known to mankind.

An approach to investigating this issue was first suggested by Unruh [32], who noted that a close analog of the Hawking effect occurs for quantized sound waves in a fluid undergoing supersonic flow. A similar blueshifting of the modes quickly brings one into a regime well outside the domain of validity of the continuum fluid equations. Unruh suggested replacing the continuum fluid equations with a more realistic model at high frequencies to see if the fluid analog of the Hawking effect would still occur. More recently, Unruh investigated models where the dispersion relation is altered at ultra-high frequencies, and he found no deviation from the Hawking prediction [33]. A variety of alternative models have been considered by other researchers [34, 35, 36, 37, 38, 39, 40]. Again, agreement with the Hawking effect prediction was found in all cases, despite significant modifications of the theory at high frequencies.

The robustness of the Hawking effect with respect to modifications of the theory at ultra-high frequency probably can be understood on the following grounds. One may view the backward-in-time propagation of modes as consisting of two stages: a first stage where the blueshifting of the mode brings it into a WKB regime but the frequencies remain well below the Planck scale, and a second stage where the continued blueshifting takes one to the Planck scale and beyond. In the first stage, the usual field theory calculations should be reliable. On the other hand, after the mode has entered a WKB regime, it seems plausible that the kinds of modifications to its propagation laws considered in [33, 34, 35, 36, 37, 38, 39, 40] should not affect its essential properties, in particular the magnitude of its negative frequency part.

Indeed, an issue closely related to the validity of the original Hawking derivation arises if one asks how a uniformly accelerating observer in Minkowski spacetime perceives the ordinary (inertial) vacuum state (see below). The outgoing modes of a given frequency ω as seen by the accelerating observer at proper time τ along his worldline correspond to modes of frequency $\sim \omega \exp(a\tau)$ in a fixed inertial frame. Therefore, at time $\tau \gg 1/a$ one might worry about field-theoretic derivations of what the accelerating observer would see. However, in this case one can appeal to Lorentz invariance to argue that what the accelerating observer sees cannot change with time. It seems likely that one could similarly argue that the Hawking effect cannot be altered by modifications of the theory at ultra-high frequencies, provided that these modifications preserve an appropriate "local Lorentz invariance" of the theory. Thus, there appears to be strong reasons for believing in the validity of the Hawking effect despite the occurrence of ultra-high-frequency modes in the derivation.

There is a second, logically independent result – namely, the Unruh effect [41] and its generalization to curved spacetime – which also gives rise to the formula (10). Although the Unruh effect is mathematically very closely related to the Hawking effect, it is important to distinguish clearly between them. In its most general form, the Unruh effect may be stated as follows (see [42, 15] for further discussion): Consider a classical spacetime (M, g_{ab}) that contains a bifurcate Killing horizon, $\mathcal{K} = \mathcal{K}_A \cup \mathcal{K}_B$, so that there is a one-parameter group of isometries whose associated Killing field, ξ^a, is normal to $\mathcal{K}$. Consider a free quantum field on this spacetime. Then there exists at most one globally nonsingular state of the field which is invariant under the isometries. Furthermore, in the "wedges" of the spacetime where the isometries have timelike orbits, this state (if it exists) is a KMS (i.e., thermal equilibrium) state at temperature (10) with respect to the isometries.

Note that in Minkowski spacetime, any one-parameter group of Lorentz boosts has an associated bifurcate Killing horizon, comprised by two intersecting null planes. The unique, globally nonsingular state which is invariant under these isometries is simply the usual ("inertial") vacuum state, $|0\rangle$. In the "right and left wedges" of Minkowski spacetime defined by the Killing horizon, the orbits of the Lorentz boost isometries are timelike, and, indeed, these orbits correspond to worldlines of uniformly accelerating observers. If we normalize the boost Killing field, b^a, so that Killing time equals proper time on an orbit with acceleration a, then the surface gravity of the Killing horizon is $\kappa = a$. An observer following this orbit would naturally use b^a to define a notion of "time translation symmetry". Consequently, by the above general result, when the field is in the inertial vacuum state, a uniformly accelerating observer would describe the field as being in a thermal equilibrium state at temperature

$$T = \frac{a}{2\pi} \tag{11}$$

as originally discovered by Unruh [41]. A mathematically rigorous proof of the Unruh effect in Minkowski spacetime was given by Bisognano and Wichmann [43] in work motivated by entirely different considerations (and done independently of and nearly simultaneously with the work of Unruh). Furthermore, the Bisognano-Wichmann theorem is formulated in the general context of axiomatic quantum field theory, thus establishing that the Unruh effect is not limited to free field theory.

Although there is a close mathematical relationship between the Unruh effect and the Hawking effect, it should be emphasized that these results refer to *different* states of the quantum field. We can divide the late time modes of the quantum field in the following manner, according to the properties that they would have in the analytically continued spacetime [14] representing the asymptotic final stationary state of the black hole: We refer to modes that would have emanated from the white hole region of the analytically continued spacetime as "UP modes" and those that would have originated from infinity as "IN modes". In the Hawking effect, the asymptotic final state of the quantum field is a state in which the UP modes of the quantum field are thermally populated at temperature (10), but the IN modes are unpopulated. This state (usually referred to as the "Unruh vacuum") would be singular on the white hole horizon in the analytically continued spacetime. On the other hand, in the Unruh effect and its generalization to curved spacetimes, the state in question (usually referred to as the "Hartle-Hawking vacuum" [44]) is globally nonsingular, and *all* modes of the quantum field in the "left and right wedges" are thermally populated.[2]

The differences between the Unruh and Hawking effects can be seen dramatically in the case of a Kerr black hole. For the Kerr black hole, it can be shown [42] that there does not exist any globally nonsingular state of the field which is invariant under the isometries associated with the Killing horizon, i.e., there does not exist a "Hartle-Hawking vacuum state" on Kerr spacetime. However, there is no difficulty with the derivation of the Hawking effect for Kerr black holes, i.e., the "Unruh vacuum state" does exist.

It should be emphasized that in the Hawking effect, the temperature (10) represents the temperature as measured by an observer near infinity. For any observer following an orbit of the Killing field, ξ^a, normal to the horizon, the locally measured temperature of the UP modes is given by

$$T = \frac{\kappa}{2\pi V} , \tag{12}$$

where $V = (-\xi^a \xi_a)^{1/2}$. In other words, the locally measured temperature of the Hawking radiation follows the Tolman law. Now, as one approaches the horizon of the black hole, the UP modes dominate over the IN modes. Taking Eq. (4) into account, we see that $T \to a/2\pi$ as the black hole horizon, $\mathcal{H}$, is approached, i.e., in this limit Eq. (12) corresponds to the flat spacetime Unruh effect.

Equation (12) shows that when quantum effects are taken into account, a black hole is surrounded by a "thermal atmosphere" whose local temperature as measured by observers following orbits of ξ^a becomes divergent as one approaches the horizon. As we shall see in the next section, this thermal atmosphere produces important physical effects on quasi-stationary bodies near the black hole. On the other hand, it should be emphasized that for a macroscopic black hole, observers who freely fall into the black hole would not notice any important quantum effects as they approach and cross the horizon.

4.   The Generalized Second Law (GSL)

In this section, we shall review some arguments for the validity of the generalized second law (GSL). We also shall review the status of several proposed entropy bounds on matter that have played a role in discussions and analyses of the GSL.

4.1   Arguments for the validity of the GSL

Even in classical general relativity, there is a serious difficulty with the ordinary second law of thermodynamics when a black hole is present, as originally emphasized by J.A. Wheeler: One can simply take some ordinary matter and drop it into a black hole, where, according to classical general relativity, it will disappear into a spacetime singularity. In this process, one loses the entropy initially present in the matter, and no compensating gain of ordinary entropy occurs, so the total entropy, S, of matter in the universe decreases. One could attempt to salvage the ordinary second law by invoking the bookkeeping rule that one must continue to count the entropy of matter dropped into a black hole as still contributing to the total entropy of the universe. However, the second law would then have the status of being observationally unverifiable.

As already mentioned in section 2, after the area theorem was proven, Bekenstein [5, 6] proposed a way out of this difficulty: Assign an entropy, S_{bh}, to a black hole given by a numerical factor of order unity times the area, A, of the black hole in Planck units. Define the *generalized entropy*, S', to be the sum of the ordinary entropy, S, of matter outside of a black hole plus the black hole entropy

$$S' \equiv S + S_{bh}. \tag{13}$$

Finally, replace the ordinary second law of thermodynamics by the *generalized second law* (GSL): The total generalized entropy of the universe never decreases with time,

$$\Delta S' \geq 0. \tag{14}$$

Although the ordinary second law will fail when matter is dropped into a black hole, such a process will tend to increase the area of the black hole, so there is a possibility that the GSL will hold.

Bekenstein's proposal of the GSL was made prior to the discovery of Hawking radiation. When Hawking radiation is taken into account, a serious problem also arises with the second law of black hole mechanics (i.e., the area theorem): Conservation of energy requires that an isolated black hole must lose mass in order to compensate for the energy radiated to infinity by the Hawking process. Indeed, if one equates the rate of mass loss of the black hole to the energy flux at infinity due to particle creation, one arrives at the startling conclusion that an isolated black hole will radiate away all of its mass within a finite time. During

this process of black hole "evaporation", A will decrease. Such an area decrease
can occur because the expected stress-energy tensor of quantum matter does
not satisfy the null energy condition – even for matter for which this condition
holds classically – in violation of a key hypothesis of the area theorem.

However, although the second law of black hole mechanics fails during the
black hole evaporation process, if we adjust the numerical factor in the definition
of S_{bh} to correspond to the identification of $\kappa/2\pi$ as temperature in the first law
of black hole mechanics – so that, as in Eq. (9) above, we have $S_{\mathrm{bh}} = A/4$ in
Planck units – then the GSL continues to hold: Although A decreases, there is at
least as much ordinary entropy generated outside the black hole by the Hawking
process. Thus, although the ordinary second law fails in the presence of black
holes and the second law of black hole mechanics fails when quantum effects
are taken into account, there is a possibility that the GSL may always hold. If
the GSL does hold, it seems clear that we must interpret S_{bh} as representing
the *physical* entropy of a black hole, and that the laws of black hole mechanics
must truly represent the ordinary laws of thermodynamics as applied to black
holes. Thus, a central issue in black hole thermodynamics is whether the GSL
holds in all processes.

It was immediately recognized by Bekenstein [5] (see also [7]) that there is a
serious difficulty with the GSL if one considers a process wherein one carefully
lowers a box containing matter with entropy S and energy E very close to the
horizon of a black hole before dropping it in. Classically, if one could lower the
box arbitrarily close to the horizon before dropping it in, one would recover all
of the energy originally in the box as "work" at infinity. No energy would be
delivered to the black hole, so by the first law of black hole mechanics, Eq. (7),
the black hole area, A, would not increase. However, one would still get rid of
all of the entropy, S, originally in the box, in violation of the GSL.

Indeed, this process makes manifest the fact that in classical general relativity, the physical temperature of a black hole is absolute zero: The above
process is, in effect, a Carnot cycle which converts "heat" into "work" with
100% efficiency [45]. The difficulty with the GSL in the above process can be
viewed as stemming from an inconsistency of this fact with the mathematical
assignment of a finite (non-zero) temperature to the black hole required by the
first law of black hole mechanics if one assigns a finite (non-infinite) entropy
to the black hole.

Bekenstein proposed a resolution of the above difficulty with the GSL in a
quasi-static lowering process by arguing [5, 6] that it would not be possible
to lower a box containing physically reasonable matter close enough to the
horizon of the black hole to violate the GSL. As will be discussed further in
the next sub-section, this proposed resolution was later refined by postulating
a universal bound on the entropy of systems with a given energy and size [46].
However, an alternate resolution was proposed in [47], based upon the idea

that, when quantum effects are taken into account, the physical temperature of a black hole is no longer absolute zero, but rather is the Hawking temperature, $\kappa/2\pi$. Since the Hawking temperature goes to zero in the limit of a large black hole, it might appear that quantum effects could not be of much relevance in this case. However, despite the fact that Hawking radiation at infinity is indeed negligible for large black holes, the effects of the quantum "thermal atmosphere" surrounding the black hole are not negligible on bodies that are quasi-statically lowered toward the black hole. The temperature gradient in the thermal atmosphere (see Eq. (12)) implies that there is a pressure gradient and, consequently, a buoyancy force on the box. This buoyancy force becomes infinitely large in the limit as the box is lowered to the horizon. As a result of this buoyancy force, the optimal place to drop the box into the black hole is no longer the horizon but rather the "floating point" of the box, where its weight is equal to the weight of the displaced thermal atmosphere. The minimum area increase given to the black hole in the process is no longer zero, but rather turns out to be an amount just sufficient to prevent any violation of the GSL from occurring in this process [47].

The analysis of [47] considered only a particular class of gedankenexperiments for violating the GSL involving the quasi-static lowering of a box near a black hole. Of course, since one does not have a general proof of the ordinary second law of thermodynamics – and, indeed, for finite systems, there should always be a nonvanishing probability of violating the ordinary second law – it would not be reasonable to expect to obtain a completely general proof of the GSL. However, general arguments within the semiclassical approximation for the validity of the GSL for arbitrary infinitesimal quasi-static processes have been given in [48, 49, 15]. These arguments crucially rely on the presence of the thermal atmosphere surrounding the black hole. Related arguments for the validity of the GSL have been given in [50, 51]. In [50], it is assumed that the incoming state is a product state of radiation originating from infinity (i.e., IN modes) and radiation that would appear to emanate from the white hole region of the analytically continued spacetime (i.e., UP modes), and it is argued that the generalized entropy must increase under unitary evolution. In [51], it is argued on quite general grounds that the (generalized) entropy of the state of the region exterior to the black hole must increase under the assumption that it undergoes autonomous evolution.

Indeed, it should be noted that if one could violate the GSL for an infinitesimal quasi-static process in a regime where the black hole can be treated semiclassically, then it also should be possible to violate the ordinary second law for a corresponding process involving a self-gravitating body. Namely, suppose that the GSL could be violated for an infinitesimal quasi-static process involving, say, a Schwarzschild black hole of mass M (with M much larger than the Planck mass). This process might involve lowering matter towards the black hole and possibly dropping the matter into it. However, an observer doing this lowering or dropping can "probe" only the region outside of the black hole, so there will be some $r_0 > 2M$ such that the detailed structure of the black hole will directly enter the analysis of the process only for $r > r_0$. Now replace the black hole by a shell of matter of mass M and radius r_0, and surround this shell with a "*real*" atmosphere of radiation in thermal equilibrium at the Hawking temperature (10) as measured by an observer at infinity. Then the ordinary second law should be violated when one performs the same process to the shell surrounded by the ("real") thermal atmosphere as one performs to violate the GSL when t

he black hole is present. Indeed, the arguments of [48, 49, 15] do not distinguish between infinitesimal quasi-static processes involving a black hole as compared with a shell surrounded by a ("real") thermal atmosphere at the Hawking temperature.

In summary, there appear to be strong grounds for believing in the validity of the GSL.

4.2   Entropy bounds

As discussed in the previous subsection, for a classical black hole the GSL would be violated if one could lower a box containing matter sufficiently close to the black hole before dropping it in. Indeed, for a Schwarzschild black hole, a simple calculation reveals that if the size of the box can be neglected, then the GSL would be violated if one lowered a box containing energy E and entropy S to within a proper distance D of the bifurcation surface of the event horizon before dropping it in, where

$$D < \frac{S}{(2\pi E)}. \tag{15}$$

(This formula holds independently of the mass, M, of the black hole.) However, it is far from clear that the finite size of the box can be neglected if one lowers a box containing physically reasonable matter this close to the black hole. If it cannot be neglected, then this proposed counterexample to the GSL would be invalidated.

As already mentioned in the previous subsection, these considerations led Bekenstein [46] to propose a universal bound on the entropy-to-energy ratio of bounded matter, given by

$$S/E \leq 2\pi R, \tag{16}$$

where R denotes the "circumscribing radius" of the body. Here "E" is normally interpreted as the energy above the ground state; otherwise, Eq. (16) would be trivially violated in cases where the Casimir energy is negative [52] – although in such cases in may still be possible to rescue Eq. (16) by postulating a suitable minimum energy of the box walls [53].

Two key questions one can ask about this bound are: (1) Does it hold in nature? (2) Is it needed for the validity of the GSL? With regard to question (1), even in Minkowski spacetime, there exist many model systems that are physically reasonable (in the sense of positive energies, causal equations of state, etc.) for which Eq. (16) fails. (For a recent discussion of such counterexamples to Eq. (16), see [54, 55, 52]; for counter-arguments to these references, see [53].) In particular it is easily seen that for a system consisting of N non-interacting species of particles with identical properties, Eq. (16) must fail when N becomes sufficiently large. However, for a system of N species of free, massless bosons or fermions, one must take N to be enormously large [56] to violate Eq. (16), so it does not appear that nature has chosen to take advantage of this possible means of violating (16). Equation (16) also is violated at sufficiently low temperatures if one defines the entropy, S, of the system via the canonical ensemble, i.e., $S(T) = -\mathrm{tr}[\rho \ln \rho]$, where ρ denotes the canonical ensemble density matrix,

$$\rho = \exp(-H/T)\mathrm{tr}[\exp(-H/T)], \tag{17}$$

where H is the Hamiltonian. However, a study of a variety of model systems [56] indicates that (16) holds at low temperatures when S is defined via the microcanonical ensemble, i.e., $S(E) = \ln n$ where n is the density of quantum states with energy E. More generally, Eq. (16) has been shown to hold for a wide variety of systems in flat spacetime [56, 57].

The status of Eq. (16) in curved spacetime is unclear; indeed, while there is some ambiguity in how "E" and "R" are defined in Minkowski spacetime [52], it is very unclear what these quantities would mean in a general, non-spherically-symmetric spacetime. (These same difficulties also plague attempts to give a mathematically rigorous formulation of the "hoop conjecture" [58].) With regard to "E", it has long been recognized that there is no meaningful local notion of gravitational energy density in general relativity. Although numerous proposals have been made to define a notion of "quasi-local mass" associated with a closed 2-surface (see, e.g., [59, 60]), none appear to have fully satisfactory properties. Although the difficulties with defining a localized notion of energy are well known, it does not seem to be as widely recognized that there also are serious difficulties in defining "R": Given any spacelike 2-surface, $\mathcal{C}$, in a 4-dimensional spacetime and given any open neighborhood, $\mathcal{O}$, of $\mathcal{C}$, there exists a spacelike 2-surface, $\mathcal{C}'$ (composed of nearly null portions) contained within $\mathcal{O}$ with arbitrarily small area and circumscribing radius. Thus, if one is given a system confined to a world tube in spacetime, it is far from clear how to define any notion of the "externally measured size" of the region unless, say, one is given a preferred slicing by spacelike hypersurfaces. Nevertheless, the fact that Eq. (16) holds for the known black hole solutions (and, indeed, is saturated by the Schwarzschild black hole) and also plausibly holds for a self-gravitating spherically symmetric body [61] provides an indication that some version of (16) may hold in curved spacetime.

With regard to question (2), in the previous section we reviewed arguments for the validity of the GSL that did not require the invocation of any entropy bounds. Thus, the answer to question (2) is "no" unless there are deficiencies in the arguments of the previous section that invalidate their conclusions. A number of such potential deficiencies have been pointed out by Bekenstein. Specifically, the analysis and conclusions of [47] have been criticized by Bekenstein on the grounds that:

  i A "thin box" approximation was made [62].

  ii It is possible to have a box whose *contents* have a greater entropy than unconfined thermal radiation of the same energy and volume [62].

  iii Under certain assumptions concerning the size/shape of the box, the nature of the thermal atmosphere, and the location of the floating point, the buoyancy force of the thermal atmosphere can be shown to be negligible and thus cannot play a role in enforcing the GSL [63].

  iv Under certain other assumptions, the box size at the floating point will be smaller than the typical wavelengths in the ambient thermal atmosphere, thus likely decreasing the magnitude of the buoyancy force [64].

Responses to criticism (i) were given in [65] and [66]; a response to criticism (ii) was given in [65]; and a response to (iii) was given in [66]. As far as I am a aware, no response to (iv) has yet been given in the literature except to note [67] that the arguments of [64] should pose similar difficulties for the ordinary second law for gedankenexperiments involving a self-gravitating body (see the end of subsection 4.1 above). Thus, my own view is that Eq. (16) is not necessary for the validity of the GSL[3]. However, this conclusion remains controversial; see [68] for a recent discussion.

More recently, an alternative entropy bound has been proposed: It has been suggested that the entropy contained within a region whose boundary has area A must satisfy [69, 70, 71]

$$S \le A/4. \tag{18}$$

This proposal is closely related to the "holographic principle", which, roughly speaking, states that the physics in any spatial region can be fully described in terms of the degrees of freedom associated with the boundary of that region. (The literature on the holographic principle is far too extensive and rapidly developing to attempt to give any review of it here.) The bound (18) would follow from (16) under the additional assumption of small self-gravitation (so that $E \lesssim R$). Thus, many of the arguments in favor of (16) are also applicable to (18). Similarly, the counterexample to (16) obtained by taking the number,

N, of particle species sufficiently large also provides a counterexample to (18), so it appears that (18) can, in principle, be violated by physically reasonable systems (although not necessarily by any systems actually occurring in nature).

Unlike Eq. (16), the bound (18) explicitly involves the gravitational constant G (although we have set $G = 1$ in all of our formulas), so there is no flat spacetime version of (18) applicable when gravity is "turned off". Also unlike (16), the bound (18) does not make reference to the energy, E, contained within the region, so the difficulty in defining E in curved spacetime does not affect the formulation of (18). However, the above difficulty in defining the "bounding area", A, of a world tube in a general, curved spacetime remains present (but see below).

The following argument has been given that the bound (18) is necessary for the validity of the GSL [71]: Suppose we had a spherically symmetric system that was not a black hole (so $R > 2E$) and which violated the bound (18), so that $S > A/4 = \pi R^2$. Now collapse a spherical shell of mass $M = R/2 - E$ onto the system. A Schwarzschild black hole of radius R should result. But the entropy of such a black hole is $A/4$, so the generalized entropy will decrease in this process.

I am not aware of any counter-argument in the literature to the argument given in the previous paragraph, so I will take the opportunity to give one here. If there were a system which violated the bound (18), then the above argument shows that it would be (generalized) entropically unfavorable to collapse that system to a black hole. I believe that the conclusion one should draw from this is that, in this circumstance, it should not be possible to form a black hole. In other words, the bound (18) should be necessary in order for black holes to be stable or metastable states, but should not be needed for the validity of the GSL.

This viewpoint is supported by a simple model calculation. Consider a massless gas composed of N species of (boson or fermion) particles confined by a spherical box of radius R. Then (neglecting self-gravitational effects and any corrections due to discreteness of modes) we have

$$S \sim N^{1/4} R^{3/4} E^{3/4}. \tag{19}$$

We wish to consider a configuration that is not already a black hole, so we need $E < R/2$. To violate (18) – and thereby threaten to violate the GSL by collapsing a shell upon the system – we need to have $S > \pi R^2$. This means that we need to consider a model with $N \gtrsim R^2$. For such a model, start with a region R containing matter with $S > \pi R^2$ but with $E < R/2$. If we try to collapse a shell upon the system to form a black hole of radius R, the collapse time will be $\gtrsim R$. But the Hawking evaporation timescale in this model is $t_H \sim R^3/N$, since the flux of Hawking radiation is proportional to N. Since $N \gtrsim R^2$, we have $t_H \lesssim R$, so the Hawking evaporation time is shorter than the collapse time! Consequently, the black hole will never actually form. Rather, at best it will merely act as a catalyst for converting the original high entropy confined state into an even higher entropy state of unconfined Hawking radiation.

As mentioned above, the proposed bound (18) is ill defined in a general (non-spherically-symmetric) curved spacetime. There also are other difficulties with (18): In a closed universe, it is not obvious what constitutes the "inside" versus the "outside" of the bounding area. In addition, (18) can be violated near cosmo-logical and other singularities, where the entropy of suitably chosen comoving volumes remains bounded away from zero but the area of the boundary of the region goes to zero. However, a reformulation of (18) which is well defined in a general curved spacetime and which avoids these difficulties has been given by Bousso [72, 73, 74]. Bousso's reformulation can be stated as follows: Let $\mathcal{L}$ be a null hypersurface such that the expansion, θ, of $\mathcal{L}$ is everywhere non-positive, $\theta \leq 0$ (or, alternatively, is everywhere non-negative, $\theta \geq 0$). In particular, $\mathcal{L}$ is not allowed to contain caustics, where θ changes sign from $-\infty$ to $+\infty$. Let B be a spacelike cross-section of $\mathcal{L}$. Bousso's reformulation conjectures that

$$S_{\mathcal{L}} \leq A_B/4, \tag{20}$$

where A_B denotes the area of B and $S_{\mathcal{L}}$ denotes the entropy flux through $\mathcal{L}$ to the future (or, respectively, the past) of B.

In [67] it was argued that the bound (21) should be valid in certain "classical regimes" (see [67]) wherein the local entropy density of matter is bounded in a suitable manner by the energy density of matter. Furthermore, the following generalization of Bousso's bound was proposed: Let $\mathcal{L}$ be a null hypersurface which starts at a cross-section, B, and terminates at a cross-section B'. Suppose further that $\mathcal{L}$ is such that its expansion, θ, is either everywhere non-negative or everywhere non-positive. Then

$$S_{\mathcal{L}} \leq |A_B - A_{B'}|/4. \tag{21}$$

Although we have argued above that the validity of the GSL should not depend upon the validity of the entropy bounds (16) or (18), there is a close relationship between the GSL and the generalized Bousso bound (21). Namely, as discussed in section 2 above, classically, the event horizon of a black hole is a null hypersurface satisfying $\theta \geq 0$. Thus, in a classical regime, the GSL itself would correspond to a special case of the generalized Bousso bound (21). This suggests the intriguing possibility that, in quantum gravity, there might be a more general formulation of the GSL – perhaps applicable to an arbitrary horizon as defined on p. 134 of [15], not merely to an event horizon of a black hole – which would reduce to (21) in a suitable classical limit.

5.   Calculations of Black Hole Entropy

The considerations of the previous sections make a compelling case for the merger of the laws of black hole mechanics with the laws of thermodynamics. In particular, they strongly suggest that S_{bh} ($= A/4$ in general relativity – see Eqs.(8) and (9) above) truly represents the physical entropy of a black hole. Now, the entropy of ordinary matter is understood to arise from the number of quantum states accessible to the matter at given values of the energy and other state parameters. One would like to obtain a similar understanding of why $A/4$ represents the entropy of a black hole in general relativity by identifying (and counting) the quantum dynamical degrees of freedom of a black hole. In order to do so, it clearly will be necessary to go beyond the classical and semiclassical considerations of the previous sections and consider black holes within a fully quantum theory of gravity. In this section, we will briefly summarize some of the main approaches that have been taken to the direct calculation of the entropy of a black hole.

The first direct quantum calculation of black hole entropy was given by Gibbons and Hawking [75] in the context of Euclidean quantum gravity. They started with a formal, functional integral expression for the canonical ensemble partition function in Euclidean quantum gravity and evaluated it for a black hole in the "zero loop" (i.e, classical) approximation. As shown in [77], the mathematical steps in this procedure are in direct correspondence with the purely classical determination of the entropy from the form of the first law of black hole mechanics. A number of other entropy calculations that have been given within the formal framework of Euclidean quantum gravity also can be shown to be equivalent to the classical derivation (see [78] for further discussion). Thus, although the derivation of [75] and other related derivations give some intriguing glimpses into possible deep relationships between black hole thermodynamics and Euclidean quantum gravity, they do not appear to provide any more insight than the classical derivation into accounting for the quantum degrees of freedom that are responsible for black hole entropy.

It should be noted that there is actually an inconsistency in the use of the canonical ensemble to derive a formula for black hole entropy, since the entropy of a black hole grows too rapidly with energy for the canonical ensemble to be defined. (Equivalently, the heat capacity of a Schwarzschild black hole is negative, so it cannot come to equilibrium with an infinite heat bath.) A derivation of black hole entropy using the microcanonical ensemble has been given in [76].

Another approach to the calculation of black hole entropy has been to attribute it to the "entanglement entropy" resulting from quantum field correlations between the exterior and interior of the black hole [79, 80, 81]. As a result of these correlations across the event horizon, the state of a quantum field when

restricted to the exterior of the black hole is mixed. Indeed, in the absence of
a short distance cutoff, the von Neumann entropy, $-\text{tr}[\rho \ln \rho]$, of any physically reasonable state would diverge. If one now inserts a short distance cutoff
of the order of the Planck scale, one obtains a von Neumann entropy of the
order of the horizon area, A. Thus, this approach provides a natural way of
accounting for why the entropy of a black hole is proportional to its surface
area. However, the constant of proportionality depends upon a cutoff and is
not (presently) calculable within this approach. (Indeed, one might argue that
in this approach, the constant of proportionality between S_{bh} and A should
depend upon the number, N, of species of particles, and thus could not equal
$1/4$ (independently of N). However, it is possible that the N-dependence in the
number of states is compensated by an N-dependent renormalization of G [82]
and, hence, of the Planck scale cutoff.) More generally, it is far from clear
why the black hole horizon should be singled out for a such special treatment
of the quantum degrees of freedom in its vicinity, since similar quantum field
correlations will exist across any other null surface. It is particularly puzzling
why the local degrees of freedom associated with the horizon should be singled
out since, as already noted in section 2 above, the black hole horizon at a given
time is defined in terms of the entire future history of the spacetime and thus
has no distinguished local significance. Finally, since the gravitational action
and field equations play no role in the above derivation, it is difficult to see how
this approach could give rise to a black hole entropy proportional to Eq. (8)
(rather than proportional to A) in a more general theory of gravity. Similar
remarks apply to approaches which attribute the relevant degrees of freedom to
the "shape" of the horizon [83] or to causal links crossing the horizon [84].

A closely related idea has been to attribute the entropy of the black hole
to the ordinary entropy of its thermal atmosphere [85]). If we assume that
the thermal atmosphere behaves like a free, massless (boson or fermion) gas,
its entropy density will be (roughly) proportional to T^3. However, since T
diverges near the horizon in the manner specified by Eq. (12), we find that the
total entropy of the thermal atmosphere near the horizon diverges. This is, in
effect, a new type of ultraviolet catastrophe. It arises because, on account of
arbitrarily large redshifts, there now are infinitely many modes – of arbitrarily
high locally measured frequency – that contribute a bounded energy as measured
at infinity. To cure this divergence, it is necessary to impose a cutoff on the
locally measured frequency of the modes. If we impose a cutoff of the order of
the Planck scale, then the thermal atmosphere contributes an entropy of order
the horizon area, A, just as in the entanglement entropy analysis. Indeed, this
calculation is really the same as the entanglement entropy calculation, since the
state of a quantum field outside of the black hole is thermal, so its von Neumann
entropy is equal to its thermodynamic entropy (see also [86]). Note that the

bulk of the entropy of the thermal atmosphere is highly localized in a "skin" surrounding the horizon, whose thickness is of order of the Planck length.

Since the attribution of black hole entropy to its thermal atmosphere is essentially equivalent to the entanglement entropy proposal, this approach has essentially the same strengths and weaknesses as the entanglement entropy approach. On one hand, it naturally accounts for a black hole entropy proportional to A. On the other hand, this result depends in an essential way on an uncalculable cutoff, and it is difficult to see how the analysis could give rise to Eq. (8) in a more general theory of gravity. The preferred status of the event horizon and the localization of the degrees of freedom responsible for black hole entropy to a "Planck length skin" surrounding the horizon also remain puzzling in this approach. To see this more graphically, consider the collapse of a massive spherical shell of matter. Then, as the shell crosses its Schwarzschild radius, the spacetime curvature outside of the shell is still negligibly small. Nevertheless, within a time of order the Planck time after the crossing of the Schwarzschild radius, the "skin" of thermal atmosphere surrounding the newly formed black hole will come to equilibrium with respect to the notion of time translation symmetry for the static Schwarzschild exterior. Thus, if an entropy is to be assigned to the thermal atmosphere in the manner suggested by this proposal, then the degrees of freedom of the thermal atmosphere – which previously were viewed as irrelevant vacuum fluctuations making no contribution to entropy – suddenly become "activated" by the passage of the shell for the purpose of counting their entropy. A momentous change in the entropy of matter in the universe has occurred, even though observers riding on or near the shell see nothing of significance occurring.

Another approach that is closely related to the entanglement entropy and thermal atmosphere approaches – and which also contains elements closely related to the Euclidean approach and the classical derivation of Eq. (8) – attempts to account for black hole entropy in the context of Sakharov's theory of induced gravity [87, 88]. In Sakharov's proposal, the dynamical aspects of gravity arise from the collective excitations of massive fields. Constraints are then placed on these massive fields to cancel divergences and ensure that the effective cosmological constant vanishes. Sakharov's proposal is not expected to provide a fundamental description of quantum gravity, but at scales below the Planck scale it may possess features in common with other more fundamental descriptions. In common with the entanglement entropy and thermal atmosphere approaches, black hole entropy is explained as arising from the quantum field degrees of freedom outside the black hole. However, in this case the formula for black hole entropy involves a subtraction of the (divergent) mode counting expression and an (equally divergent) expression for the Noether charge operator, so that, in effect, only the massive fields contribute to black hole entropy. The result of this subtraction yields Eq. (9).

More recently, another approach to the calculation of black hole entropy has been developed in the framework of quantum geometry [89, 90]. In this approach, if one considers a spacetime containing an isolated horizon (see section 2 above), the classical symplectic form and classical Hamiltonian each acquire an additional boundary term arising from the isolated horizon [26]. (It should be noted that the phase space [91] considered here incorporates the isolated horizon boundary conditions, i.e., only field variations that preserve the isolated horizon structure are admitted.) These additional terms are identical in form to that of a Chern-Simons theory defined on the isolated horizon. Classically, the fields on the isolated horizon are determined by continuity from the fields in the "bulk" and do not represent additional degrees of freedom. However, in the quantum theory – where distributional fields are allowed – these fields are interpreted as providing additional, independent degrees of freedom associated with the isolated horizon. One then counts the "surface states" of these fields on the isolated horizon subject to a boundary condition relating the surface states to "volume states" and subject to the condition that the area of the isolated horizon (as determined by the volume state) lies within a squared Planck length of the value A. This state counting yields an entropy proportional to A for black holes much larger than the Planck scale. Unlike the entanglement entropy and thermal atmosphere calculations, the state counting here yields finite results and no cutoff need be introduced. However, the formula for entropy contains a free parameter (the "Immirzi parameter"), which arises from an ambiguity in the loop quantization procedure, so the constant of proportionality between S and A is not calculable.

The most quantitatively successful calculations of black hole entropy to date are ones arising from string theory. It is believed that at "low energies", string theory should reduce to a 10-dimensional supergravity theory (see [92] for considerable further discussion of the relationship between string theory and 10-dimensional and 11-dimensional supergravity). If one treats this supergravity theory as a classical theory involving a spacetime metric, g_{ab}, and other classical fields, one can find solutions describing black holes. On the other hand, one also can consider a "weak coupling" limit of string theory, wherein the states are treated perturbatively. In the weak coupling limit, there is no literal notion of a black hole, just as there is no notion of a black hole in linearized general relativity. Nevertheless, certain weak coupling states can be identified with certain black hole solutions of the low energy limit of the theory by a correspondence of their energy and charges. (Here, it is necessary to introduce "D-branes" into string perturbation theory in order to obtain weak coupling states with the desired charges.) Now, the weak coupling states are, in essence, ordinary quantum dynamical degrees of freedom, so their entropy can be computed by the usual methods of statistical physics. Remarkably, for certain classes of extremal and nearly extremal black holes, the ordinary entropy of the weak coupling states agrees exactly with the expression for $A/4$ for the corresponding classical black hole states; see [93] and [94] for reviews of these results. Recently, it also has been shown [95] that for certain black holes, subleading corrections to the state counting formula for entropy correspond to higher order string corrections to the effective gravitational action, in precise agreement with Eq. (8).

Since the formula for entropy has a nontrivial functional dependence on energy and charges, it is hard to imagine that the agreement between the ordinary entropy of the weak coupling states and black hole entropy could be the result of a random coincidence. Furthermore, for low energy scattering, the absorption/emission coefficients ("gray body factors") of the corresponding weak coupling states and black holes also agree [96]. This suggests that there may be a close physical association between the weak coupling states and black holes, and that the dynamical degrees of freedom of the weak coupling states are likely to at least be closely related to the dynamical degrees of freedom responsible for black hole entropy. However, it remains a challenge to understand in what sense the weak coupling states could be giving an accurate picture of the local physics occurring near (and within) the region classically described as a black hole.

The relevant degrees of freedom responsible for entropy in the weak coupling string theory models are associated with conformal field theories. Recently Carlip [97, 98] has attempted to obtain a direct relationship between the string theory state counting results for black hole entropy and the classical Poisson bracket algebra of general relativity. After imposing certain boundary conditions corresponding to the presence of a local Killing horizon, Carlip chooses a particular subgroup of spacetime diffeomorphisms, generated by vector fields ξ^a. The transformations on the phase space of classical general relativity corresponding to these diffeomorphisms are generated by Hamiltonians H_ξ. However, the Poisson bracket algebra of these Hamiltonians is not isomorphic to the Lie bracket algebra of the vector fields ξ^a but rather corresponds to a central extension of this algebra. A Virasoro algebra is thereby obtained. Now, it is known that the asymptotic density of states in a conformal field theory based upon a Virasoro algebra is given by a universal expression (the "Cardy formula") that depends only on the Virasoro algebra. For the Virasoro algebra obtained by Carlip, the Cardy formula yields an entropy in agreement with Eq. (9). Since the Hamiltonians, H_ξ, are closely related to the corresponding Noether currents and charges occurring in the derivation of Eqs. (8) and (9), Carlip's approach holds out the possibility of providing a direct, general explanation of the remarkable agreement between the string theory state counting results and the classical formula for the entropy of a black hole.

6.   Open Issues

The results described in the previous sections provide a remarkably compelling case that stationary black holes are localized thermal equilibrium states of the quantum gravitational field, and that the laws of black hole mechanics are simply the ordinary laws of thermodynamics applied to a system containing a black hole. Although no results on black hole thermodynamics have been subject to any experimental or observational tests, the theoretical foundation of black hole thermodynamics appears to be sufficiently firm as to provide a solid basis for further research and speculation on the nature of quantum gravitational phenomena. In this section, I will briefly discuss two key unresolved issues in black hole thermodynamics which may shed considerable further light upon quantum gravitational physics.

6.1   Does a pure quantum state evolve to a mixed state in the process of black hole formation and evaporation?

In classical general relativity, the matter responsible for the formation of a black hole propagates into a singularity lying within the deep interior of the black hole. Suppose that the matter which forms a black hole possesses quantum correlations with matter that remains far outside of the black hole. Then it is hard to imagine how these correlations could be restored during the process of black hole evaporation unless gross violations of causality occur. In fact, the semiclassical analyses of the Hawking process show that, on the contrary, correlations between the exterior and interior of the black hole are continually built up as it evaporates (see [15] for further discussion). Indeed, these correlations play an essential role in giving the Hawking radiation an exactly thermal character [30].

As already mentioned in subsection 4.1 above, an isolated black hole will "evaporate" completely via the Hawking process within a finite time. If the correlations between the inside and outside of the black hole are not restored during the evaporation process, then by the time that the black hole has evaporated completely, an initial pure state will have evolved to a mixed state, i.e., "information" will have been lost. In a semiclassical analysis of the evaporation process, such information loss does occur and is ascribable to the propagation of the quantum correlations into the singularity within the black hole. A key unresolved issue in black hole thermodynamics is whether this conclusion continues to hold in a complete quantum theory of gravity. On one hand, arguments can be given [15] that alternatives to information loss – such as the formation of a high entropy "remnant" or the gradual restoration of correlations during the late stages of the evaporation process – seem highly implausible. On the other hand, it is commonly asserted that the evolution of an initial pure state to a final mixed state is in conflict with quantum mechanics. For this reason, the issue of whether a pure state can evolve to a mixed state in the process of black hole formation and evaporation is usually referred to as the "*black hole information paradox*".

There appear to be two logically independent grounds for the claim that the evolution of an initial pure state to a final mixed state is in conflict with quantum mechanics:

i Such evolution is asserted to be incompatible with the fundamental principles of quantum theory, which postulates a unitary time evolution of a state vector in a Hilbert space.

ii Such evolution necessarily gives rise to violations of causality and/or energy-momentum conservation and, if it occurred in the black hole formation and evaporation process, there would be large violations of causality and/or energy-momentum (via processes involving "virtual black holes") in ordinary laboratory physics.

With regard to (1), within the semiclassical framework, the evolution of an initial pure state to a final mixed state in the process of black hole formation and evaporation can be attributed to the fact that the final time slice fails to be a Cauchy surface for the spacetime [15]. No violation of any of the local laws of quantum field theory occurs. In fact, a closely analogous evolution of an initial pure state to a final mixed state occurs for a free, massless field in Minkowski spacetime if one chooses the final "time" to be a hyperboloid rather than a hyperplane [15]. (Here, the "information loss" occurring during the time evolution results from radiation to infinity rather than into a black hole.) Indeed, the evolution of an initial pure state to a final mixed state is naturally accommodated within the framework of the algebraic approach to quantum theory [15] as well as in the framework of generalized quantum theory [99].

The main arguments for (2) were given in [100] (see also [101]). However, these arguments assume that the effective evolution law governing laboratory physics has a "Markovian" character, so that it is purely local in time. As pointed out in [102], one would expect a black hole to retain a "memory" (stored in its external gravitational field) of its energy-momentum, so it is far from clear that an effective evolution law modeling the process of black hole formation and evaporation should be Markovian in nature. Furthermore, even within the Markovian context, it is not difficult to construct models where rapid information loss occurs at the Planck scale, but negligible deviations from ordinary dynamics occur at laboratory scales [102].

For the above reasons, I do not feel that the issue of whether a pure state evolves to a mixed state in the process of black hole formation and evaporation should be referred to as a "paradox". Nevertheless, the resolution of this issue is of great importance: If pure states remain pure, then our basic understanding of black holes in classical and semiclassical gravity will have to undergo significant revision in quantum gravity. On the other hand, if pure states evolve to mixed states in a fully quantum treatment of the gravitational field, then at least the aspect of the classical singularity as a place where "information can get lost" must continue to remain present in quantum gravity. In that case, rather than "smooth out" the singularities of classical general relativity, one might expect singularities to play a fundamental role in the formulation of quantum gravity [103]. Thus, the resolution of this issue would tell us a great deal about both the nature of black holes and the existence of singularities in quantum gravity.

6.2   What (and where) are the degrees of freedom responsible for black hole entropy? ·

The calculations described in section 5 yield a seemingly contradictory picture of the degrees of freedom responsible for black hole entropy. In the entanglement entropy and thermal atmosphere approaches, the relevant degrees of freedom are those associated with the ordinary degrees of freedom of quantum fields outside of the black hole. However, the dominant contribution to these degrees of freedom comes from (nearly) Planck scale modes localized to (nearly) a Planck length of the black hole, so, effectively, the relevant degrees of freedom are associated with the horizon. In the quantum geometry approach, the relevant degrees of freedom are also associated with the horizon but appear to have a different character in that they reside directly on the horizon (although they are constrained by the exterior state). Finally the string theory calculations involve weak coupling states, so it is not clear what the degrees of freedom of these weak coupling states would correspond to in a low energy limit where these states may admit a black hole interpretation. However, there is no indication in the calculations that these degrees of freedom should be viewed as being localized near the black hole horizon.

The above calculations are not necessarily in conflict with each other, since it is possible that they each could represent a complementary aspect of the same physical degrees of freedom. Nevertheless, it seems far from clear as to whether we should think of these degrees of freedom as residing outside of the black hole (e.g., in the thermal atmosphere), on the horizon (e.g., in Chern-Simons states), or inside the black hole (e.g., in degrees of freedom associated with what classically corresponds to the singularity deep within the black hole).

The following puzzle [104] may help bring into focus some of the issues related to the degrees of freedom responsible for black hole entropy and, indeed, the meaning of entropy in quantum gravitational physics. As we have already discussed, one proposal for accounting for black hole entropy is to attribute it to the ordinary entropy of its thermal atmosphere. If one does so, then, as previously mentioned in section 5 above, one has the major puzzle of explaining why the quantum field degrees of freedom near the horizon contribute enormously to entropy, whereas the similar degrees of freedom that are present throughout the universe – and are locally indistinguishable from the thermal atmosphere – are treated as mere "vacuum fluctuations" which do not contribute to entropy. But perhaps an even greater puzzle arises if we assign a negligible entropy to the thermal atmosphere (as compared with the black hole area, A), as would be necessary if we wished to attribute black hole entropy to other degrees of freedom. Consider a black hole enclosed in a reflecting cavity which has come to equilibrium with its Hawking radiation. Surely, far from the black hole, the thermal atmosphere in the cavity must contribute an entropy given by the usual formula for a thermal gas in (nearly) flat spacetime. However, if the thermal atmosphere is to contribute a negligible total entropy (as compared with A), then at some proper distance D from the horizon much greater than the Planck length, the thermal atmosphere must contribute to the entropy an amount that is much less than the usual result ($\propto T^3$) that would be obtained by a naive counting of modes. If that is the case, then consider a box of ordinary thermal matter at infinity whose energy is chosen so that its floating point would be less than this distance D from the horizon. Let us now slowly lower the box to its floating point. By the time it reaches its floating point, the contents of the box are indistinguishable from the thermal atmosphere, so the entropy within the box also must be less than what would be obtained by usual mode counting arguments. It follows that the entropy within the box must have decreased during the lowering process, despite the fact that an observer inside the box still sees it filled with thermal radiation and would view the lowering process as having been adiabatic. Furthermore, suppose one lowers (or, more accurately, pushes) an empty box to the same distance from the black hole. The entropy difference between the empty box and the box filled with radiation should still be given by the usual mode counting formulas. Therefore, the empty box would have to be assigned a negative entropy.

I believe that in order to gain a better understanding of the degrees of freedom responsible for black hole entropy, it will be necessary to achieve a deeper understanding of the notion of entropy itself. Even in flat spacetime, there is far from universal agreement as to the meaning of entropy – particularly in quantum theory – and as to the nature of the second law of thermodynamics. The situation in general relativity is considerably murkier [105], as, for example, there is no unique, rigid notion of "time translations" and classical general relativistic dynamics appears to be incompatible with any notion of "ergodicity". It seems likely that a new conceptual framework will be required in order to have a proper understanding of entropy in quantum gravitational physics.

Acknowledgments

This research was supported in part by NSF grant PHY 95-14726 to the University of Chicago.

Notes

1. If the black hole is rotating, the the spectrum seen by an observer at infinity corresponds to what would emerge from a "rotating black body".

2. The state in which none of the modes in the region exterior to the black hole are populated is usually referred to as the "Boulware vacuum". The Boulware vacuum is singular on both the black hole and white hole horizons.

3. It is worth noting that if the buoyancy effects of the thermal atmosphere were negligible, the bound (16) also would not be sufficient to ensure the validity of the GSL for non-spherical bodies: The bound (16) is formulated in terms of the "circumscribing radius", i.e., the *largest* linear dimension, whereas if buoyancy effects were negligible, then to enforce the GSL one would need a bound of the form (16) with R being the *smallest* linear dimension.

References

[1] R.M. Wald, *General Relativity*, University of Chicago Press (Chicago, 1984).

[2] S.W. Hawking and G.F.R. Ellis, *The Large Scale Structure of Spacetime*, Cambridge University Press (Cambridge, 1973).

[3] P. T. Chrusciel, E. Delay. G.J. Galloway, and R. Howard, "The Area Theorem", gr-qc/0001003.

[4] S.W. Hawking, "Gravitational Radiation from Colliding Black Holes", Phys. Rev. Lett. **26**, 1344-1346 (1971).

[5] J.D. Bekenstein, "Black Holes and Entropy", Phys. Rev. **D7**, 2333-2346 (1973).

[6] J.D. Bekenstein, "Generalized Second Law of Thermodynamics in Black-Hole Physics", Phys. Rev. **D9**, 3292-3300 (1974).

[7] J.M. Bardeen, B. Carter, and S.W. Hawking, "The Four Laws of Black Hole Mechanics" Commun. Math. Phys. **31**, 161-170 (1973).

[8] M. Heusler, *Black Hole Uniqueness Theorems*, Cambridge University Press (Cambridge, 1996).

[9] B. Carter, "Black Hole Equilibrium States" in *Black Holes*, ed. by C. DeWitt and B.S. DeWitt, 57-214, Gordon and Breach (New York, 1973).

[10] H. Friedrich, I. Racz, and R.M. Wald, "On the Rigidity Theorem for Spacetimes with a Stationary Event Horizon or a Compact Cauchy Horizon", Commun. Math. Phys. **204**, 691-707 (1999); gr-qc/9811021.

[11] D. Sudarsky and R.M. Wald, "Extrema of Mass, Stationarity and Staticity, and Solutions to the Einstein-Yang-Mills Equations" Phys. Rev. **D46**, 1453-1474 (1992).

[12] D. Sudarsky and R.M. Wald, "Mass Formulas for Stationary Einstein-Yang-Mills Black Holes and a Simple Proof of Two Staticity Theorems" Phys. Rev. **D47**, R5209-R5213 (1993).

[13] P.T. Chrusciel and R.M. Wald, "Maximal Hypersurfaces in Stationary Asymptotically Flat Spacetimes" Commun. Math Phys. **163**, 561-604 (1994).

[14] I. Racz and R.M. Wald, "Global Extensions of Spacetimes Describing Asymptotic Final States of Black Holes" Class. Quant. Grav. **13**, 539-552 (1996); gr-qc/9507055.

[15] R.M. Wald, *Quantum Field Theory in Curved Spacetime and Black Hole Thermodynamics*, University of Chicago Press (Chicago, 1994).

[16] V. Iyer and R.M. Wald, "Some Properties of Noether Charge and a Proposal for Dynamical Black Hole Entropy", Phys. Rev. **D50**, 846-864 (1994).

[17] R. Sorkin, "Two Topics Concerning Black Holes: Extremality of the Energy, Fractality of the Horizon" in *Proceedings of the Conference on Heat Kernel Techniques and Quantum Gravity*, ed. by S.A. Fulling, 387-407, University of Texas Press, (Austin, 1995); gr-qc/9508002.

[18] W. Israel, "Third Law of Black-Hole Dynamics: a Formulation and Proof", Phys. Rev. Lett. **57**, 397-399 (1986).

[19] M. Aizenman and E.H. Lieb, "The Third Law of Thermodynamics and the Degeneracy of the Ground State for Lattice Systems", J. Stat. Phys. **24**, 279-297 (1981).

[20] R.M. Wald, "'Nernst Theorem' and Black Hole Thermodynamics", Phys. Rev. **D56**, 6467-6474 (1997); gr-qc/9704008.

[21] A. Ashtekar, C. Beetle, O. Dreyer, S. Fairhurst, B. Krishnan, J. Lewandowski, and J. Wisniewski, "Generic Isolated Horizons and Their Applications", gr-qc/0006006.

[22] A. Ashtekar, C. Beetle, and S. Fairhurst, "Isolated Horizons: A Generalization of Black Hole Mechanics", Class. Quant. Grav. **16**, L1-L7 (1999); gr-qc/9812065.

[23] A. Ashtekar, C. Beetle, and S. Fairhurst, "Mechanics of Isolated Horizons", Class. Quant. Grav. **17**, 253-298 (2000); gr-qc/9907068.

[24] A. Ashtekar and A. Corichi, "Laws Governing Isolated Horizons: Inclusion of Dilaton Couplings", Class. Quant. Grav. **17**, 1317-1332 (2000); gr-qc/9910068.

[25] J. Lewandowski, "Spacetimes Admitting Isolated Horizons", Class. Quant. Grav. **17**, L53-L59 (2000); gr-qc/9907058.

[26] A. Ashtekar, S. Fairhurst, and B. Krishnan, "Isolated Horizons: Hamiltonian Evolution and the First Law", gr-qc/0005083.

[27] A. Corichi, U. Nucamendi, and D. Sudarsky, "Einstein-Yang-Mills Isolated Horizons: Phase Space, Mechanics, Hair and Conjectures" Phys. Rev. **D62**, 044046 (19 pages) (2000); gr-qc/0002078.

[28] S.W. Hawking, "Particle Creation by Black Holes", Commun. Math. Phys. **43**, 199-220 (1975).

[29] L. Parker, "Quantized Fields and Particle Creation in Expanding Universes", Phys. Rev. **183**, 1057-1068 (1969).

[30] R.M. Wald, "On Particle Creation by Black Holes", Commun. Math. Phys. **45**, 9-34 (1975).

[31] K. Fredenhagen and R. Haag, "On the Derivation of the Hawking Radiation Associated with the Formation of a Black Hole", Commun. Math. Phys. **127**, 273-284 (1990).

[32] W.G. Unruh, "Experimental Black-Hole Evaporation?", Phys. Rev. Lett. **46**, 1351-1353 (1981).

[33] W.G. Unruh, "Dumb Holes and the Effects of High Frequencies on Black Hole Evaporation" Phys. Rev. **D51**, 2827-2838 (1995); gr-qc/9409008.

[34] R. Brout, S. Massar, R. Parentani, and Ph. Spindel, "Hawking Radiation Without Transplanckian Frequencies", Phys. Rev. **D52**, 4559-4568 (1995); hep-th/9506121.

[35] S. Corley and T. Jacobson, "Hawking Spectrum and High Frequency Dispersion" Phys. Rev. **D54**, 1568-1586 (1996); hep-th/9601073.

[36] T. Jacobson, "On the Origin of the Outgoing Black Hole Modes" Phys. Rev. **D53**, 7082-7088 (1996); hep-th/9601064.

[37] B. Reznik, "Trans-Planckian Tail in a Theory with a Cutoff", Phys. Rev. **D55**, 2152-2158 (1997); gr-qc/9606083.

[38] M. Visser, "Hawking radiation without black hole entropy", Phys. Rev. Lett. **80**, 3436-3439 (1998); gr-qc/9712016.

[39] S. Corley and T. Jacobson, "Lattice Black Holes", Phys. Rev. **D57**, 6269-6279 (1998); hep-th/9709166.

[40] T. Jacobson and D. Mattingly, "Hawking radiation on a falling lattice", Phys. Rev. **D61** 024017 (10 pages) (2000); hep-th/9908099.

[41] W.G. Unruh, "Notes on Black Hole Evaporation", Phys. Rev. **D14**, 870-892 (1976).

[42] B.S. Kay and R.M. Wald, "Theorems on the Uniqueness and Thermal Properties of Stationary, Nonsingular, Quasifree States on Spacetimes with a Bifurcate Killing Horizon", Phys. Rep. **207**, 49-136 (1991).

[43] J.J. Bisognano and E.H. Wichmann, "On the Duality Condition for Quantum Fields", J. Math. Phys. **17**, 303-321 (1976).

[44] J.B. Hartle and S.W. Hawking, "Path Integral Derivation of Black Hole Radiance", Phys. Rev. **D13**, 2188-2203 (1976).

[45] R. Geroch, colloquium given at Princeton University, December, 1971 (unpublished).

[46] J.D. Bekenstein, "Universal Upper Bound on the Entropy-to-Energy Ratio for Bounded Systems", Phys. Rev. **D23**, 287-298 (1981).

[47] W.G. Unruh and R.M. Wald, "Acceleration Radiation and the Generalized Second Law of Thermodynamics", Phys. Rev. **D25**, 942-958 (1982).

[48] W.H. Zurek and K.S. Thorne, "Statistical Mechanical Origin of the Entropy of a Rotating, Charged Black Hole", Phys. Rev. Lett. **54**, 2171-2175 (1986).

[49] K.S. Thorne, W.H. Zurek, and R.H. Price, "The Thermal Atmosphere of a Black Hole", in *Black Holes: The Membrane Paradigm*, ed. by K.S. Thorne, R.H. Price, and D.A. Macdonald, 280-340, Yale University Press (New Haven, 1986).

[50] V.P. Frolov and D.N. Page, "Proof of the Generalized Second Law for Quasistatic, Semiclassical Black Holes", Phys. Rev. Lett. **71**, 3902-3905 (1993).

[51] R.D. Sorkin, "The Statistical Mechanics of Black Hole Thermodynamics", in *Black Holes and Relativistic Stars*, ed. by R.M. Wald, 177-194, University of Chicago Press (Chicago, 1998); gr-qc/9705006.

[52] D.N. Page, "Defining Entropy Bounds", hep-th/0007238.

[53] J.D. Bekenstein, "On Page's Examples Challenging the Entropy Bound", gr-qc/0006003.

[54] D.N. Page, "Huge Violations of Bekenstein's Entropy Bound", gr-qc/0005111.

[55] D.N. Page, "Subsystem Entropy Exceeding Bekenstein's Bound", hep-th/0007237.

[56] J.D. Bekenstein, "Entropy Content and Information Flow in Systems with Limited Energy", Phys. Rev. **D30**, 1669-1679 (1984).

[57] J.D. Bekenstein and M. Schiffer, "Quantum Limitations on the Storage and Transmission of Information", Int. J. Mod. Phys. **C1**, 355 (1990).

[58] C.W. Misner, K.S. Thorne, and J.A. Wheeler, *Gravitation*, Freeman (San Francisco, 1973).

[59] R. Penrose, "Quasi-Local Mass and Angular Momentum ", Proc. Roy. Soc. Lond. **A381**, 53-63 (1982).

[60] J.D. Brown and J.W. York, "Quasilocal Energy and Conserved Charges Derived from the Gravitational Action", Phys. Rev. **D47**, 1407-1419 (1993).

[61] R.D. Sorkin, R.M. Wald, and Z.J. Zhang, "Entropy of Self-Gravitating Radiation", Gen. Rel. Grav. **13**, 1127-1146 (1981).

[62] J.D. Bekenstein, "Entropy Bounds and the Second Law for Black Holes", Phys. Rev. **D27**, 2262-2270 (1983).

[63] J.D. Bekenstein, "Entropy Bounds and Black Hole Remnants", Phys. Rev. **D49**, 1912-1921 (1994).

[64] J.D. Bekenstein, "Non-Archimedian Character of Quantum Buoyancy and the Generalized Second Law of Thermodynamics", Phys. Rev. **D60**, 124010 (9 pages) (1999); gr-qc/9906058.

[65] W.G. Unruh and R.M. Wald, "Entropy Bounds, Acceleration Radiation and the Generalized Second Law", Phys. Rev. **D27**, 2271-2276 (1983).

[66] M.A. Pelath and R.M.Wald, "Comment on Entropy Bounds and the Generalized Second Law", Phys. Rev. **D60**, 104009 (4 pages) (1999); gr-qc/9901032.

[67] E.E. Flanagan, D. Marolf, and R.M. Wald, "Proof of Classical Versions of the Bousso Entropy Bound and of the Generalized Second Law" Phys. Rev. **D62**, 084035 (11 pages) (2000); hep-th/9909373

[68] W. Anderson, "Does the GSL Imply and Entropy Bound?", in *Matters of Gravity*, ed. by J. Pullin, gr-qc/9909022.

[69] G. 't Hooft, "On the Quantization of Space and Time", in *Quantum Gravity*, ed. by M.A. Markov, V.A. Berezin, and V.P. Frolov, 551-567, World Scientific Press (Singapore, 1988).

[70] J.D. Bekenstein, "Do We Understand Black Hole Entropy ?", in *Proceedings of the VII Marcel Grossman Meeting*, 39-58, World Scientific Press (Singapore, 1996); gr-qc/9409015.

[71] L. Susskind, "The World as a Hologram", J. Math. Phys. **36**, 6377-6396 (1995); hep-th/9409089.

[72] R. Bousso, "A Covariant Entropy Conjecture", JHEP **07**, 004 (1999); hep-th/9905177.

[73] R. Bousso, "Holography in General Space-times", JHEP **06**, 028 (1999); hep-th/9906022.

[74] R. Bousso, "The Holographic Principle for General Backgrounds", hep-th/9911002.

[75] G. Gibbons and S.W. Hawking "Action Integrals and Partition Functions in Quantum Gravity", Phys. Rev. **D15**, 2752-2756 (1977).

[76] J.D. Brown and J.W. York, "Microcanonical Functional Integral for the Gravitational Field", Phys. Rev. **D47**, 1420-1431 (1993).

[77] R.M. Wald, "Black Hole Entropy is the Noether Charge", Phys. Rev. **D48**, R3427-R3431 (1993).

[78] V. Iyer and R.M. Wald, "A Comparison of Noether Charge and Euclidean Methods for Computing the Entropy of Stationary Black Holes", Phys. Rev. **D52**, 4430-4439 (1995); gr-qc/9503052.

[79] L. Bombelli, R.K. Koul, J. Lee, and R. Sorkin, "Quantum Source of Entropy for Black Holes" Phys. Rev. **D34**, 373-383 (1986).

[80] C. Callen and F. Wilzcek, "On Geometric Entropy", Phys. Lett **B333**, 55-61 (1994).

[81] C. Holzhey, F. Larsen, and F. Wilzcek, "Geometric and Renormalized Entropy in Conformal Field Theory", Nucl. Phys. **B424**, 443-467 (1994).

[82] L. Susskind and J. Uglam, "Black Hole Entropy in Canonical Quantum Gravity and Superstring Theory", Phys. Rev. **D50**, 2700-2711 (1994).

[83] R. Sorkin, "How Wrinkled is the Surface of a Black Hole?" in *Proceedings of the First Australasian Conference on General Relativity and Gravitation*, ed. by D. Wiltshire, 163-174, University of Adelaide Press, (Adelaide, 1996); gr-qc/9701056.

[84] D. Dou, "Causal Sets, a Possible Interpretation for the Black Hole Entropy, and Related Topics", Ph.D. thesis (SISSA, Trieste, 1999).

[85] G. 't Hooft, "On the Quantum Structure of a Black Hole", Nucl. Phys. **B256**, 727-745 (1985).

[86] S. Mukohyama, "Aspects of Black Hole Entropy", gr-qc/9912103.

[87] V.P. Frolov, D.V. Fursaev and A.I. Zelnikov, "Statistical Origin of Black Hole Entropy in Induced Gravity", Nucl. Phys. **B486**, 339-352 (1997); hep-th/9607104.

[88] V.P. Frolov and D.V. Fursaev, "Mechanism of the Generation of Black Hole Entropy in Sakharov's Induced Gravity", Phys. Rev. **D56**, 2212-2225 (1997); hep-th/9703178.

[89] A. Ashtekar, J. Baez, A. Corichi, and K. Krasnov, "Quantum Geometry and Black Hole Entropy", Phys. Rev. Lett. **80**, 904-907 (1998); gr-qc/9710007.

[90] A. Ashtekar and K. Krasnov, "Quantum Geometry and Black Holes", in *Black Holes, Gravitational Radiation, and the Universe*, ed. by B.R. Iyer and B. Bhawal, 149-170, Kluwer Academic Publishers (Dordrecht, 1999); gr-qc/9804039.

[91] A. Ashtekar, A. Corichi, and K. Krasnov, "Isolated Horizons: the Classical Phase Space", gr-qc/9905089.

[92] D. Marolf, "String/M-branes for Relativists", gr-qc/9908045.

[93] G. Horowitz, "Quantum States of Black Holes", in *Black Holes and Relativistic Stars*, ed. by R.M. Wald, 241-266, University of Chicago Press (Chicago, 1998); gr-qc/9704072.

[94] A. Peet, "TASI Lectures on Black Holes in String Theory", hep-th/0008241.

[95] G.L. Cardoso, B. de Wit, and T. Mohaupt, "Area Law Corrections from State Counting and Supergravity", Class. Quant. Grav. **17** 1007-1015 (2000); hep-th/9910179.

[96] J.M. Maldacena and A. Strominger, "Black Hole Greybody Factors and D-Brane Spectroscopy", Phys.Rev. **D55**, 861-870 (1997); hep-th/9609026.

[97] S. Carlip, "Entropy from Conformal Field Theory at Killing Horizons", Class. Quant. Grav. **16**, 3327-3348 (1999); gr-qc/9906126.

[98] S. Carlip, "Black Hole Entropy from Horizon Conformal Field Theory", gr-qc/9912118.

[99] J. Hartle, "Generalized Quantum Theory in Evaporating Black Hole Spacetimes", in *Black Holes and Relativistic Stars*, ed. by R.M. Wald, 195-219, University of Chicago Press (Chicago, 1998); gr-qc/9705022.

[100] T. Banks, L. Susskind, and M.E. Peskin, "Difficulties for the Evolution of Pure States into Mixed States", Nucl. Phys. **B244**, 125-134 (1984).

[101] J. Ellis, J.S. Hagelin, D.V. Nanopoulos, and M. Srednicki, "Search for Violations of Quantum Mechanics", Nucl. Phys. **B241**, 381-405 (1984).

[102] W.G. Unruh and R.M. Wald, "Evolution Laws Taking Pure States to Mixed States in Quantum Field Theory", Phys. Rev. **D52**, 2176-2182 (1995); hep-th/9503024.

[103] R. Penrose, "Singularities and Time-Asymmetry" in *General Relativity, an Einstein Centennary Survey*, ed. by S.W. Hawking and W. Israel, 581-638, Cambridge University Press (Cambridge, 1979).

[104] R.M. Wald, "Gravitation, Thermodynamics, and Quantum Theory", Class. Quant. Grav. **16**, A177-A190 (1999); gr-qc/9901033.

[105] R.M. Wald, "Black Holes and Thermodynamics", in *Black Holes and Relativistic Stars*, ed. by R.M. Wald, 155-176, University of Chicago Press (Chicago, 1998); gr-qc/9702022.

INTRODUCTION TO QUANTUM FIELDS IN CURVED SPACETIME AND THE HAWKING EFFECT

Ted Jacobson

Department of Physics, University of Maryland

College Park, MD 20742-4111

jacobson@physics.umd.edu

Abstract   These notes introduce the subject of quantum field theory in curved spacetime and some of its applications and the questions they raise. Topics include particle creation in time-dependent metrics, quantum origin of primordial perturbations, Hawking effect, the trans-Planckian question, and Hawking radiation on a lattice.

1.   Introduction

Quantum gravity remains an outstanding problem of fundamental physics. The bottom line is we don't even know the nature of the system that should be quantized. The spacetime metric may well be just a collective description of some more basic stuff. The fact[1] that the semi-classical Einstein equation can be derived by demanding that the first law of thermodynamics hold for local causal horizons, assuming the proportionality of entropy and area, leads one to suspect that the metric is only meaningful in the thermodynamic limit of something else. This led me at first to suggest that the metric shouldn't be quantized at all. However I think this is wrong. Condensed matter physics abounds with examples of collective modes that become meaningless at short length scales, and which are nevertheless accurately treated as quantum fields within the appropriate domain. (Consider for example the sound field in a Bose-Einstein condensate of atoms, which loses meaning at scales below the so-called "healing length", which is still several orders of magnitude longer than the atomic size of the fundamental constituents.) Similarly, there exists a perfectly good perturbative approach to quantum gravity in the framework of low energy effective field theory[2]. However, this is not regarded as a solution to the problem of quantum gravity, since the most pressing questions are non-perturbative in nature: the nature and fate of spacetime singularities, the fate of

Cauchy horizons, the nature of the microstates counted by black hole entropy, and the possible unification of gravity with other interactions.

At a shallower level, the perturbative approach of effective field theory is nevertheless relevant both for its indications about the deeper questions and for its application to physics phenomena in their own right. It leads in particular to the subject of quantum field theory in curved spacetime backgrounds, and the "back-reaction" of the quantum fields on such backgrounds. Some of the most prominent of these applications are the Hawking radiation by black holes, primordial density perturbations, and early universe phase transitions. It also fits into the larger category of quantum field theory (qft) in inhomogeneous and/or time-dependent backgrounds of other fields or matter media, and is also intimately tied to non-inertial effects in flat space qft such as the Unruh effect. The pertubative approach is also known as "semi-classical quantum gravity", which refers to the setting where there is a well-defined classical background geometry about which the quantum fluctuations are occuring.

The present notes are an introduction to some of the essentials and phenomena of quantum field theory in curved spacetime. Familiarity with quantum mechanics and general relativity are assumed. Where computational steps are omitted I expect that the reader can fill these in as an exercise.

Given the importance of the subject, it is curious that there are not very many books dedicated to it. The standard reference by Birrell and Davies[3] was published twenty years ago, and another monograph by Grib, Mamaev, and Mostapanenko[4], half of which addresses strong background field effects in flat spacetime, was published two years earlier originally in Russian and then in English ten years ago. Two books with a somewhat more limited scope focusing on fundamentals with a mathematically rigorous point of view are those by Fulling[5] and Wald[6]. This year DeWitt[7] published a comprehensive two volume treatise with a much wider scope but including much material on quantum fields in curved spacetime. A number of review articles (see e.g. [8, I76, 10, 11, 12, 13]) and many shorter introductory lecture notes (see e.g. [14, 15, 16]) are also available. For more information on topics not explicitly referenced in the text of these notes the above references should be consulted.

In these notes the units are chosen with $c = 1$ but $\hbar$ and G are kept explicit. The spacetime signature is $(+---)$. Please send corrections if you find things here that are wrong.

2.   Planck length and black hole thermodynamics

Thanks to a scale separation it is useful to distinguish quantum field theory in a curved background spacetime (qftcs) from true quantum gravity (qg). Before launching into the qftcs formalism, it seems worthwhile to have a quick look at some of the interesting issues that qftcs is concerned with.

2.1   Planck length

It is usually presumed that the length scale of quantum gravity is the Planck length $L_P = (\hbar G/c^3)^{1/2} \approx 10^{-33}$ cm. The corresponding energy scale is 10^{19} GeV. Recent "braneworld scenarios", in which our 4d world is a hypersurface in a higher dimensional spacetime, put the scale of quantum gravity much lower, at around a TeV, corresponding to $L_{\text{TeV}} = 10^{16} L_P \approx 10^{-17}$ cm. In either case, there is plenty of room for applicability of qftcs. (On the other hand, we are much closer to seeing true qg effects in TeV scale qg. For example we might see black hole creation and evaporation in cosmic rays or accelerators.)

Here I will assume Planck scale qg, and look at some dimensional analysis to give a feel for the phenomena. First, how should we think of the Planck scale? The Hilbert-Einstein action is $S_{HE} = (\hbar/16\pi L_P^2) \int d^4x |g|^{1/2} R$. For a spacetime region with radius of curvature L and 4-volume L^4 the action is $\sim \hbar (L/L_P)^2$. This suggests that quantum curvature fluctuations with radius less than the Planck length $L \lesssim L_P$ are unsuppressed.

Another way to view the significance of the Planck length is as the minimum localization length Δx, in the sense that if $\Delta x < L_P$ a black hole swallows the Δx. To see this, note that the uncertainty relation $\Delta x \Delta p \geq \hbar/2$ implies $\Delta p \gtrsim \hbar/\Delta x$ which implies $\Delta E \gtrsim \hbar c/\Delta x$. Associated with this uncertain energy is a Schwarzschild radius $R_s(\Delta x) = 2G\Delta M/c^2 = 2G\Delta E/c^4$, hence quantum mechanics and gravity imply $R_s(\Delta x) \gtrsim L_P^2/\Delta x$. The uncertain $R_s(\Delta x)$ is less than Δx only if $\Delta x \gtrsim L_P$.

2.2   Hawking effect

Before Hawking, spherically symmetric, static black holes were assumed to be completely inert. In fact, it seems more natural that they can decay, since there is no conservation law preventing that. The decay is quantum mechanical, and thermal: Hawking found that a black hole radiates at a temperature proportional to $\hbar$, $T_H = (\hbar/2\pi)\kappa$, where κ is the surface gravity. The fact that the radiation is thermal is even natural, for *what else could it be*? The very nature of the horizon is a causal barrier to information, and yet the Hawking radiation emerges from just outside the horizon. Hence there can be no information in the Hawking radiation, save for the mass of the black hole which is visible on the outside, so it must be a maximum entropy state, i.e. a thermal state with a temperature determined by the black hole mass.

For a Schwarzschild black hole $\kappa = 1/4GM = 1/2R_s$, so the Hawking temperature is inversely proportional to the mass. This implies a thermal wavelength $\lambda_H = 8\pi^2 R_s$, a purely geometrical relationship which indicates two things. First, although the emission process involves quantum mechanics, its "kinematics" is somehow classical. Second, as long as the Schwarzschild radius is much longer than the Planck length, it should be possible to understand the Hawking effect using only qftcs, i.e. semi-classical qg. (Actually one must also require that the back-reaction is small, in the sense that the change in the Hawking temperature due to the emission of a single Hawking quantum is small. This fails to hold for for a very nearly extremal black hole[17].)

A Planck mass ($\sim 10^{-5}$ gm) black hole—if it could be treated semi-classically— would have a Schwarzschild radius of order the Planck length ($\sim 10^{-33}$ cm) and a Hawking temperature of order the Planck energy ($\sim 10^{19}$ GeV). From this the Hawking temperatures for other black holes can be found by scaling. A solar mass black hole has a Schwarzschild radius $R_s \sim 3$ km hence a Hawking temperature $\sim 10^{-38}$ times smaller than the Planck energy, i.e. 10^{-19} GeV. Evaluated more carefully it works out to $T_H \sim 10^{-7}$ K. For a mini black hole of mass $M = 10^{15}$ gm one has $R_s \sim 10^{-13}$ cm and $T_H \sim 10^{11}$K ~ 10 MeV.

The "back reaction", i.e. the response of the spacetime metric to the Hawking process, should be well approximated by the semi-classical Einstein equation $G_{\mu\nu} = 8\pi G \langle T_{\mu\nu} \rangle$ provided it is a small effect. To assess the size, we can compare the stress tensor to the background curvature near the horizon (not to $G_{\mu\nu}$, since that vanishes in the background). The background Riemann tensor components are $\sim 1/R_s^2$ (in suitable freely falling reference frame), while for example the energy density is $\sim T_H^4/\hbar^3 \sim \hbar/R_s^4$. Hence $G\langle T_{\mu\nu} \rangle \sim \hbar G/R_s^4 = (L_P/R_s)^2 R_s^{-2}$, which is much less than the background curvature provided $R_s \gg L_P$, i.e. provided the black hole is large compared to the Planck length.

Although a tiny effect for astrophysical black holes, the Hawking process has a profound implication: black holes can "evaporate". How long does one take to evaporate? It emits roughly one Hawking quantum per light crossing time, hence $dM/dt \sim T_H/R_S \sim \hbar/R_S^2 \sim \hbar/G^2 M^2$. In Planck units $\hbar = c = G = 1$, we have $dM/dt \sim M^{-2}$. Integration yields the lifetime $\sim M^3 \sim (R_s/L_P)^2 R_S$. For the 10^{15} gm, 10^{-13} cm black hole mentioned earlier we have $(10^{20})^2 10^{-13}$cm $= 10^{27}$ cm, which is the age of the universe. Hence a black hole of that mass in the early universe would be explosively ending its life now. These are called "primordial black holes" (pbh's). None have been knowingly observed so far, nor their detritus, which puts limits on the present density of pbh's (see for example the references in[18]). However, it has been suggested[18] that they might nevertheless be the source of the highest energy ($\sim 3 \times 10^{20}$ eV) cosmic rays. Note that even if pbh's were copiously produced in the early universe, their initial number density could easily have been inflated away if they formed before inflation.

2.3   Black hole entropy

How about the black hole entropy? The thermodynamic relation $dM = T_H dS = (\hbar/8\pi GM)dS$ (for a black hole with no angular momentum or charge) implies $S_{BH} = 4\pi GM^2/\hbar = A_H/4L_P^2$, where $A_H = 4\pi R_S^2$ is the black hole horizon area. What is this huge entropy? What microstates does it count?

Whatever the microstates may be, S_{BH} is the lower bound for the entropy emitted to the outside world as the black hole evaporates, i.e. the minimal missing information brought about by the presence of the black hole. By sending energy into a black hole one could make it last forever, emitting an arbitrarily large amount of entropy, so it only makes sense to talk about the minimal entropy. This lower bound is attained when the black hole evaporates reversibly into a thermal bath at a temperature infinitesimally below T_H. (Evaporation into vacuum is irreversible, so the total entropy of the outside increases even more[19], $S_{\text{emitted}} \sim (4/3)S_{BH}$.) This amounts to a huge entropy increase. The only way out of this conclusion would be if the semi-classical analysis breaks down for some reason...which is suspected by many, not including me.

The so-called "information paradox" refers to the loss of information associated with this entropy increase when a black hole evaporates, as well as to the loss of any other information that falls into the black hole. I consider it no paradox at all, but many think it is a problem to be avoided at all costs. I think this viewpoint results from missing the strongly non-perturbative role of quantum gravity in evolving the spacetime into and beyond the classically singular region, whether by generating baby universes or otherwise (see section 6.3.12 for more discussion of this point). Unfortunately it appears unlikely that semi-classical qg can "prove" that information is or is not lost, though people have tried hard. For it not to be lost there would have to be subtle effects not captured by semi-classical qg, yet in a regime where semi-classical description "should" be the whole story at leading order.

3.   Harmonic oscillator

One can study a lot of interesting issues with free fields in curved spacetime, and I will restrict entirely to that case except for a brief discussion. Free fields in curved spacetime are similar to collections of harmonic oscillators with time-dependent frequencies. I will therefore begin by developing the properties of a one-dimensional quantum harmonic oscillator in a formalism parallel to that used in quantum field theory.

The action for a particle of mass m moving in a time dependent potential $V(x, t)$ in one dimension takes the form

$$S = \int dt\, L \qquad\qquad L = \frac{1}{2}m\dot{x}^2 - V(x, t) \qquad\qquad (1)$$

from which follows the equation of motion $m\ddot{x} = -\partial_x V(x, t)$. Canonical quantization proceeds by (i) defining the momentum conjugate to x, $p = \partial L/\partial\dot{x} = m\dot{x}$, (ii) replacing x and p by operators $\hat{x}$ and $\hat{p}$, and (iii) imposing the canonical commutation relations $[\hat{x}, \hat{p}] = i\hbar$. The operators are represented as hermitian linear operators on a Hilbert space, the hermiticity ensuring that their spectrum is real as befits a quantity whose classical correspondent is a real number.[1] In the Schrödinger picture the state is time-dependent, and the operators are time-independent. In the position representation for example the momentum operator is given by $\hat{p} = -i\hbar\partial_x$. In the Heisenberg picture the state is time-independent while the operators are time-dependent. The commutation relation then should hold at each time, but this is still really only one commutation relation since the the equation of motion implies that if it holds at one initial time it will hold at all times. In terms of the position and velocity, the commutation relation(s) in the Heisenberg picture take the form

$$[x(t), \dot{x}(t)] = i\hbar/m. \tag{2}$$

Here and from here on the hats distinguishing numbers from operators are dropped.

Specializing now to a harmonic oscillator potential $V(x,t) = \frac{1}{2}m\omega^2(t)x^2$ the equation of motion takes the form

$$\ddot{x} + \omega^2(t)x = 0. \tag{3}$$

Consider now any operator solution $x(t)$ to this equation. Since the equation is second order the solution is determined by the two hermitian operators $x(0)$ and $\dot{x}(0)$, and since the equation is linear the solution is linear in these operators. It is convenient to trade the pair $x(0)$ and $\dot{x}(0)$ for a single time-independent non-hermitian operator a, in terms of which the solution is written as

$$x(t) = f(t)a + \bar{f}(t)a^\dagger, \tag{4}$$

where $f(t)$ is a complex function satisfying the classical equation of motion,

$$\ddot{f} + \omega^2(t)f = 0, \tag{5}$$

$\bar{f}$ is the complex conjugate of f, and $a^\dagger$ is the hermitian conjugate of a. The commutation relations (2) take the form

$$\langle f, f \rangle [a, a^\dagger] = 1, \tag{6}$$

where the bracket notation is defined by

$$\langle f, g \rangle = (im/\hbar)\left(\bar{f}\partial_t g - (\partial_t \bar{f})g\right). \tag{7}$$

If the functions f and g are solutions to the harmonic oscillator equation (5), then the bracket (7) is independent of the time t at which the right hand side is evaluated, which is consistent with the assumed time independence of a.

Let us now assume that the solution f is chosen so that the real number $\langle f, f \rangle$ is positive. Then by rescaling f we can arrange to have

$$\langle f, f \rangle = 1. \tag{8}$$

In this case the commutation relation (6) becomes

$$[a, a^\dagger] = 1, \tag{9}$$

the standard relation for the harmonic oscillator raising and lowering operators. Using the bracket with the operator x we can pluck out the raising and lowering operators from the position operator,

$$a = \langle f, x \rangle, \qquad\qquad a^\dagger = -\langle \bar{f}, x \rangle. \tag{10}$$

Since both f and x satisfy the equation of motion, the brackets in (10) are time independent as they must be.

A Hilbert space representation of the operators can be built by introducing a state $|0\rangle$ defined to be normalized and satisfying $a|0\rangle = 0$. For each n, the state $|n\rangle = (1/\sqrt{n!})(a^\dagger)^n|0\rangle$ is a normalized eigenstate of the number operator $N = a^\dagger a$ with eigenvalue n. The span of all these states defines a Hilbert space of "excitations" above the state $|0\rangle$.

So far the solution $f(t)$ is arbitrary, except for the normalization condition (8). A change in $f(t)$ could be accompanied by a change in a that keeps the solution $x(t)$ unchanged. In the special case of a constant frequency $w(t) = w$ however, the energy is conserved, and a special choice of $f(t)$ is selected if we require that the state $|0\rangle$ be the ground state of the Hamiltonian. Let us see how this comes about.

For a general f we have

$$
\begin{aligned}
H &= \frac{1}{2}m\dot{x}^2 + \frac{1}{2}mw^2 x^2 \tag{11} \\
&= \frac{1}{2}m\Big[(\dot{f}^2 + w^2 f^2)aa + (\dot{f}^2 + w^2 f^2)^* a^\dagger a^\dagger \\
&\quad + (|\dot{f}|^2 + w^2|f|^2)(aa^\dagger + a^\dagger a)\Big]. \tag{12}
\end{aligned}
$$

Thus

$$H|0\rangle = \frac{1}{2}m(\dot{f}^2 + w^2 f^2)^* a^\dagger a^\dagger|0\rangle + (|\dot{f}|^2 + w^2|f|^2)|0\rangle, \tag{13}$$

where the commutation relation (9) was used in the last term. If $|0\rangle$ is to be an eigenstate of H, the first term must vanish, which requires

$$\dot{f} = \pm i w f. \tag{14}$$

For such an f the norm is

$$\langle f, f \rangle = \mp \frac{2m\omega}{\hbar} |f|^2, \tag{15}$$

hence the positivity of the normalization condition (8) selects from (14) the minus sign. This yields what is called the normalized *positive frequency* solution to the equation of motion, defined by

$$f(t) = \sqrt{\frac{\hbar}{2m\omega}} e^{-i\omega t} \tag{16}$$

up to an arbitrary constant phase factor.

With f given by (16) the Hamiltonian (12) becomes

$$H = \frac{1}{2}\hbar\omega(aa^\dagger + a^\dagger a) \tag{17}$$

$$= \hbar\omega(N + \frac{1}{2}), \tag{18}$$

where the commutation relation (9) was used in the last step. The spectrum of the number operator is the non-negative integers, hence the minimum energy state is the one with $N = 0$, and "zero-point energy" $\hbar\omega/2$. This is just the state $|0\rangle$ annihilated by a as defined above. If any function other than (16) is chosen to expand the position operator as in (4), the state annihilated by a is not the ground state of the oscillator.

Note that although the mean value of the position is zero in the ground state, the mean of its square is

$$\langle 0|x^2|0\rangle = \hbar/2m\omega. \tag{19}$$

This characterizes the "zero-point fluctuations" of the position in the ground state.

4.   Quantum scalar field in curved spacetime

Much of interest can be done with a scalar field, so it suffices for an introduction. The basic concepts and methods extend straightforwardly to tensor and spinor fields. To being with let's take a spacetime of arbitrary dimension D, with a metric $g_{\mu\nu}$ of signature $(+ - \cdots -)$. The action for the scalar field φ is

$$S = \int d^D x \sqrt{|g|} \frac{1}{2} \left(g^{\mu\nu}\partial_\mu\varphi\partial_\nu\varphi - (m^2 + \xi R)\varphi^2 \right), \tag{20}$$

for which the equation of motion is

$$(\Box + m^2 + \xi R)\,\varphi = 0, \qquad \Box = |g|^{-1/2}\partial_\mu|g|^{1/2}g^{\mu\nu}\partial_\nu. \tag{21}$$

(With $\hbar$ explicit, the mass m should be replaced by $m/\hbar$, however we'll leave $\hbar$ implicit here.) The case where the coupling ξ to the Ricci scalar R vanishes is referred to as "minimal coupling", and that equation is called the *Klein-Gordon* (KG) equation. If also the mass m vanishes it is called the "massless, minimally coupled scalar". Another special case of interest is "conformal coupling" with $m = 0$ and $\xi = (D-2)/4(D-1)$.

4.1 Conformal coupling

Let me pause briefly to explain the meaning of conformal coupling since it comes up often in discussions of quantum fields in curved spacetime, primarily either because Robertson-Walker metrics are conformally flat or because all two-dimensional metrics are conformally flat. Consider making a position dependent conformal transformation of the metric:

$$\widetilde{g}_{\mu\nu} = \Omega^2(x)g_{\mu\nu}, \tag{22}$$

which induces the changes

$$\widetilde{g}^{\mu\nu} = \Omega^{-2}(x)g^{\mu\nu}, \qquad |\widetilde{g}|^{1/2} = \Omega^D(x)|g|^{1/2}, \tag{23}$$

$$|\widetilde{g}|^{1/2}\widetilde{g}^{\mu\nu} = \Omega^{D-2}(x)|g|^{1/2}g^{\mu\nu}, \tag{24}$$

$$\begin{aligned}
\widetilde{R} &= \widetilde{g}^{\mu\nu}\widetilde{R}_{\mu\nu} \\
&= \Omega^{-2}\Big(R - 2(D-1)\Box\ln\Omega \\
&\quad -(D-1)(D-2)g^{\alpha\beta}(\ln\Omega)_{,\alpha}(\ln\Omega)_{,\beta}\Big).
\end{aligned} \tag{25}$$

In $D = 2$ dimensions, the action is simply invariant in the massless, minimally coupled case without any change of the scalar field: $S[\varphi, g] = S[\varphi, \widetilde{g}]$. In any other dimension, the kinetic term is invariant under a conformal transformation with a *constant* Ω if we accompany the metric change (22) with a change of the scalar field, $\widetilde{\varphi} = \Omega^{(2-D)/2}\varphi$. (This scaling relation corresponds to the fact that the scalar field has dimension $[\text{length}]^{(2-D)/2}$ since the action must be dimensionless after factoring out an overall $\hbar$.) For a non-constant Ω the derivatives in the kinetic term ruin the invariance in general. However it can be shown that the action is invariant (up to a boundary term) if the coupling constant ξ is chosen to have the special value given in the previous paragraph, i.e. $S[\varphi, g] = S[\widetilde{\varphi}, \widetilde{g}]$. In $D = 4$ dimensions that value is $\xi = 1/6$.

4.2   Canonical quantization

To canonically quantize we first pass to the Hamiltonian description. Separating out a time coordinate x^0, $x^\mu = (x^0, x^i)$, we can write the action as

$$S = \int dx^0\, L, \qquad\qquad L = \int d^{D-1}x\, \mathcal{L}. \qquad (26)$$

The canonical momentum at a time x^0 is given by

$$\pi(\underline{x}) = \frac{\delta L}{\delta\left(\partial_0 \varphi(\underline{x})\right)} = |g|^{1/2} g^{\mu 0} \partial_\mu \varphi(\underline{x}) = |h|^{1/2} n^\mu \partial_\mu \varphi(\underline{x}). \qquad (27)$$

Here $\underline{x}$ labels a point on a surface of constant x^0, the x^0 argument of φ is suppressed, n^μ is the unit normal to the surface, and h is the determinant of the induced spatial metric h_{ij}. To quantize, the field φ and its conjugate momentum φ are now promoted to hermitian operators[2] and required to satisfy the canonical commutation relation,

$$[\varphi(\underline{x}), \pi(\underline{y})] = i\hbar \delta^{D-1}(\underline{x}, \underline{y}) \qquad (28)$$

It is worth noting that, being a variational derivative, the conjugate momentum is a density of weight one, and hence the Dirac delta function on the right hand side of (28) is a density of weight one in the second argument. It is defined by the property $\int d^{D-1}y\, \delta^{D-1}(\underline{x}, \underline{y}) f(\underline{y}) = f(\underline{x})$ for any scalar function f, without the use of a metric volume element.

In analogy with the bracket (7) defined for the case of the harmonic oscillator, one can form a conserved bracket from two complex solutions to the scalar wave equation (21),

$$\langle f, g\rangle = \int_\Sigma d\Sigma_\mu\, j^\mu, \qquad\qquad j^\mu(f, g) = (i/\hbar)|g|^{1/2} g^{\mu\nu}\left(\overline{f}\partial_\nu g - (\partial_\nu \overline{f})g\right). \qquad (29)$$

This bracket is sometimes called the *Klein-Gordon inner product*, and $\langle f, f\rangle$ the *Klein-Gordon norm* of f. The current density $j^\mu(f, g)$ is divergenceless $(\partial_\mu j^\mu = 0)$ when the functions f and g satisfy the KG equation (21), hence the value of the integral in (29) is independent of the spacelike surface Σ over which it is evaluated, provided the functions vanish at spatial infinity. The KG inner product satisfies the relations

$$\overline{\langle f, g\rangle} = -\langle \overline{f}, \overline{g}\rangle = \langle g, f\rangle, \qquad\qquad \langle f, \overline{f}\rangle = 0 \qquad (30)$$

Note that it is not positive definite.

4.3　　Hilbert space

At this point it is common to expand the field operator in modes and to associate annihilation and creation operators with modes, in close analogy with the harmonic oscillator (4), however instead I will begin with individual wave packet solutions. My reason is that in some situations there is no particularly natural set of modes, and none is needed to make physical predictions from the theory. (An illustration of this statement will be given in our treatment of the Hawking effect.) A mode decomposition is a basis in the space of solutions, and has no fundamental status.

In analogy with the harmonic oscillator case (10), we define the annihilation operator associated with a complex classical solution f by the bracket of f with the field operator φ:

$$a(f) = \langle f, \varphi \rangle \tag{31}$$

Since both f and φ satisfy the wave equation, $a(f)$ is well-defined, independent of the surface on which the bracket integral is evaluated. It follows from the above definition and the hermiticity of φ that the hermitian conjugate of $a(f)$ is given by

$$a^\dagger(f) = -a(\overline{f}). \tag{32}$$

The canonical commutation relation (28) together with the definition of the momentum (27) imply that

$$[a(f), a^\dagger(g)] = \langle f, g \rangle. \tag{33}$$

The converse is also true, in the sense that if (33) holds for all solutions f and g, then the canonical commutation relation holds. Using (32), we immediately obtain the similar relations

$$[a(f), a(g)] = -\langle f, \overline{g} \rangle, \qquad\qquad [a^\dagger(f), a^\dagger(g)] = -\langle \overline{f}, g \rangle \tag{34}$$

If f is a positive norm solution with unit norm $\langle f, f \rangle = 1$, then $a(f)$ and $a^\dagger(f)$ satisfy the usual commutation relation for the raising and lowering operators for a harmonic oscillator, $[a(f), a^\dagger(f)] = 1$. Suppose now that $|\Psi\rangle$ is a normalized quantum state satisfying $a(f)|\Psi\rangle = 0$. Of course this condition does not specify the state, but rather only one aspect of the state. Nevertheless, for each n, the state $|n, \Psi\rangle = (1/\sqrt{n!})(a^\dagger(f))^n |\Psi\rangle$ is a normalized eigenstate of the number operator $N(f) = a^\dagger(f)a(f)$ with eigenvalue n. The span of all these states defines a Fock space of f-wavepacket "n-particle excitations" above the state $|\Psi\rangle$.

If we want to construct the *full* Hilbert space of the field theory, how can we proceed? We should find a decomposition of the space of complex solutions to the wave equation $\mathcal{S}$ into a direct sum of a positive norm subspace $\mathcal{S}_p$ and its complex conjugate $\overline{\mathcal{S}_p}$, such that all brackets between solutions from the two subspaces vanish. That is, we must find a direct sum decomposition

$$\mathcal{S} = \mathcal{S}_p \oplus \overline{\mathcal{S}_p} \tag{35}$$

such that

$$\langle f, f \rangle \; > \; 0 \qquad \forall f \in \mathcal{S}_p \tag{36}$$
$$\langle f, \overline{g} \rangle \; = \; 0 \qquad \forall f, g \in \mathcal{S}_p. \tag{37}$$

The first condition implies that each f in $\mathcal{S}_p$ can be scaled to define its own harmonic oscillator sub-albegra as in the previous paragraph. The second condition implies, according to (34), that the annihilators and creators for f and g in the subspace $\mathcal{S}_p$ commute amongst themselves: $[a(f), a(g)] = 0 = [a^\dagger(f), a^\dagger(g)]$.

Given such a decompostion a total Hilbert space for the field theory can be defined as the space of finite norm sums of possibly infinitely many states of the form $a^\dagger(f_1) \cdots a^\dagger(f_n)|0\rangle$, where $|0\rangle$ is a state such that $a(f)|0\rangle = 0$ for *all* f in $\mathcal{S}_p$, and all $f_1, \ldots, f_n$ are in $\mathcal{S}_p$. The state $|0\rangle$ is called a *Fock vacuum*. It depends on the decomposition (35), and in general is not the ground state (which is not even well defined unless the background metric is globally static). The representation of the field operator on this Fock space is hermitian and satisfies the canonical commutation relations.

4.4   Flat spacetime

Now let's apply the above generalities to the case of a massive scalar field in flat spacetime. In this setting a natural decomposition of the space of solutions is defined by positive and negative frequency with respect to a Minkowski time translation, and the corresponding Fock vacuum is the ground state. I summarize briefly since this is standard flat spacetime quantum field theory.

Because of the infinite volume of space, plane wave solutions are of course not normalizable. To keep the physics straight and the language simple it is helpful to introduce periodic boundary conditions, so that space becomes a large three-dimensional torus with circumferences L and volume $V = L^3$. The allowed wave vectors are then $\mathbf{k} = (2\pi/L)\mathbf{n}$, where the components of the vector $\mathbf{n}$ are integers. In the end we can always take the limit $L \to \infty$ to obtain results for local quantities that are insensitive to this formal compactification.

A complete set of solutions ("modes") to the classical wave equation (21) with $\square$ the flat space d'Alembertian and $R = 0$ is given by

$$f_{\mathbf{k}}(t, \mathbf{x}) = \sqrt{\frac{\hbar}{2V\omega(\mathbf{k})}} e^{-i\omega(\mathbf{k})t} e^{i\mathbf{k}\cdot\mathbf{x}} \tag{38}$$

where

$$\omega(\mathbf{k}) = \sqrt{\mathbf{k}^2 + m^2}, \tag{39}$$

together with the solutions obtained by replacing the positive frequency $\omega(\mathbf{k})$ by its negative, $-\omega(\mathbf{k})$. The brackets between these solutions satisfy

$$\langle f_{\mathbf{k}}, f_{\mathbf{l}} \rangle = \delta_{\mathbf{k},\mathbf{l}} \tag{40}$$

$$\langle \overline{f}_{\mathbf{k}}, \overline{f}_{\mathbf{l}} \rangle = -\delta_{\mathbf{k},\mathbf{l}} \tag{41}$$

$$\langle f_{\mathbf{k}}, \overline{f}_{\mathbf{l}} \rangle = 0, \tag{42}$$

so they provide an orthogonal decomposition of the solution space into positive norm solutions and their conjugates as in (37), with S_p the space spanned by the positive frequency modes $f_{\mathbf{k}}$. As described in the previous subsection this provides a Fock space representation.

If we define the annihilation operator associated to $f_{\mathbf{k}}$ by

$$a_{\mathbf{k}} = \langle f_{\mathbf{k}}, \varphi \rangle, \tag{43}$$

then the field operator has the expansion

$$\varphi = \sum_{\mathbf{k}} \left(f_{\mathbf{k}} \, a_{\mathbf{k}} + \overline{f}_{\mathbf{k}} \, a_{\mathbf{k}}^{\dagger} \right). \tag{44}$$

Since the individual solutions $f_{\mathbf{k}}$ have positive frequency, and the Hamiltonian is a sum over the contributions from each $\mathbf{k}$ value, our previous discussion of the single oscillator shows that the vacuum state defined by

$$a_{\mathbf{k}} |0\rangle = 0 \tag{45}$$

for all $\mathbf{k}$ is in fact the ground state of the Hamiltonian. The states

$$a_{\mathbf{k}}^{\dagger} |0\rangle \tag{46}$$

have momentum $\hbar\mathbf{k}$ and energy $\hbar\omega(\mathbf{k})$, and are interpreted as single particle states. States of the form $a_{\mathbf{k}_1}^{\dagger} \cdots a_{\mathbf{k}_n}^{\dagger} |0\rangle$ are interpreted as n-particle states.

Note that although the field Fourier component $\varphi_{\mathbf{k}} = f_{\mathbf{k}} a_{\mathbf{k}} + \overline{f}_{-\mathbf{k}} a_{-\mathbf{k}}^{\dagger}$ has zero mean in the vacuum state, like the harmonic oscillator position it undergoes "zero-point fluctuations" characterized by

$$\langle 0 | \varphi_{\mathbf{k}}^{\dagger} \varphi_{\mathbf{k}} | 0 \rangle = |f_{-\mathbf{k}}|^2 = \frac{\hbar}{2V\omega(\mathbf{k})}, \tag{47}$$

which is entirely analogous to the oscillator result (19).

4.5   Curved spacetime, "particles", and stress tensor

In a general curved spacetime setting there is no analog of the preferred Minkowski vacuum and definition of particle states. However, it is clear that we can import these notions locally in an approximate sense if the wavevector and frequency are high enough compared to the inverse radius of curvature. Slightly more precisely, we can expand the metric in Riemann normal coordinates about any point x_0:

$$g_{\mu\nu}(x) = \eta_{\mu\nu} + \frac{1}{3}R_{\mu\nu\alpha\beta}(x_0)(x - x_0)^\alpha(x - x_0)^\beta + O((x - x_0)^3). \quad (48)$$

If $\mathbf{k}^2$ and $\omega^2(\mathbf{k})$ are much larger than any component of the Riemann tensor $R_{\mu\nu\alpha\beta}(x_0)$ in this coordinate system then it is clear that the flat space interpretation of the corresponding part of Fock space will hold to a good approximation, and in particular a particle detector will respond to the Fock states as it would in flat spacetime. This notion is useful at high enough wave vectors locally in any spacetime, and for essentially all wave vectors asymptotically in spacetimes that are asymptotically flat in the past or the future or both.

More generally, however, the notion of a "particle" is ambiguous in curved spacetime, and one should use field observables to characterize states. One such observable determines how a "particle detector" coupled to the field would respond were it following some particular worldline in spacetime and the coupling were adiabatically turned on and off at prescribed times. For example the transition probability of a point monopole detector is determined in lowest order perturbation theory by the two-point function $\langle\Psi|\varphi(x)\varphi(x')|\Psi\rangle$ evaluated along the worldline[21, 11] of the detector. (For a careful discussion of the regularization required in the case of a point detector see [22].) Alternatively this quantity—along with the higher order correlation functions—is itself a probe of the state of the field.

Another example of a field observable is the expectation value of the stress energy tensor, which is the source term in the semi-classical Einstein equation

$$G_{\mu\nu} = 8\pi G\langle\Psi|T_{\mu\nu}(x)|\Psi\rangle. \quad (49)$$

This quantity is infinite because it contains the product of field operators at the same point. Physically, the infinity is due to the fluctuations of the infinitely many ultraviolet field modes. For example, the leading order divergence of the energy density can be attributed to the zero-point energy of the field fluctuations, but there are subleading divergences as well. We have no time here to properly go into this subject, but rather settle for a few brief comments.

One way to make sense of the expectation value is via the *difference* between its values in two different states. This difference is well-defined (with a suitable regulator) and finite for any two states sharing the same singular short distance behavior. The result depends of course upon the comparison state. Since the divergence is associated with the very short wavelength modes, it might seem that to uniquely define the expectation value at a point x it should suffice to just subtract the infinities for a state defined as the vacuum in the local flat spacetime approximation at x. This subtraction is ambiguous however, even after making it as local as possible and ensuring local conservation of energy ($\nabla^\mu \langle T_{\mu\nu} \rangle = 0$). It defines the expectation value only up to a tensor $H_{\mu\nu}$ constructed locally from the background metric with four or fewer derivatives and satisfying the identity $\nabla^\mu H_{\mu\nu} = 0$. The general such tensor is the variation with respect to the metric of the invariant functional

$$\int d^D x \sqrt{|g|} \left(c_0 + c_1 R + c_2 R^2 + c_3 R^{\mu\nu} R_{\mu\nu} + c_4 R^{\mu\nu\rho\sigma} R_{\mu\nu\rho\sigma} \right). \quad (50)$$

(In four spacetime dimensions the last term can be rewritten as a combination of the first two and a total divergence.) Thus $H_{\mu\nu}$ is a combination of $g_{\mu\nu}$, the Einstein tensor $G_{\mu\nu}$, and curvature squared terms. In effect, the ambiguity $-H_{\mu\nu}$ is added to the metric side of the semi-classical field equation, where it renormalizes the cosmological constant and Newton's constant, and introduces curvature squared terms.

A different approach is to define the expectation value of the stress tensor via the metric variation of the renormalized effective action, which possesses ambiguities of the same form as (50). Hence the two approaches agree.

4.6   Remarks

4.6.1   Continuum normalization of modes.
Instead of the "box normalization" used above we could "normalize" the solutions $f_{\mathbf{k}}$ (38) with the factor $V^{-1/2}$ replaced by $(2\pi)^{-3/2}$. Then the Kronecker δ's in (40, 41) would be replaced by Dirac δ-functions and the discrete sum over momenta in (44) would be an integral over $\mathbf{k}$. In this case the annihilation and creation operators would satisfy $[a_{\mathbf{k}}, a_{\mathbf{l}}^\dagger] = \delta^3(\mathbf{k}, \mathbf{l})$.

4.6.2   Massless minimally coupled zero mode.
The massless minimally coupled case $m = 0$, $\xi = 0$ has a peculiar feature. The spatially constant function $f(t, \mathbf{x}) = c_0 + c_1 t$ is a solution to the wave equation that is not included among the positive frequency solutions $f_{\mathbf{k}}$ or their conjugates. This "zero mode" must be quantized as well, but it behaves like a free particle rather than like a harmonic oscillator. In particular, the state of lowest energy would have vanishing conjugate field momentum and hence would be described by a Schrödinger wave function $\psi(\varphi_0)$ that is totally delocalized in the field amplitude φ_0. Such a wave function would be non-normalizable as a quantum state, just as any momentum eigenstate of a non-relativistic particle is non-normalizable. Any normalized state would be described by a Schrödinger wavepacket that would spread in φ_0 like a free particle spreads in position space, and would have an expectation value $\langle \varphi_0 \rangle$ growing linearly in time. This suggests that no time-independent state exists. That is indeed true in the case where the spatial directions are compactified on a torus for example, so

that the modes are discrete and the zero mode carries as much weight as any other mode. In non-compact space one must look more closely. It turns out that in $1 + 1$ dimensions the zero mode continues to preclude a time independent state (see e.g. [23], although the connection to the behavior of the zero mode is not made there), however in higher dimensions it does not, presumably because of the extra factors of k in the measure $k^{D-2}dk$. A version of the same issue arises in deSitter space, where no deSitter invariant state exists for the massless, minimally coupled field [24]. That is true in higher dimensions as well, however, which may be related to the fact that the spatial sections of deSitter space are compact.

5.   Particle creation

We turn now to the subject of particle creation in curved spacetime (which would more appropriately be called "field excitation", but we use the standard term). The main applications are to the case of expanding cosmological space-times and to the Hawking effect for black holes. To begin with I will discuss the analogous effect for a single harmonic oscillator, which already contains the essential elements of the more complicated cases. The results will then be carried over to the cosmological setting. The following section then takes up the subject of the Hawking effect.

5.1   Parametric excitation of a harmonic oscillator

A quantum field in a time-dependent background spacetime can be modeled in a simple way by a harmonic oscillator whose frequency $\omega(t)$ is a given function of time. The equation of motion is then (3),

$$\ddot{x} + \omega^2(t)\, x = 0. \tag{51}$$

We consider the situation where the frequency is asymptotically constant, approaching ω_{in} in the past and ω_{out} in the future. The question to be answered is this: if the oscillator starts out in the ground state $|0_{in}\rangle$ appropriate to ω_{in} as $t \to -\infty$, what is the state as $t \to +\infty$? More precisely, in the Heisenberg picture the state does not evolve, so what we are really asking is how is the state $|0_{in}\rangle$ expressed as a Fock state in the out-Hilbert space appropriate to ω_{out}? It is evidently not the same as the ground state $|0_{out}\rangle$ of the Hamiltonian in the asymptotic future.

To answer this question we need only relate the annihilation and creation operators associated with the *in* and *out* normalized positive frequency modes

$f_{\substack{in \\ out}}(t)$, which are solutions to (51) with the asymptotic behavior

$$f_{\substack{in \\ out}}(t) \overset{t\to\mp\infty}{\longrightarrow} \sqrt{\frac{\hbar}{2m\omega_{\substack{in \\ out}}}} \exp(-i\omega_{\substack{in \\ out}} t). \tag{52}$$

Since the equation of motion is second order in time derivatives it admits a two-parameter family of solutions, hence there must exist complex constants α and β such that

$$f_{out} = \alpha f_{in} + \beta \bar{f}_{in}. \tag{53}$$

The normalization condition $\langle f_{out}, f_{out}\rangle = 1$ implies that

$$|\alpha|^2 - |\beta|^2 = 1. \tag{54}$$

The *out* annihilation operator is given (see (10)) by

$$
\begin{aligned}
a_{out} &= \langle f_{out}, x \rangle \tag{55}\\
&= \langle \alpha f_{in} + \beta \bar{f}_{in}, x \rangle \tag{56}\\
&= \alpha a_{in} - \bar{\beta} a_{in}^\dagger. \tag{57}
\end{aligned}
$$

This sort of linear relation between two sets of annihilation and creation operators (or the corresponding solutions) is called a *Bogoliubov transformation*, and the coefficients are the *Bogoliubov coefficients*. The mean value of the *out* number operator $N_{out} = a_{out}^\dagger a_{out}$ is nonzero in the state $|0_{in}\rangle$:

$$\langle 0_{in} | N_{out} | 0_{in} \rangle = |\beta|^2. \tag{58}$$

In this sense the time dependence of $\omega(t)$ excites the oscillator, and $|\beta|^2$ characterizes the excitation number.

To get a feel for the Bogoliubov coefficient β let us consider two extreme cases, adiabatic and sudden.

5.1.1   Adiabatic transitions and ground state.   The adiabatic case corresponds to a situation in which the frequency is changing very slowly compared to the period of oscillation,

$$\frac{\dot{\omega}}{\omega} \ll \omega. \tag{59}$$

In this case there is almost no excitation, so $|\beta| \ll 1$. Generically in the adiabatic case the Bogoliubov coefficient is exponentially small, $\beta \sim \exp(-\omega_0 T)$, where ω_0 is a typical frequency and T characterizes the time scale for the variations in the frequency.

In the Schrödinger picture, one can say that during an adiabatic change of $\omega(t)$ the state continually adjusts to remain close to the instantaneous *adiabatic ground state*. This state at time t_0 is the one annihilated by the lowering operator $a(f_{t_0})$ defined by the solution $f_{t_0}(t)$ satisfying the initial conditions at t_0 corresponding to the "instantaneous positive frequency solution",

$$f_{t_0}(t_0) = \sqrt{\hbar/2m\omega(t_0)} \tag{60}$$

$$\dot{f}_{t_0}(t_0) = -i\omega(t_0)f_{t_0}(t_0). \tag{61}$$

The instantaneous adiabatic ground state is sometimes called the "lowest order adiabatic ground state at t_0". One can also consider higher order adiabatic ground states as follows (see e.g. [3, 5]). A function of the WKB form $(\hbar/2mW(t))^{1/2} \exp(-i \int^t W(t')\,dt')$ is a normalized solution to (51) provided $W(t)$ satisfies a certain second order differential equation. That equation can be solved iteratively, yielding an expansion $W(t) = \omega(t) + \cdots$, where the subsequent terms involve time derivatives of $\omega(t)$. The lowest order adiabatic ground state at t_0 is defined using the solution whose initial conditions (61) match the lowest order truncation of the expansion for $W(t)$. A higher order adiabatic ground state is similarly defined using a higher order truncation.

5.1.2   Sudden transitions.

The opposite extreme is the sudden one, in which ω changes instantaneously from ω_{in} to ω_{out} at some time t_0. We can then find the Bogoliubov coefficients using (53) and its first derivative at t_0. For $t_0 = 0$ the result is

$$\alpha = \frac{1}{2}\left(\sqrt{\frac{\omega_{in}}{\omega_{out}}} + \sqrt{\frac{\omega_{out}}{\omega_{in}}}\right) \tag{62}$$

$$\beta = \frac{1}{2}\left(\sqrt{\frac{\omega_{in}}{\omega_{out}}} - \sqrt{\frac{\omega_{out}}{\omega_{in}}}\right). \tag{63}$$

(For $t_0 \neq 0$ there are extra phase factors in these solutions.) Interestingly, the amount of excitation is precisely the same if the roles of ω_{in} and ω_{out} are interchanged. For an example, consider the case where $\omega_{out} = 4\omega_{in}$, for which $\alpha = 5/4$ and $\beta = -3/4$. In this case the expectation value (58) of the out number operator is $9/16$, so there is about "half an excitation".

5.1.3   Relation between *in* and *out* ground states & the squeeze operator.

The expectation value of N_{out} is only one number characterizing the relation between $|O_{in}\rangle$ and the out states. We shall now determine the complete relation

$$|O_{in}\rangle = \sum_n c_n |n\rangle_{out} \tag{64}$$

where the states $|n\rangle_{out}$ are eigenstates of N_{out} and the c_n are constants.

One can find a_{in} in terms of a_{out} and $a_{out}^\dagger$ by combining (53) with its complex conjugate to solve for f_{in} in terms of f_{out} and $\bar{f}_{out}$. In analogy with (57) one then finds

$$a_{in} = \alpha a_{out} + \bar{\beta} a_{out}^\dagger. \tag{65}$$

Thus the defining condition $a_{in}|0_{in}\rangle = 0$ implies

$$a_{out}|0_{in}\rangle = -\frac{\bar{\beta}}{\alpha} a_{out}^\dagger |0_{in}\rangle. \tag{66}$$

A transparent way to solve this is to note that the commutation relation $[a_{out}, a_{out}^\dagger] = 1$ suggests the formal analogy $a_{out} = \partial/\partial a_{out}^\dagger$. This casts (66) as a first order ordinary differential equation, with solution

$$\begin{aligned} |0_{in}\rangle &= \mathcal{N} \exp\left[-\left(\frac{\bar{\beta}}{2\alpha}\right) a_{out}^\dagger a_{out}^\dagger\right] |0_{out}\rangle \tag{67} \\ &= \mathcal{N} \sum_n \frac{\sqrt{2n!}}{n!} \left(-\frac{\bar{\beta}}{2\alpha}\right)^n |2n\rangle_{out}. \tag{68} \end{aligned}$$

Note that the state $|0_{out}\rangle$ on which the exponential operator acts is annihilated by a_{out}, so there is no extra term from $a_{out}^\dagger$-dependence.

The state $|0_{in}\rangle$ thus contains only even numbered excitations when expressed in terms of the out number eigenstates. The normalization constant $\mathcal{N}$ is given by

$$|\mathcal{N}|^{-2} = \sum_n \frac{2n!}{(n!)^2} \left|\frac{\bar{\beta}}{2\alpha}\right|^{2n}. \tag{69}$$

For large n the summand approaches $|\bar{\beta}/\alpha|^{2n} = \left[|\beta|^2/(|\beta|^2 + 1)\right]^n$, where the relation (54) was used in the last step. The sum therefore converges, and can be evaluated[3] to yield

$$|\mathcal{N}| = \left(1 - |\beta/\alpha|^2\right)^{1/4} = |\alpha|^{-1/2}. \tag{70}$$

An alternate way of describing $|0_{in}\rangle$ in the out Hilbert space is via the *squeeze operator*

$$S = \exp\left[\frac{z}{2} a^\dagger a^\dagger - \frac{\bar{z}}{2} aa\right]. \tag{71}$$

Since its exponent is anti-hermitian, S is unitary. Conjugating a by S yields

$$S^\dagger a S = \cosh|z|\, a + \sinh|z| \frac{z}{|z|} a^\dagger. \tag{72}$$

This has the form of the Bogoliubov transformation (65) with $\alpha = \cosh|z|$ and $\beta = \sinh|z|(z/|z|)$. With a_{out} in place of a in S this gives

$$a_{in} = S^\dagger a_{out} S. \tag{73}$$

The condition $a_{in}|0_{in}\rangle = 0$ thus implies $a_{out}S|0_{in}\rangle = 0$, so evidently

$$|0_{in}\rangle = S^\dagger|0_{out}\rangle \tag{74}$$

up to a constant phase factor. That is, the *in* and *out* ground states are related by the action of the squeeze operator S. Since S is unitary, the right hand side of (74) is manifestly normalized.

5.2   Cosmological particle creation

We now apply the ideas just developed to a free scalar quantum field satisfying the KG equation (21) in a homogeneous isotropic spacetime. The case when the spatial sections are flat is slightly simpler, and it is already quite applicable, hence we restrict to that case here.

The spatially flat Robertson-Walker (RW) line element takes the form:

$$ds^2 = dt^2 - a^2(t)dx^i dx^i = a^2(\eta)(d\eta^2 - dx^i dx^i) \tag{75}$$

It is conformally flat, as are all RW metrics. The coordinate $\eta = \int dt/a(t)$ is called the *conformal time*, to distinguish it from the proper time t of the isotropic observers. The d'Alembertian $\Box$ for this metric is given by

$$\Box = \partial_t^2 + (3\dot{a}/a)\partial_t - a^{-2}\partial_{x^i}^2, \tag{76}$$

where the dot stands for $\partial/\partial t$. The spatial translation symmetry allows the spatial dependence to be separated from the time dependence. A field

$$u_k(\mathbf{x}, t) = \zeta_{\mathbf{k}}(t)e^{i\mathbf{k}\cdot\mathbf{x}} \tag{77}$$

satisfies the field equation (21) provided $\zeta_{\mathbf{k}}(t)$ satisfies an equation similar to that of a damped harmonic oscillator but with time-dependent damping coefficient $3\dot{a}/a$ and time-dependent frequency $a^{-2}k^2 + m^2 + \xi R$. It is worth emphasizing that in spite of the "damping", the field equation is Hamiltonian, and the Klein-Gordon norm (29) of any solution is conserved in the evolution.

The field equation can be put into the form of an undamped oscillator with time-dependent frequency by using the conformal time η instead of t and factoring out an appropriate power of the conformal factor $a^2(\eta)$:

$$\zeta_{\mathbf{k}} = a^{-1}\chi_{\mathbf{k}}. \tag{78}$$

The function $u_{\mathbf{k}}$ satisfies the field equation if and only if

$$\chi_{\mathbf{k}}'' + \omega^2(\eta)\chi_{\mathbf{k}} = 0, \tag{79}$$

where the prime stands for $d/d\eta$ and

$$\omega^2(\eta) = k^2 + m^2 a^2 - (1 - 6\xi)(a''/a). \tag{80}$$

(In the special case of conformal coupling $m = 0$ and $\xi = 1/6$, this becomes the time-independent harmonic oscillator, so that case is just like flat spacetime. All effects of the curvature are then incorporated by the prefactor $a(\eta)^{-1}$ in (78).)

The u_k are orthogonal in the Klein-Gordon inner product (29), and they are normalized[4] provided the χ_k have unit norm:

$$\langle u_k, u_l \rangle = \delta_{k,l} \iff (iV/\hbar)(\bar{\chi}_k \chi'_k - \bar{\chi}'_k \chi_k) = 1. \tag{81}$$

Note that the relevant norm for the χ_k differs from that for the harmonic oscillator (7) only by the replacement $m \to V$, where V is the x^i-coordinate volume of the constant t surfaces. We also have $\langle u_k, \bar{u}_l \rangle = 0$ for all k, l, hence these modes provide an orthogonal postive/negative norm decomposition of the space of complex solutions. As discussed in section 4.3, this yields a corresponding Fock space representation for the field operators. The field operator can be expanded in terms of the corresponding annihilation and creation operators:

$$\varphi(\mathbf{x}, t) = \sum_k \left(u_k(\mathbf{x}, t) a_k + \bar{u}_k(\mathbf{x}, t) a_k^\dagger \right). \tag{82}$$

Consider now the special case where there is no time dependence in the past and future, $a(\eta) \to$ constant. The *in* and *out* "vacua" are the ground states of the Hamiltonian at early and late times, and are the states annihilated by the a_k associated with the $u_k^{in,out}$ constructed with early and late time positive frequency modes $\chi_k^{in,out}$, as explained in section 4.4:

$$\chi_k^{\substack{in\\out}}(\eta) \overset{\eta \to \mp\infty}{\longrightarrow} \sqrt{\frac{\hbar}{2V\omega_{\substack{in\\out}}}} \exp(-i\omega_{\substack{in\\out}} \eta). \tag{83}$$

The Bogoliubov transformation now takes the form

$$u_k^{out} = \sum_{k'} \left(\alpha_{kk'} u_{k'}^{in} + \beta_{kk'} \bar{u}_{k'}^{in} \right). \tag{84}$$

Matching the coefficients of $\exp(i\mathbf{k} \cdot \mathbf{x})$, we see that

$$\alpha_{kk'} = \alpha_k \delta_{k,k'}, \qquad \beta_{kk'} = \beta_k \delta_{k,-k'}, \tag{85}$$

i.e. the Bogoliubov coefficients mix only modes of wave vectors $\mathbf{k}$ and $-\mathbf{k}$, and they depend only upon the magnitude of the wavevector on account of rotational symmetry (eqn. (79) for χ_k does not depend on the direction of $\mathbf{k}$). The normalization condition on u_k^{out} implies

$$|\alpha_k|^2 - |\beta_k|^2 = 1. \tag{86}$$

As in the harmonic oscillator example (58), if the state is the *in*-vacuum, then the expected excitation level of the $\mathbf{k}$ out-mode, i.e. the average number of particles in that mode, is given by

$$\langle 0_{in}|N_{\mathbf{k}}^{out}|0_{in}\rangle = |\beta_k|^2. \tag{87}$$

To convert this statement into one about particle density, we sum over $\mathbf{k}$ and divide by the physical spatial volume $V_{phys} = a^3 V$, which yields the number density of particles. Alternatively one can work with the continuum normalized modes. The relation between the discrete and continuous sums is

$$\frac{1}{V_{phys}} \sum_{\mathbf{k}} \longleftrightarrow \frac{1}{(2\pi a)^3} \int d^3\mathbf{k}. \tag{88}$$

the number density of out-particles is thus

$$n^{out} = \frac{1}{(2\pi a)^3} \int d^3\mathbf{k} |\beta_k|^2. \tag{89}$$

The mean particle number characterizes only certain aspects of the state. As in the oscillator example, a full description of the in-vacuum in the out-Fock space is obtained from the Bogoliubov relation between the corresponding annihilation and creation operators. From (84) we can solve for u^{in} and thence find

$$a_{\mathbf{k}}^{in} = \alpha_k \, a_{\mathbf{k}}^{out} + \bar{\beta}_k \, a_{-\mathbf{k}}^{\dagger \, out}, \tag{90}$$

whence

$$a_{\mathbf{k}}^{out}|0_{in}\rangle = -\frac{\bar{\beta}_k}{\alpha_k} \, a_{-\mathbf{k}}^{\dagger \, out}|0_{in}\rangle, \tag{91}$$

which can be solved to find

$$|0_{in}\rangle = \left(\prod_{\mathbf{k}'} \mathcal{N}_{\mathbf{k}}'\right) \exp\left[-\sum_{\mathbf{k}} \left(\frac{\bar{\beta}_k}{2\alpha_k}\right) a_{\mathbf{k}}^{\dagger \, out} a_{-\mathbf{k}}^{\dagger \, out}\right] |0_{out}\rangle \tag{92}$$

where the $\mathcal{N}_{\mathbf{k}}'$ are normalization factors. This solution is similar to the corresponding expression (68) for the harmonic oscillator and it can be found by a similar method. Although the operators $a_{\mathbf{k}}^{\dagger \, out}$ and $a_{-\mathbf{k}}^{\dagger \, out}$ are distinct, each product appears twice in the sum, once for $\mathbf{k}$ and once for $-\mathbf{k}$, hence the factor of 2 in the denominator of the exponent is required. Using the two-mode analog of the squeeze operator (71) the state (92) can also be written in a manifestly normalized fashion analogous to (74). It is sometimes called a *squeezed vacuum*.

5.3   Remarks

5.3.1   Momentum correlations in the squeezed vacuum.
The state (92) can be expressed as a sum of terms each of which has equal numbers of $\mathbf{k}$ and $-\mathbf{k}$ excitations. These degrees of freedom are thus entangled in the state, in such a way as to ensure zero total momentum. This is required by translation invariance of the states $|0_{in}\rangle$ and $|0_{out}\rangle$, since momentum is the generator of space translations.

5.3.2   Normalization of the squeezed vacuum.
The norm sum for the part of the state (92) involving $\mathbf{k}$ and $-\mathbf{k}$ is a standard geometric series, which evaluates to $|\alpha_k|^2$ using (86). Hence to normalize the state one can set $\mathcal{N}_{\mathbf{k}} = |\alpha_k|^{-1/2}$ for all $\mathbf{k}$, including $\mathbf{k} = 0$ as in (70). The overall normalization factor is a product of infinitely many numbers less than unity. Unless those numbers converge rapidly enough to unity the state is not normalizable. The condition for normalizability is easily seen to be $\sum_k |\beta_k|^2 < \infty$, i.e. according to (87) the average total number of excitations must be finite. If it is not, the state $|0_{in}\rangle$ does not lie in the Fock space built on the the state $|0_{out}\rangle$. Note that, although formally unitary, the squeeze operator does not act unitarily on the *out* Fock space if the corresponding state is not in fact normalizable.

5.3.3   Energy density.
If the scale factor a changes by over a time interval $\Delta\tau$, then for a massless field dimensional analysis indicates that the in vacuum has a resulting energy density $\rho \sim \hbar(\Delta\tau)^{-4}$ after the change. To see how the formalism produces this, according to (89) we have $\rho = (2\pi a)^{-3} \int d^3k |\beta_k|^2 (\hbar\omega/a)$. The Bogoliubov coefficent β_k is of order unity around $k_c/a \sim 1/\Delta\tau$ and decays exponentially above that. The integral is dominated by the upper limit and hence yields the above mentioned result.

5.3.4   Adiabatic vacuum.
Modes with frequency much larger than a'/a see the change of the scale factor as adiabatic, hence they remain relatively unexcited. The state that corresponds to the instantaneously defined ground state, in analogy with (61) for the single harmonic oscillator, is called the *adiabatic vacuum* at a given time.

5.4   de Sitter space

The special case of de Sitter space is of interest for various reasons. The first is just its high degree of symmetry, which makes it a convenient arena for the study of qft in curved space. It is the maximally symmetric Lorentzian space with (constant) positive curvature. Maximal symmetry refers to the number of Killing fields, which is the same as for flat spacetime. The Euclidean version of de Sitter space is just the sphere.

de Sitter (dS) space has hypersurface-orthogonal timelike Killing fields, hence is locally static, which further simplifies matters, but not to the point of triviality. The reason is that all such Killing fields have Killing horizons, null surfaces to which they are tangent, and beyond which they are spacelike. Hence dS space serves as a highly symmetric analog of a black hole spacetime. In particular, a symmetric variant of the Hawking effect takes place in de Sitter space, as first noticed by Gibbons and Hawking[25]. See [26] for a recent review.

Inflationary cosmology provides another important use of deSitter space, since during the period of exponential expansion the spacetime metric is well described by dS space. In this application the dS line element is usually written using spatially flat RW coordinates:

$$ds^2 = dt^2 - e^{2Ht}dx^i dx^i. \tag{93}$$

These coordinates cover only half of the global dS space, and they do not make the existence of a time translation symmetry manifest. This takes the conformal form (75) with $\eta = -H^{-1}\exp(-Ht)$ and $a(\eta) = -1/H\eta$. The range of t is $(-\infty, \infty)$ while that of η is $(-\infty, 0)$.

The flat patch of de Sitter space is asymptotically static with respect to conformal time η in the past, since $a'/a = -1/\eta \to 0$ as $\eta \to -\infty$. Therefore in the asymptotic past the adiabatic vacuum (with respect to positive η-frequency) defines a natural initial state. This is the initial state used in cosmology. In fact it happens to define a deSitter invariant state, also known as the Euclidean vacuum or the Bunch-Davies vacuum.

5.4.1   Primordial perturbations from zero point fluctuations.

Observations of the Cosmic Microwave Background radiation support the notion that the origin of primordial perturbations lies in the quantum fluctuations of scalar and tensor metric modes (see [27] for a recent review and [28] for a classic reference.) The scalar modes arise from (and indeed are entirely determined by, since the metric has no independent scalar degree of freedom) coupling to matter.[5] Let's briefly discuss how this works for a massless minimally coupled scalar field, which is just how these perturbations are described. First I'll describe the scenario in words, then add a few equations.

Consider a field mode with a frequency high compared to the expansion rate $\dot{a}/a$ during the early universe. To be specific let us assume this rate to be a constant H, i.e. de Sitter inflation. Such a mode was presumably in its ground state, as the prior expansion would have redshifted away any initial excitation. As the universe expanded the frequency redshifted until it became comparable to the expansion rate, at which point the oscillations ceased and the field amplitude approached a time-independent value. Just before it stopped oscillating the field had quantum zero point fluctuations of its amplitude, which were then preserved during the further expansion. Since the amplitude was frozen when the mode had the fixed proper wavenumber H, it is the same for all modes apart from the proper volume factor in the mode normalization which varies with the cosmological time of freezeout. Finally after inflation ended the expansion rate dropped faster than the wavenumber, hence eventually the mode could begin oscillating again when its wavelength became shorter than the Hubble length H^{-1}. This provided the seeds for density perturbations that would then grow by gravitational interactions. On account of the particular wavevector dependence of the amplitude of the frozen spatial fluctuations, the spectrum of these perturbations turns out to be scale-invariant (when appropriately defined).

More explicitly, the field equation for a massless minimally coupled field is given by (79), with

$$\omega^2(\eta) = k^2 - a''/a = k^2 - 2H^2 a^2, \tag{94}$$

where the last equality holds in de Sitter space. In terms of proper frequency $\omega_p = \omega/a$ and proper wavenumber $k_p = k/a$ we have $\omega_p^2 = k_p^2 - 2H^2$. The first term is the usual flat space one that produces oscillations, while the second term tends to oppose the oscillations. For proper wavenumbers much higher than H the second term is negligible. The field oscillates while the proper wavenumber redshifts exponentially. Eventually the two terms cancel, and the mode stops oscillating. This happens when

$$k_p = \sqrt{2}H. \tag{95}$$

As the wavenumber continues to redshift into the region $k_p \ll H$, to a good approximation the amplitude satisfies the equation $\chi_k'' - (a''/a)\chi_k = 0$, which has a growing and a decaying solution[6]. The growing solution is $\chi_k \propto a$, which implies that the field mode u_k (77,78) is constant in time. (This conclusion is also evident directly from the fact that the last term of the wave operator (76) vanishes as a grows.)

The squared amplitude of the fluctuations when frozen is, according to (47),

$$\langle 0|\varphi_k^\dagger \varphi_k|0\rangle = |u_{-k}|^2 \sim \frac{\hbar}{k_p V_p} = \frac{\hbar H^2}{V k^3}, \tag{96}$$

where the proper values are used for consistent matching to the previous flat space result. This gives rise to the scale invariant spectrum of density perturbations.

6.   Black hole evaporation

The vacuum of a quantum field is unstable to particle emission in the presence of a black hole event horizon. This instability is called the *Hawking effect*. Unlike the cosmological particle creation discussed in the previous section,

this effect is not the result of time dependence of the metric exciting the field
oscillators. Rather it is more like pair creation in an external electric field
[13]. (For more discussion of the role of time dependence see section 6.3.13.)
A general introduction to the Hawking effect was given in section 2.2. The
present section is devoted to a derivation and discussion related topics.

The historical roots of the Hawking effect lie in the classical Penrose process
for extracting energy from a rotating black hole. We first review that process
and indicate how it led to Hawking's discovery. Then we turn to the qft analysis.

6.1   Historical sketch

The Kerr metric for a rotating black hole is stationary, but the asymptotic
time translation Killing vector χ becomes spacelike *outside* the event horizon.
The region where it is spacelike is called the *ergoregion*. The conserved Killing
energy for a particle with four-momentum p is $E = \chi \cdot p$. Physical particles
have future pointing timelike 4-momenta, hence E is positive provided χ is
also future timelike. Where χ is spacelike however, some physical 4-momenta
have negative Killing energy.

In the Penrose process, a particle of energy $E_0 > 0$ is sent into the ergoregion
of a rotating black hole where it breaks up into two pieces with Killing energies
E_1 and E_2, so that $E_0 = E_1 + E_2$. If E_2 is arranged to be negative, then
$E_1 > E_0$, that is, more energy comes out than entered. The black hole absorbs
the negative energy E_2 and thus loses mass. It also loses angular momentum,
hence the process in effect extracts the rotational energy of the black hole.

The Penrose process is maximally efficient and reversible if the horizon area
is unchanged. That condition is achievable in the limit that the absorbed particle enters the black hole on a trajectory tangent to one of the null generators
of the horizon. This role of the horizon area in governing efficiency of energy extraction exhibits the analogy between area and entropy. Together with
Bekenstein's information theoretic arguments, it gave birth to the subject of
black hole thermodynamics.

When a field scatters from a rotating black hole a version of the Penrose
process called superradiant scattering can occur. The analogy with stimulated
emission suggests that quantum fields should exhibit spontaneous emission
from a rotating black hole. To calculate this emission rate from an "eternal"
spinning black hole one must specify a condition on the state of the quantum
field that determines what emerges from the past horizon. In order to avoid this
unphysical specification Hawking considered instead a black hole that forms
from collapse, which has no past horizon. In this case only initial conditions
before the collapse need be specified.

Much to his surprise, Hawking found that for any initial state, even a *nonrotating* black hole will spontaneously emit radiation. This *Hawking effect* is

a pair creation process in which one member of the pair lies in the ergoregion inside the horizon and has negative energy, while the other member lies outside and escapes to infinity with positive energy. The Hawking radiation emerges in a steady flux with a thermal spectrum at the temperature $T_H = \hbar\kappa/2\pi$. The surface gravity κ had already been seen to play the role of temperature in the classical first law of black hole mechanics, which thus rather remarkably presaged the quantum Hawking effect.

6.2   The Hawking effect

Two different notions of "frequency" are relevant to this discussion. One is the "Killing frequency", which refers to time dependence with respect to the time-translation symmetry of the background black hole spacetime. In the asymptotically flat region at infinity the Killing frequency agrees with the usual frequency defined by the Minkowski observers at rest with respect to the black hole. The other notion is "free-fall frequency" defined by an observer falling across the event horizon. Since the Killing flow is tangent to the horizon, Killing frequency there is very different from free-fall frequency, and that distinction lies at the heart of the Hawking effect.

6.2.1   Average number of outgoing particles.   The question to be answered is this: if a black hole forms from collapse with the quantum field in any 'regular' state $|\Psi\rangle$, then at late times, long after the collapse, what will be the average particle number and other observables for an outgoing positive Killing frequency wavepacket P of a quantum field far from the black hole? We address this question here for the case of a noninteracting scalar field and a static (non-rotating) black hole At the end we make some brief remarks about generalizations. Figure 1 depicts the various ingredients in the following discussion.

To begin with we evaluate the expectation value $\langle\Psi|N(P)|\Psi\rangle$ of the number operator $N(P) = a^\dagger(P)a(P)$ in the quantum state $|\Psi\rangle$ of the field. This does not fully characterize the state, but it will lead directly to considerations that do.

The annihilation operator (31) corresponding to a normalized wavepacket P is given by

$$a(P) = \langle P, \varphi\rangle_{\Sigma_f}, \tag{97}$$

where the Klein-Gordon inner product (29) is evaluated on the "final" spacelike slice Σ_f. To evaluate the expectation value of $N(P)$ we use the field equation satisfied by φ to relate $N(P)$ to an observable on an earlier slice Σ_i on which we know enough about the quantum state. Specifically, we assume there are no incoming excitations long after the black hole forms, and we assume that the state looks like the vacuum at very short distances (or high frequencies) as seen

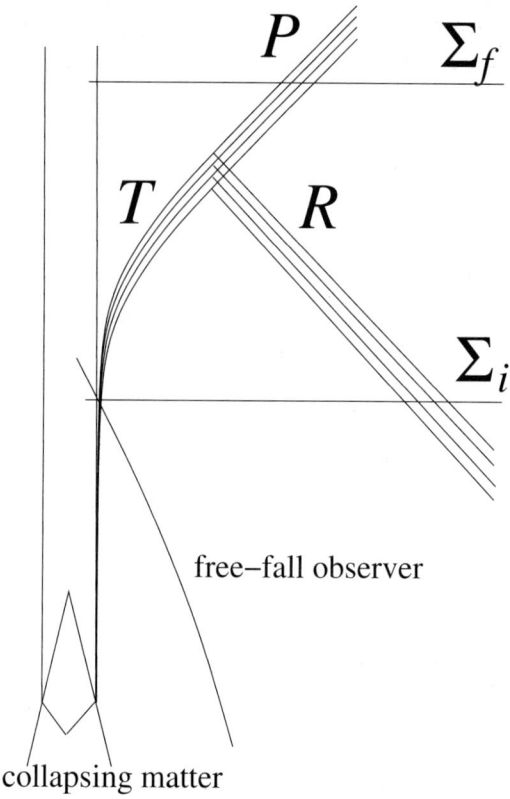

Figure 1. Spacetime diagram of black hole formed by collapsing matter. The outgoing wavepacket P splits into the transmitted part T and reflected part R when propagated backwards in time. The two surfaces $\Sigma_{f,i}$ are employed for evaluating the Klein-Gordon inner products between the wavepacket and the field operator. Although P, and hence R and T, have purely positive Killing frequency, the free-fall observer crossing T just outside the horizon sees both positive and negative frequency components with respect to his proper time.

by observers falling across the event horizon. Hawking originally propagated the field through the time-dependent collapsing part of the metric and back out all the way to spatial infinity, where he assumed $|\Psi\rangle$ to be the incoming vacuum at very high frequencies. As pointed out by Unruh[21] (see also [29, 30]) the result can be obtained without propagating all the way back, but rather stopping on a spacelike surface Σ_i far to the past of Σ_f but still after the formation of the black hole. This is important since propagation back out to infinity invokes arbitrarily high frequency modes whose behavior may not be given by the standard relativistic free field theory.

If Σ_i lies far enough to the past of Σ_f, the wavepacket P propagated backwards by the Klein-Gordon equation breaks up into two distinct parts,

$$P = R + T. \tag{98}$$

(See Fig. 1.) R is the "reflected" part that scatters from the black hole and returns to large radii, while T is the "transmitted" part that approaches the horizon. R has support only at large radii, and T has support only in a very small region just outside the event horizon where it oscillates very rapidly due to the backwards gravitational blueshift. Since both the wavepacket P and the field operator φ satisfy the Klein-Gordon equation, the Klein-Gordon inner product in (97) can be evaluated on Σ_i instead of Σ_f without changing $a(P)$. This yields a corresponding decomposition for the annihilation operator,

$$a(P) = a(R) + a(T). \tag{99}$$

Thus we have

$$\langle\Psi|N(P)|\Psi\rangle = \langle\Psi|(a^\dagger(R) + a^\dagger(T))(a(R) + a(T))|\Psi\rangle. \tag{100}$$

Now it follows from the stationarity of the black hole metric that the Killing frequencies in a solution of the KG equation are conserved. Hence the wavepackets R and T both have the same, purely positive Killing, frequency components as P. As R lies far from the black hole in the nearly flat region, this means that it has purely positive asymptotic Minkowski frequencies, hence the operator $a(R)$ is a *bona fide* annihilation operator—or rather $\langle R, R\rangle^{1/2}$ times the annihilation operator—for incoming excitations. Assuming that long after the black hole forms there are no such incoming excitations, we have $a(R)|\Psi\rangle = 0$. Equation (100) then becomes

$$\langle\Psi|N(P)|\Psi\rangle = \langle\Psi|a^\dagger(T)a(T)|\Psi\rangle. \tag{101}$$

If $a(T)|\Psi\rangle = 0$ as well, then no P-particles are emitted at all. The state with this property is called the "Boulware vacuum". It is the state with no positive Killing frequency excitations anywhere, including at the horizon.

The Boulware vacuum does not follow from collapse however. The reason is that the wavepacket T does not have purely positive frequency with respect to the time of a free fall observer crossing the horizon, and it is this latter frequency that matches to the local Minkowski frequency in a neighborhood of the horizon small compared with the radius of curvature of the spacetime.

More precisely, consider a free-fall observer (i.e. a timelike geodesic $x(\tau)$) with proper time τ who falls across the horizon at $\tau = 0$ at a point where the slice Σ_i meets the horizon (see Fig. 1). For this observer the wavepacket T has time dependence $T(\tau)$ (i.e. $T(x(\tau))$) that vanishes for $\tau > 0$ since the wavepacket has no support behind the horizon. Such a function cannot possibly have purely positive frequency components. To see why, recall that if a function vanishes on a continuous arc in a domain of analyticity, then it vanishes everywhere in that domain (since its power series vanishes identically on the arc and hence by analytic continuation everywhere). Any positive frequency function

$$h(\tau) = \int_0^\infty d\omega \, e^{-i\omega\tau} \, \tilde{h}(\omega) \qquad (102)$$

is analytic in the lower half τ plane, since the addition of a negative imaginary part to τ leaves the integral convergent. The positive real τ axis is the limit of an arc in the lower half plane, hence if $h(\tau)$ were to vanish for $\tau > 0$ it would necessarily vanish also for $\tau < 0$. (Conversely, a function that is analytic on the lower half-plane and does not blow up exponentially as $|\tau| \to \infty$ must contain only positive frequency components, since $\exp(-i\omega\tau)$ does blow up exponentially as $|\tau| \to \infty$ when ω is negative.)

The wavepacket T can be decomposed into its positive and negative frequency parts with respect to the free fall time τ,

$$T = T^+ + T^-, \qquad (103)$$

which yields the corresponding decomposition of the annihilation operator

$$\begin{aligned} a(T) &= a(T^+) + a(T^-) & (104) \\ &= a(T^+) - a^\dagger(\overline{T^-}). & (105) \end{aligned}$$

Since T^- has negative KG norm, Eqn. (32) has been used in the last line to trade $a(T^-)$ for the *bona fide* creation operator $a^\dagger(\overline{T^-})$. The τ-dependence of T consists of very rapid oscillations for $\tau < 0$, so the wavepackets T^+ and $\overline{T^-}$ have very high energy in the free-fall frame.

A free fall observer crossing the horizon long after the black hole forms would presumably see the ground state of the field at short distances, that is, such an observer would see no very high positive free-fall frequency excitations. The reason is that the collapse process occurs on the much longer time scale of the Schwarzschild radius r_s, so the modes with frequency much higher than $1/r_s$ should remain in their ground state. We therefore assume that the wavepackets T^+ and $\overline{T^-}$ are in their ground states,

$$a(T^+)|\Psi\rangle = 0, \qquad a(\overline{T^-})|\Psi\rangle = 0. \qquad (106)$$

For further discussion of this assumption see section 7.

Using (105) the number expectation value (101) can be evaluated as

$$\begin{aligned} \langle\Psi|N(P)|\Psi\rangle &= \langle\Psi|a(\overline{T^-})a^\dagger(\overline{T^-})|\Psi\rangle & (107) \\ &= \langle\Psi|[a(\overline{T^-}), a^\dagger(\overline{T^-})]|\Psi\rangle & (108) \\ &= \langle\overline{T^-}, \overline{T^-}\rangle_{\Sigma_i} & (109) \\ &= -\langle T^-, T^-\rangle_{\Sigma_i}, & (110) \end{aligned}$$

where (106) is used in the first and second lines, (33) is used in the third line, and (30) is used in the last step. The problem has thus been reduced to the computation of the Klein-Gordon norm of the negative frequency part of the transmitted wavepacket T. This requires that we be more explicit about the form of the wavepacket.

6.2.2   Norm of the negative frequency part & thermal flux.   For definiteness we consider a spherically symmetric vacuum black hole in 3+1 dimensions, that is a *Schwarzschild* black hole. The Schwarzschild line element can variously be expressed as

$$ds^2 \ = \ (1 - \frac{r_s}{r})dt^2 - (1 - \frac{r_s}{r})^{-1}dr^2 - r^2(d\theta^2 + \sin^2\theta\, d\varphi^2) \quad (111)$$

$$ = \ (1 - \frac{r_s}{r})(dt^2 - dr_*^2) - r^2(d\theta^2 + \sin^2\theta\, d\varphi^2) \quad (112)$$

$$ = \ (1 - \frac{r_s}{r})du\, dv - r^2(d\theta^2 + \sin^2\theta\, d\varphi^2). \quad (113)$$

The first form is in "Schwarzschild coordinates" and $r_s = 2GM$ is the Schwarzschild radius. The second form uses the "tortoise coordinate" r_*, defined by $dr_* = dr/(1 - r_s/r)$ or $r_* = r + r_s \ln(r/r_s - 1)$, which goes to $-\infty$ at the horizon. The third form uses the retarded and advanced time coordinates $u = t - r_*$ and $v = t + r_*$, which are also called outgoing and ingoing null coordinates respectively.

A scalar field satisfiying the Klein-Gordon equation $(\Box + m^2)\varphi = 0$ can be decomposed into spherical harmonics

$$\varphi(t, r, \theta, \phi) = \sum_{lm} \frac{\varphi_{lm}(t, r)}{r} Y_{lm}(\theta, \phi), \quad (114)$$

where $\varphi_{lm}(t, r)$ satisfies the 1+1 dimensional equation

$$(\partial_t^2 - \partial_{r_*}^2 + V_{lm})\varphi_{lm} = 0 \quad (115)$$

with the effective potential

$$V_{lm}(r) = \left(1 - \frac{r_s}{r}\right)\left(\frac{r_s}{r^3} + \frac{l(l+1)}{r^2} + m^2\right). \quad (116)$$

As $r \to \infty$ the potential goes to m^2. As $r \to r_s$, the factor $(r - r_s)$ approaches zero exponentially as $\exp(r_*/r_s)$ with respect to r_*. Near the horizon $\varphi_{lm}(t, r_*)$ therefore satisfies the massless wave equation, hence has the general form $f(u) + g(v)$.

Since the wavepacket $P = \sum_{lm} P_{lm}(t, r)Y_{lm}(\theta, \phi)$ is purely outgoing with support only at large radii at late times, near the horizon $P_{lm}(t, r)$ must be only a function of the 'retarded time $u = t - r_*$. That is, there can be no ingoing component. Since the metric is static, i.e. invariant with respect to t-translations, we can decompose any solution into components with a fixed t-frequency ω. A positive frequency outgoing mode at infinity has t-dependence $\exp(-i\omega t)$, hence its form near the horizon must be $\exp(-i\omega u)$.

Consider now a late time outgoing positive frequency wavepacket P that is narrowly peaked in Killing frequency ω. Propagating backwards in time, T is the part of the wavepacket that is squeezed up against the horizon, and its Y_{lm} component has the form $T_{lm} \sim \exp(-i\omega u)$ for all l, m. The coordinate u diverges as the horizon is approached. It is related to the proper time τ of a free-fall observer crossing the horizon at $\tau = 0$ via $\tau \simeq -\tau_0 \exp(-\kappa u)$, where $\kappa = 1/2r_s$ is the surface gravity of the black hole and the constant τ_0 depends on the velocity of the free-fall observer.[7] Hence the τ-dependence of the wavepacket along the free-fall worldline is

$$T \sim \exp\left(i\frac{\omega}{\kappa}\ln(-\tau)\right) \tag{117}$$

for $\tau < 0$, and it vanishes for $\tau > 0$.

To find the positive frequency part we use a method introduced by Unruh [31], which exploits the fact that a function analytic and bounded as $|\tau| \to \infty$ in the lower half complex τ plane has purely positive frequency (see the discussion after Eqn. (102)). The positive frequency extension of $T(\tau)$ from $\tau < 0$ to $\tau > 0$ is thus obtained by analytic continuation of $\ln(-\tau)$ in the lower half complex τ-plane. This continuation is given by $\ln \tau + i\pi$, provided the branch cut of $\ln \tau$ is taken in the upper half-plane. The positive frequency extension of $T(\tau)$ to $\tau > 0$ is therefore obtained by replacing $\ln(-\tau)$ with $\ln \tau + i\pi$ in (117), which yields $T(-\tau)\exp(-\pi\omega/\kappa)$ for $\tau > 0$. Similarly, the negative frequency extension of $\ln(-\tau)$ is given by $\ln \tau - i\pi$, provided the branch cut of $\ln \tau$ is taken instead in the lower half-plane. The negative frequency extension of $T(\tau)$ to $\tau > 0$ is therefore $T(-\tau)\exp(+\pi\omega/\kappa)$. Knowing these two extensions, we proceed as follows.

Define a new wavepacket $\widetilde{T}$, with support only inside the horizon, by "flipping" the wavepacket $T(u)$ across the horizon (see Fig. 2). That is, $\widetilde{T}$ vanishes outside the horizon and inside is constant on the outgoing null lines, with $\widetilde{T}(\tau) = T(-\tau)$ for $\tau > 0$. The above argument shows that the wavepackets

$$T^+ = c_+(T + e^{-\pi\omega/\kappa}\widetilde{T}) \tag{118}$$

$$T^- = c_-(T + e^{+\pi\omega/\kappa}\widetilde{T}) \tag{119}$$

have positive and negative free-fall frequency respectively. The two constants $c_\pm$ can be chosen so that $T^+ + T^-$ agrees with T outside the horizon and vanishes (as does T) inside the horizon. This yields $c_- = (1 - e^{2\pi\omega/\kappa})^{-1}$ and $c+/c_- = -e^{2\pi\omega/\kappa}$.

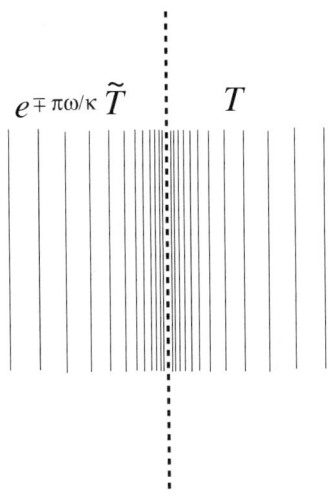

Figure 2. Spacetime sketch of phase contours of the transmitted wavepacket T and its flipped version $\widetilde{T}$ on either side of the horizon (dashed line). The upper and lower signs in the exponent of the factor $\exp(\mp\pi\omega/\kappa)$ yield the positive and negative free-fall frequency extensions of T.

Now $\langle T, \widetilde{T} \rangle = 0$ (since the two wavepackets do not overlap) and $\langle \widetilde{T}, \widetilde{T} \rangle = -\langle T, T \rangle$ (since the flipped wavepacket has the reverse τ-dependence), so using (119) one finds

$$\langle T^-, T^- \rangle = \frac{\langle T, T \rangle}{1 - e^{2\pi\omega/\kappa}}. \tag{120}$$

Inserting this in the expression (110) for the number operator expectation value yields Hawking's result,

$$\langle \Psi | N(P) | \Psi \rangle = \frac{\langle T, T \rangle}{e^{2\pi\omega/\kappa} - 1}. \tag{121}$$

The number expectation value (121) corresponds to the result for a thermal state at the Hakwing temperature $T_H = \hbar\kappa/2\pi$, multiplied by the so-called "greybody factor"

$$\Gamma = \langle T, T \rangle. \tag{122}$$

This factor is the probability for an excitation described by the the wavepacket P to pile up just outside the event horizon when propagated backwards in time, rather than being scattered back out to infinity. Equivalently, Γ is the probability for the excitation with wavepacket P to fall across the horizon when sent in forwards in time. This means that the black hole would be in detailed balance with a thermal bath at the Hawking temperature. Yet another interpretation of Γ is the probability for an excitation originating close to the horizon with normalized wavepacket $T/\langle T, T\rangle^{1/2}$ to escape to infinity rather than scattering back and falling into the black hole.

6.2.3   The quantum state.

The local free-fall vacuum condition (106) can be used to find the quantum state of the near horizon modes. We can do that here just "pair by pair", since the field is noninteracting so the state is the tensor product of the states for each outside/inside pair.

Using (119) the vacuum conditions become

$$a(T^+)|\Psi\rangle \quad \propto \quad \left[a(T) - e^{-\pi\omega/\kappa}a^\dagger(\tilde{T}^*)\right]|\Psi\rangle = 0 \qquad (123)$$

$$a(\overline{T^-})|\Psi\rangle \quad \propto \quad \left[-a^\dagger(T) + e^{+\pi\omega/\kappa}a(\tilde{T}^*)\right]|\Psi\rangle = 0 \qquad (124)$$

(here we use $*$ instead of "bar" for complex conjugation of $\tilde{T}$ for typographical reasons). Note that since $\overline{T}$ and $\tilde{T}$ have negative norm (as explained before (120)), we have replaced the corresponding annihilation operators by minus the creation operators of their conjugates. These equations define what is called the *Unruh vacuum* $|U\rangle$ for these wavepacket modes.

Now let $|B\rangle$ denote the quantum state of the T and $\tilde{T}^*$ modes such that

$$a(T)|B\rangle \;\; = \;\; 0 \qquad (125)$$

$$a(\tilde{T}^*)|B\rangle \;\; = \;\; 0. \qquad (126)$$

This state is called the *Boulware vacuum* for these modes. In analogy with (66) and (67) the vacuum conditions (123,124) imply that the Unruh and Boulware vacua are related by

$$|U\rangle \propto \exp\left[e^{-\pi\omega/\kappa}a^\dagger(\hat{T})a^\dagger(\hat{\tilde{T}}^*)\right]|B\rangle \qquad (127)$$

where the hats denote the corresponding normalized wavepackets.

The Unruh vacuum is thus a two-mode squeezed state, analogous to the one (92) found at each wavevector $\mathbf{k}$ in the case of cosmological pair creation. There each pair had zero total momentum since the background was space translation invariant. In the present case, each pair has zero total Killing energy, since the background is time translation invariant. The mode $\tilde{T}$ has the same positive Killing frequency as T has (because the Killing flow is symmetric under the flipping across the horizon operation that defines $\tilde{T}$), however its conjugate has negative Killing frequency, and therefore negative Killing energy.

The Unruh vacuum is a pure state, but because of its entangled structure it becomes mixed when restricted to the exterior. To find that mixed state we expand the exponential in a series. Denoting by $|n_{R,L}\rangle$ the level-n excitations of the modes T and $\widetilde{T}^*$ respectively, we have

$$|U\rangle \propto \sum_n e^{-n\pi\omega/\kappa}|n_L\rangle|n_R\rangle. \tag{128}$$

The reduced density matrix is thus

$$Tr_L|U\rangle\langle U| \propto \sum_n e^{-2n\pi\omega/\kappa}|n_R\rangle\langle n_R|, \tag{129}$$

a thermal canonical ensemble at the Hawking temperature.

The essence of the Hawking effect is the correlated structure of the local vacuum state at short distances near the horizon and its thermal character outside the horizon. Given this, the outgoing flux at infinity is just a consequence of the propagation of a fraction Γ (122) of each outgoing wavepacket from the horizon to infinity.

6.3   Remarks

In this subsection we make a large number of brief remarks about related topics that we have no time or space to go into deeply. Where no references are given see the sources listed at the end of the Introduction.

6.3.1   Local temperature.
The Hawking temperature refers to the Killing energy, or, since the Schwarzschild Killing vector is normalized at infinity, to the energy defined by a static observer at infinity. A static observer at finite radius will perceive the thermal state to have the blueshifted temperature $T_{loc} = T_H/|\xi|$, where $|\xi|$ is the local norm of the Schwarzschild time translation Killing vector. At infinity this is just the Hawking temperature, whereas it diverges as the horizon is approached. This divergence is due to the infinite acceleration of the static observer at the horizon and it occurs even for an accelerated observer in the Minkowski vacuum of flat spacetime (see section 6.3.4 below). A freely falling observer sees nothing divergent.

6.3.2   Equilibrium state: Hartle-Hawking vacuum.
A black hole will be in equilibrium with an incoming thermal flux at the Hawking temperature. The state that includes this incoming flux is called the Hartle-Hawking vacuum.

6.3.3   Stimulated emission.
Suppose that the field is not in the free-fall vacuum at the horizon (106), but rather that there are n excitations in the mode T^+, so that $a^\dagger(T^+)a(T^+)|\Psi\rangle = n\langle T^+, T^+\rangle|\Psi\rangle$. Then instead of (110) the expectation value of the number operator will be

$$\langle\Psi|N(P)|\Psi\rangle = n\langle T, T\rangle + (n+1)\langle\overline{T^-}, \overline{T^-}\rangle. \tag{130}$$

That is, if n quanta are present to begin with in the T^+ mode, the observer at infinity will observe in the P mode $n + 1$ times the usual number of Hawking quanta, in addition to n times the greybody factor (122). To produce a state in which the T^+ mode is occupied in standard physics one would have to send in particles of enormous energy just before the black hole formed[32]. As explained in section 7.4.2, however, trans-Planckian considerations could in principle allow stimulated emission at times long after the black hole formed. (This has nothing to do with the standard late time stimulated emission of the super-radiant modes of a rotating black hole[33, 34].)

6.3.4   Unruh effect.   The argument given above for the structure of the vacuum near a black hole horizon applies equally well to the Minkowski vacuum near an acceleration horizon in flat spacetime, where it is known as the Unruh effect. From a logical point of view it might be better to introduce the Unruh effect first, and then export it to the neighborhood of a black hole horizon to infer the Hawking effect. However, I chose here to go in the other direction

In the Unruh effect the boost Killing field $\xi_B = x\partial_t + t\partial_x$ (which generates hyperbolic rotations) plays the role of the Schwarzschild time translation, and the corresponding "temperature" is $\hbar/2\pi$. The Minkowski vacuum is the analog of the Hartle-Hawking equilibrium state, rather than the Unruh evaporating state. A uniformly accelerated observer following a hyperbolic orbit of the Killing field will perceive the Minkowski vacuum as a thermal state with temperature $\hbar/2\pi|\xi_B|$. The norm $|\xi_B|$ is just $(x^2 - t^2)^{1/2}$, which is also the inverse of the acceleration of the orbit, hence the local temperature is the *Unruh temperature* $T_U = \hbar a/2\pi$. As $(x^2 - t^2)^{1/2} \to \infty$ this temperature is redshifted to zero, so a Killing observer at infinity sees only the zero temperature vacuum. As the acceleration horizon $x = \pm t$ is approached a Killing observer sees a diverging temperature. The same temperature divergence is seen by a static observer approaching the horizon of a black hole in the Unruh or Hartle-Hawking states (*cf.* section 6.3.1 above).

6.3.5   Rotating black hole.   A small portion of the event horizon of a rotating black hole is indistinguishable from that of a Schwarzschild black hole, so the Hawking effect carries over to that case as well. The frequency ω in (121) should be replaced by the frequency with respect to the horizon generating Killing field $\partial_t + \Omega_H\partial_\phi$, where ∂_t and ∂_ϕ are the asymptotic time translation and rotation Killing vectors, and Ω_H is the angular velocity of the horizon. Thus ω is replaced by $\omega - m\Omega_H$, where m denotes the angular momentum. In effect there is a chemical potential $m\Omega_H$. See e.g. [35] for a discussion of the quantization of the "super-radiant" modes with $\omega - m\Omega_H < 0$.

6.3.6   de Sitter space.   The reasoning in the black hole case applies *mutatis mutandis* to de Sitter spacetime, where an observer is surrounded by a horizon that is locally indistinguishable from a black hole horizon. This leads to the temperature of deSitter spacetime[25, 26].

6.3.7   Higher spin fields.   The Hawking effect occurs also for higher spin fields, the only difference being (1) the greybody factors are different, and (2) for half-integer spin fields the Fermi distribution rather than the Bose distribution arises for the Hawking emission.

6.3.8   Interacting fields.   Our discussion here exploited the free field equation of motion, but the Hawking effect occurs for interacting fields as well. The essence, as in the free field case, is the Unruh effect, the interacting version of which can easily be established with the help of a Euclidean functional integral representation of the Minkowski vacuum[36]. (This result was found a decade earlier, at about the same time as the original Unruh effect, via a theorem [37] in the context of axiomatic quantum field theory, although the interpretation in terms of the thermal observations of uniformly accelerated observers was not noted until later[38].) The direct analog for the Hawking effect involves a Euclidean functional integral expression for the interacting Hartle-Hawking equilibrium state (for an introduction see [39, 40] and references therein).

More directly, in an asymptotically free theory one can presumably use the free field analysis to discover the structure of the vacuum near the horizon as in the free field case. The propagation of the field from that point on will involve the interactions. If the Hawking temperature is much higher than the scale Λ of asymptotic freedom then free particles will stream away from the black hole and subsequently be "dressed" by the interactions and fragment into asymptotic states[41, 42]. If the Hawking temperature is much lower than Λ then it is not so clear (to me at least) how to determine what is emitted.

6.3.9   Stress-energy tensor.   The Unruh and Hartle-Hawking states are "regular" on the horizon, i.e. they look like the Minkowski vacuum at short distances. (Recall that it is precisely the local vacuum property (106) or (123,124) that determines the thermal state of the outgoing modes at infinity.) Hence the mean value of the stress energy tensor is finite. The Boulware state $|B\rangle$ referred to above is obtained by removing the $T, \widetilde{T}^*$ pair excitations. This produces a state with a negative mean energy density that diverges as the horizon is approached. In fact, if even just *one* Hawking quantum is removed from the Unruh state $|U\rangle$ to obtain the Boulware state for that mode, a negative energy density divergence will be produced at the horizon. This can be viewed as the result of the infinite blueshift of the negative energy "hole". That is quite odd in the context of flat space, since the Minkowski vacuum should be the lowest energy state, hence any other state should have higher energy. The explanation is that there is a positive energy density divergence *on* the horizon that more than compensates the negative energy off the horizon[43].

6.3.10   Back-reaction.   As previously noted the Unruh state corresponds to an entangled state of positive and negative Killing energy excitations. As the positive energy excitations escape to infinity, there must be a corresponding negative energy flux into the black hole. Studies of the mean value of the stress tensor confirm this. Turning on the gravitational dynamics, this would lead to a mass loss for the black hole via the Einstein equation. The backreaction driven by the mean value is called the "semi-classical" evolution. There are quantum fluctuations about this mean evolution on a time and length scale of the Schwarzschild radius, unless a large number of matter fields is invoked to justify a large N limit that suppresses the quantum fluctuations.

6.3.11   Statistical entropy.   The entangled structure (128) of the Unruh state leads to a mixed state (129) when observations are restricted to the region outside the horizon. The "entanglement entropy" $-Tr\rho\ln\rho$ of this state is the same as the thermal entropy of the canonical ensemble (129). Summing over all modes this entropy diverges due to the infinite density of modes. To characterize the divergence one can use the thermodynamic entropy density $s \propto T^3$ of a bath of radiation at temperature T. The local temperature measured by a static observer (*cf.* section 6.3.1) is given by

$$T_{loc} = T_H/|\xi| \simeq T_H/\kappa\ell = 1/2\pi\ell, \tag{131}$$

where ℓ is the proper distance to the horizon on a surface of constant t, and the relation $\kappa = (d|\xi|/d\ell)_H$ has been used. Thus the entropy diverges like

$$S = \int s \, dv \sim \int T_{loc}^3 \, d\ell d^2 A \sim A/\ell_c^2, \tag{132}$$

where ℓ_c is a cutoff length above the horizon.[8]

What is the meaning of this entropy? It seems clear on the one hand that it must be included in the black hole entropy, but on the other hand it must somehow be meaningfully cut off. It is natural to try to understand the scaling of the black hole entropy with area in this way, but this is only the contribution from one quantum field, and there is also the classical contribution from the gravitational field itself that emerges from the partition function for quantum gravity[47]. The apparent dependence on the number of fields is perhaps removed by the corresponding renormalization of Newton's constant[48]. (The last reference in [48] is a review.) However this is regularization dependent and hence difficult to interpret physically, and moreover at least in dimensional regularization vector fields and some non-minimally coupled scalars contribute *negatively* to renormalizing G, which does not seem to match the entanglement entropy. There has been much work in this area but it remains to be fully understood.

6.3.12   **Information loss.**   Two types of potential information loss occur in black hole physics. First, when something falls into a black hole any information it carries is apparently lost to the outside world. Second, when a black hole radiates Hawking quanta, each radiated quantum is entangled with a partner lying inside the horizon, as Eqn. (128) shows. As long as the black hole does not evaporate completely, all information remains available on a spacelike surface that crosses the horizon and enters the black hole. If on the other hand the black hole evaporates completely, then no single spacelike surface stretching to infinity and filling "all space" can capture all information, according to the semi-classical analysis of the Hawking effect. Rather a disconnected surface behind the horizon must be included. The information on this disconnected surface flows into the strong curvature region at the singularity, where its fate is not yet understood.

This situation has generated much discussion. Some researchers (myself included) see the information loss to the outside world not as a sign of the breakdown of quantum mechanics but just as a consequence of the mutability of spatial topology in quantum gravity. When a black hole is about to evaporate completely it looks very small to the world outside the horizon. However this outside smallness has absolutely nothing to do with the size of the region inside available for storing information. Let us look into this a bit more.

Consider for example the spacelike singularity at $r = 0$ inside the Schwarzschild black hole. The metric in Eddington-Finkelstein coordinates is

$$ds^2 = (1 - r_s/r)dv^2 - 2dvdr - r^2(d\theta^2 + \sin^2\theta d\phi^2), \qquad (133)$$

where $v = t + r_*$ is the advanced time coordinate defined below (113). Inside the horizon, where $r < r_s$, a line of constant r, θ, ϕ is spacelike, and has a proper length $L(r, \Delta v) = (r_s/r - 1)^{1/2}\Delta v$. This goes to infinity as the singularity is approached, *for any interval of advanced time Δv.* Hence there is no dearth of space inside. On the other hand, the transverse angular dimensions go to zero size, and also we do not know how to describe the spacetime too near to the singularity. Therefore let's stop at the radius where the curvature $\sim r_s/r^3$ is equal to the Planck curvature, i.e at $r \sim r_s^{1/3}$ in Planck units. Then then proper length goes as $L(r_s^{1/3}, \Delta v) \sim r_s^{1/3}\Delta v$. For a solar mass black hole we have $r_s \sim 1\,\mathrm{km} \sim 10^{38}$ in Planck units, so $r_s^{1/3} \sim 10^{13}$. That means that for an external advanced time of $\Delta v = 1$ second, the proper length inside is one million light years. After a day or so, the length is the size of the universe, and so on over the Hawking lifetime $r_s^3 \sim 10^{114}$.

What happens to the future of this *spacelike* $r = r_s^{1/3}$ cylinder inside the black hole is governed by quantum gravity. We don't know what form the evolution takes. It is conceivable that time just stops running, like a frozen engine, producing a boundary of spacetime. It seems much more likely however that spacetime persists beyond, either into "quantum foam" or into a plump baby universe (see e.g. [49] and references therein) or universes[50]. In any of these scenarios, the tiny outside size of a black hole at the end stage of Hawking evaporation is no indication of the information carrying capacity of the interior.

Others believe that the validity of quantum mechanics requires that all information winds up available on the exterior slice after the evaporation. Results from string theory are often invoked to support this viewpoint. (For a critique of these arguments see [51].) If true, it would require some breakdown of the semi-classical description where none seems to be otherwise called for. The role of ultra high frequencies in the Hawking effect is often brought up in this context, however they are irrelevant since the derivation of the correlated structure of the vacuum and the Hawking effect does not need to access those frequencies. For more on this see the discussion of the trans-Planckian question in section 7.

6.3.13   Role of the black hole collapse.   The Hawking flux continues in a steady state long after the collapse that forms a black hole, which suggests that the collapse phase has nothing to do with the Hawking effect. Indeed the derivation given above makes use of only the free-fall vacuum conditions near the horizon, and the collapse phase plays no role. On the other hand if, as Hawking originally did, we follow the positive Killing frequency wavepacket T all the way backwards in time, it would go through the collapse phase and back out to infinity where Killing frequency and free-fall frequency coincide. Were it not for the time dependence of the background during the collapse, there would be no change of the Killing frequency, so there would be no negative frequency part of the ingoing wavepacket and there would be no particle creation. As discussed in the following section this propagation through the collapse phase invokes ultra high frequency field modes, and therefore may not even be physically relevant. However it seems clear is that whatever physics delivers the outgoing vacuum near the horizon, it must involve some violation of the time translation symmetry of the classical black hole background, even if it does not involve the collapse phase. In the lattice model of section 7.3 the violation arises from microscopic time dependence of the lattice spacing. In quantum gravity it may come just from the quantum gravitational fluctuations (see section 7.4.3).

7.   The trans-Planckian question

A deep question arises on account of the infinite redshift at the black hole horizon: do the outgoing modes that carry the Hawking radiation really emerge from a reservoir of modes with frequency arbitrarily far beyond the Planck frequency just outside the horizon, or is there another possibility? Does the existence or properties of the Hawking effect depend on the existence of such a *trans-Planckian reservoir*?

The reasoning leading to the expression (128) for the Unruh state in terms of positive Killing frequency modes is largely shielded from this question. The essential input is the free-fall vacuum conditions (106), which can be applied at any length or time scale much shorter than the Schwarzschild radius and inverse surface gravity (which are roughly the same unless the black hole is near extremally rotating or charged). There is no need to appeal to Planckian or trans-Planckian frequencies.

From this persepctive it is clear that, as far as the derivation of the Hawking effect is concerned, the only question is whether or not the free-fall vacuum in fact arises at short distances near the horizon from the initial conditions before collapse. As mentioned earlier, the modes with frequencies much higher than the inverse of the collapse time scale would be expected to remain unexcited. Nevertheless, in standard relativistic field theory these modes arise from trans-Planckian modes.

Consider an outgoing wavepacket near the horizon and peaked around frequency ω_1^{ff} as measured by a free-fall observer crossing the horizon at the advanced time v_1. At an earlier time $v_2 = v_1 - \Delta v$, the wavepacket would be blueshifted and squeezed closer to the horizon, with exponentially higher free-fall frequency,

$$\omega_2^{\text{ff}}/\omega_1^{\text{ff}} \sim e^{\kappa \Delta v} = e^{\Delta v/2R_s}. \tag{134}$$

For a solar mass black hole and $\Delta v = 2$ seconds, the ratio is $\exp(10^5)$.

To *predict* the state of the positive free-fall frequency modes T^+ and $\overline{T^-}$ from the initial state thus seems to require trans-Planckian physics. This is a breakdown of the usual separation of scales invoked in the application of effective field theory and it leaves some room for doubt[52, 30, 53] about the existence of the Hawking effect.

While the physical arguments for the Hawking effect do seem quite plausible, the trans-Planckian question is nevertheless pressing. After all, there are reasons to suspect that the trans-Planckian modes do not even *exist*. They imply an infinite contribution to black hole entanglement entropy from quantum fields, and they produce other divergences in quantum field theory that are not desirable in a fundamental theory.

The trans-Planckian question is really two-fold:

i Is the Hawking effect *universal*, i.e. insensitive to short distance physics, or at least can it be reliably derived in a quantum gravity theory with acceptable short distance behavior?

ii If there is no trans-Planckian reservoir, from where do the outgoing black hole modes arise?

7.1   String theory viewpoint

String theory has made impressive progress towards answering the first question, at least for some special black holes. In particular[54, 55], some near-extremal black holes become well-understood D-brane configurations in the weak coupling limit, and supersymmetry links the weak coupling to strong coupling results. Thus the Hawking effect can be reliably analyzed in a full quantum gravity theory. By a rather remarkable and unexpected correspondence the computations yield agreement with the semi-classical predictions, at least in the long wavelength limit. Moreover, the D-brane entropy is understood in terms of the counting of microstates, and agrees with the corresponding black hole entropy at strong coupling, just as the supersymmetry reasoning says it should. From yet another angle, the AdS/CFT duality in string theory offers other support[56]. There the Hawking effect and black hole entropy are interpreted in terms of a thermal state of the CFT (a conformally invariant super-Yang-Mills theory). However, neither of these approaches from string theory has so far been exploited to address the origin of the outgoing modes, since a local spacetime picture of the black hole horizon is lacking. This seems to be a question worth pursuing.

7.2   Condensed matter analogy

Condensed matter physics provides an analogy for effective field theory with a fundamental cutoff, hence it can be used to explore the consequences of a missing trans-Planckian reservoir. (For a review of these ideas see [57], and for a very brief summary see [58].) The first such black hole analog was Unruh's sonic black hole, which consists of a fluid with an inhomogeneous flow exceeding the speed of sound at a sonic horizon. A molecular fluid does not support wavelengths shorter than the intermolecular spacing, hence the sonic horizon has no "trans-molecular" reservoir of outgoing modes. Unruh found that nevertheless outgoing modes are produced, by a process of "mode conversion" from ingoing to outgoing modes. This phenomenon comes about already because of the alteration of the dispersion relation for the sound waves. It has been studied in various field theoretic models, however none are fully satisfactory since the unphysical short distance behavior of the field is always eventually called into play. The model that most closely mirrors the fluid analogy is a falling lattice[59] which has sensible short distance physics. The mode conversion on the lattice involves what is known as a Bloch oscillation in the condensed matter context. Here I will briefly explain how it works.

7.3   Hawking effect on a falling lattice

The model is 2d field theory on a lattice of points falling freely into a black hole. We begin with the line element in Gaussian normal coordinates,

$$ds^2 = dt^2 - a^2(t, z)\, dz^2. \tag{135}$$

A line of constant z is an infalling geodesic, at rest at infinity, and the "local scale factor" $a(t, z)$ satisfies

$$a(t, z \to \infty) = 1 \tag{136}$$

$$a(0, z) = 1 \tag{137}$$

$$a(t, z_H(t)) \sim \kappa t \qquad \text{for } \kappa t \gtrsim 1, \tag{138}$$

expressing the facts that the metric is asymptotically flat, the coordinate z measures proper distance on the $t = 0$ time slice, and at the horizon $z_H(t)$ the time scale for variations of the local scale factor is the surface gravity κ. The specific form of $a(t, z)$ is not required for the present discussion. (For details see [59].) A scalar field on this spacetime is governed by the action

$$S = \frac{1}{2} \int d^2x \, \sqrt{-g} g^{\mu\nu} \, \partial_\mu \varphi \, \partial_\nu \varphi \tag{139}$$

$$= \frac{1}{2} \int dt dz \left[a(z, t)(\partial_t \varphi)^2 - \frac{1}{a(t, z)} (\partial_z \varphi)^2 \right]. \tag{140}$$

Now we discretize the z coordinate with spacing δ:

$$z \to z_m = m\delta, \qquad \partial_z \varphi \to D\varphi = \frac{\varphi_{m+1} - \varphi_m}{\delta}, \tag{141}$$

where $\varphi_m = \varphi(z_m)$. The proper distance between the lattice points z_m and z_{m+1} on a constant t slice is approximately $a(t, z_m)\delta$. At $t = 0$ this is just δ everywhere, that is the points start out equidistant. However, since they are on free fall trajectories at different distances from the horizon, they spread out as time goes on. In particular the lattice spacing at the horizon grows with time like $\sim \kappa t$.

The discretized action is

$$S_{\text{lattice}} = \frac{1}{2} \int dt \sum_m \left[a_m(t)(\partial_t \varphi_m)^2 - \frac{2(D\varphi_m(t))^2}{a_{m+1}(t) + a_m(t)} \right]. \tag{142}$$

The discrete field equations produce the dispersion relation

$$\omega^{\text{ff}}(k) = \pm \frac{2}{a(z, t)\delta} \sin(k\delta/2) \tag{143}$$

for a mode of the form $\exp(-i\omega^{\text{ff}}t + ikz)$, provided $\partial_t a \ll \omega^{\text{ff}}$ and $\partial_z a \ll k$. (See Fig. 3.) For small wave numbers the lattice dispersion agrees with

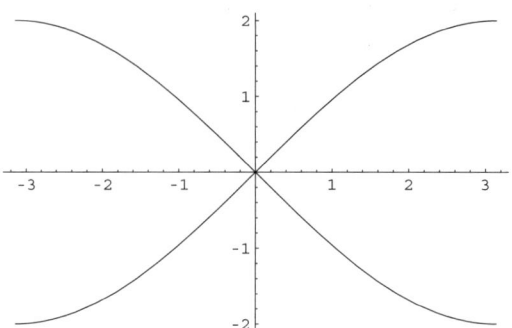

Figure 3.   Dispersion relation $\omega\delta = \pm 2\sin(k\delta/2)$ plotted vs. $k\delta$. Wavevectors differing by $2\pi/\delta$ are equivalent. Only the Brillouin zone $|k| \leq \pi/\delta$ is shown.

the continuum, however it is periodic in translation of k by $2\pi/\delta$. At the wavenumber $k = \pi/\delta$ there is a maximum frequency $2/a\delta$ and vanishing group velocity $d\omega/dk$. Beyond that wavenumber the group velocity reverses, and k is equivalent to $k - 2\pi/\delta$ which lies in the Brillouin zone $|k| \leq \pi/\delta$.

On the lattice the trans-Planckian redshift cannot take place, because of the lattice cutoff. Hence the outgoing modes—provided they exist—must come from ingoing modes. Both WKB "eikonal" trajectory and numerical evolution of the discrete wave equation confirm that indeed this occurs. The behavior of a typical wavepacket throughout the process of bouncing off the horizon is illustrated in Fig. 4. The real part of the wavepacket is plotted vs. the static coordinate at several different times. Following backwards in time, the packet starts to squeeze up against the horizon and then a trailing dip freezes and develops oscillations that grow until they balloon out, forming into a compact high frequency wavepacket that propagates neatly away from the horizon backwards in time.

The mode conversion can be understood as follows, following a wavepacket peaked around a long wavelength $\lambda \gg \delta$ backwards in time. The wavepacket blueshifts as it approaches the horizon, eventually enough for the lattice structure and therefore the curvature of the dispersion relation to be felt. At this point its group velocity begins to drop. In the WKB calculation the wavepacket motion reverses direction at a turning point outside the horizon. This occurs before its group velocity in the falling lattice frame is negative, so it is falling in at that stage because its outward velocity is not great enough to overcome the infalling of the lattice.

As the wavepapcket continues backwards in time now away from the black hole its wavevector continues to grow until it goes past the edge of the Brillouin

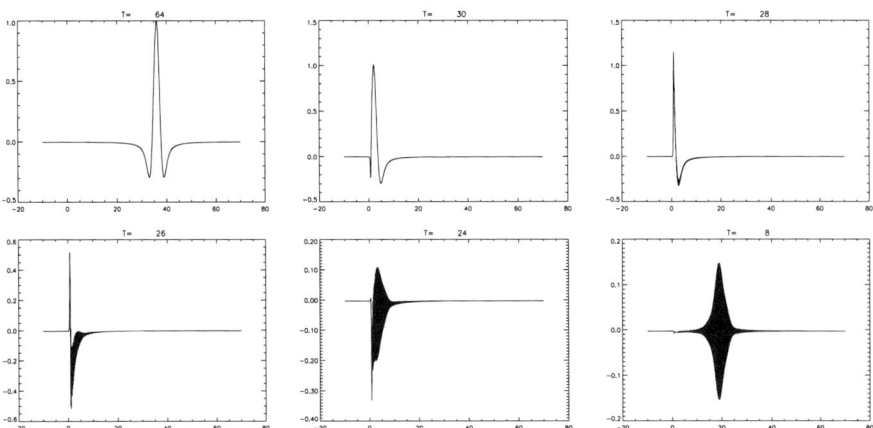

Figure 4. A typical wavepacket evolution on the lattice. The oscillations of the incoming wavepacket are too dense to resolve in the plots.

zone, thus becoming an ingoing mode also in the lattice frame. This reversal of group velocity is precisely what happens in a "Bloch oscillation" when a quantum particle in a periodic potential is accelerated. As the turnaround is occuring, another equally important effect is that the time-dependence of the underlying lattice is felt. Thus, unlike in the continuum limit of this stationary background, *the Killing frequency of the wavepacket is no longer conserved.*

Following the wavepacket all the way backwards in time out to the asymptotic region, it winds up with a short wavelength of order δ and a large frequency of order $1/\delta$. This frequency shift is absolutely critical to the existence of the outgoing modes, since an ingoing low frequency mode would simply fall across the horizon. Only the "exotic" modes with sufficiently high frequency will undergo the mode conversion process.

The Hawking flux is determined by the negative frequency part of the ingoing wavepacket. The eikonal approximation just described does not capture the negative frequency mixing which occurs during the turnaround at the horizon. Just as in the continuum, the wavepacket squeezed against the horizon has both positive and negative free-fall frequency components. As these components propagate backwards in time away from the horizon, their frequency slowly shifts, but their relative amplitude remains fixed. Hence the norm of the negative freqency part of the incoming wavepacket turns out to be just what the continuum Hawking effect indicates, with small lattice corrections.

Put differently, the infalling vacuum is adiabatically modified by the underlying microscopic time dependence of the lattice in such a way that the Unruh conditions (106) on the state of the outgoing modes at the horizon are satisfied.

7.4   Remarks

7.4.1   Finite Entanglement entropy.   Since the lattice has a short distance cutoff the entanglement entropy between the modes just inside and outside of the horizon is finite at any time. As illustrated in Figure 5, any given entangled pair of vacuum modes began in the past outside the horizon, propagated towards the horizon where it was "split", and then separated, with one half falling in and the other converted into an outgoing mode. As time goes on, new pairs continually propagate in and maintain a constant entanglement entropy.

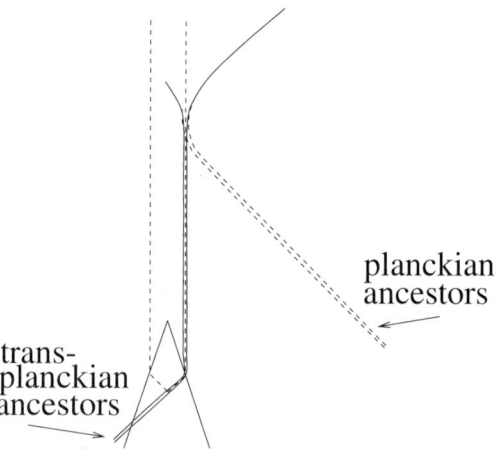

planckian ancestors

trans- planckian ancestors

Figure 5.   The ancestors of a Hawking quantum and its negative energy partner. In standard relativistic field theory the ancestors are trans-Planckian and pass through the collapsing matter at the moment of horizon formation. On the lattice the ancestors ar Planckian and propagate in towards the black hole at late times.

7.4.2   Stimulated emission of Hawking radiation at late times.   As discussed in section 6.3.3, if the Unruh vacuum conditions (106) do not hold at the horizon then stimulated emission of Hawking radiation will occur. In the falling lattice model, these horizon modes arise from ingoing modes long after the black hole formed as shown in Figure 5. Thus it is possible to stimulate the emission of Hawking radiation by sending in radiation at late times, in contrast to the usual continuum case of a static black hole. This seems a generic feature of theories with a cutoff, for which the outgoing modes must arise from modes that are ingoing after the collapse. (Something like it should happen also in string theory if, as many suppose, the trans-Planckian reservoir at the horizon is also eliminated there.) Note however that the linear model described here is surely a gross oversimplification. Turning on the gravitational interactions between the modes and the background, one is led to a picture in which the modes "dissipate" when propagated backwards in time to the Planckian regime. Hence what really produces the outgoing Hawking quantum must be a complicated collective mode of the interacting vacuum that "anti-dissipates" as it approaches the horizon and turns around. Calculations exploring this in quantum gravity were carried out in [60].

7.4.3   Lattice time dependence and geometry fluctuations.   The microscopic time dependence of the lattice, i.e. the slow spreading of the lattice points, plays a critical role in transforming an ingoing mode with high Killing frequency to an outgoing mode with low Killing frequency, and in allowing for mixing of positive and negative frequencies despite the stationarity of the continuum black hole background. This suggests the conjecture that in quantum gravity the underlying quantum fluctuations of the geometry that do not share the stationarity of the black hole background metric might play this role. A step towards understanding this might be provided by a two or three dimensional version of the lattice model, in which the density of lattice points remains fixed but their microscopic positions fluctuate. This is precisely what happens with an inhomogeneous flow of a real molecular fluid. The quantum gravity analysis of [60] lends some support to this conjecture, though it is not clear whether the Lorentz violation seen there is just an artifact of a non-covariant cutoff or a feature introduced by the global geometry of the black hole spacetime.

7.5   Trans-Planckian question in cosmology

The modes producing the inflationary perturbation spectrum (*cf.* section 5.4.1) redshift exponentially from their trans-Planckian origins. It has been suggested that this might leave a visible imprint on the perturbation spectrum, via a modified high frequency dispersion relation and/or a modified initial quantum state for the field modes. As illustrated by the Hawking process on the lattice however, as long as the redshifting is adiabatic on the timescale of the modes, they would remain in their ground states. If the Hubble rate H during inflation were much less than the Planck mass M_P (or whatever scale the modified dispersion sets in) the modes could be treated in the standard relativistic manner by the time the perturbation spectrum is determined. At most one might expect an effect of order H/M_P, and some analyses suggest the effect will be even smaller. This all depends on what *state* the modes are in. They cannot be too far from the vacuum, since otherwise running them backwards in time they would develop an exponentially growing energy density which would be incompatible with the inflationary dynamics. Hence it seems the best one can say at present is that there may be room for noticable deviations from the inflationary predictions, if H/M_P is large enough. (See for example [61] and references therein for discussions of these issues.)

Acknowledgments

I would like to thank the organizers of the CECS School on Quantum Gravity both for a stimulating, enjoyable experience, and for extraordinary patience. I'm also grateful to Brendan Foster, Bei-Lok Hu, and Albert Roura for helpful corrections, suggestions, and questions about these notes. This work was supported in part by the NSF under grants PHY-9800967 and PHY-0300710 at the University of Maryland.

Notes

1. In a more abstract algebraic approach, one does not require at this stage a representation but rather requires that the quantum variables are elements of an algebra equipped with a star operation satisfying certain axioms. For quantum mechanics the algebraic approach is no different from the concrete representation approach, however for quantum fields the more general algebraic approach turns out to be necessary to have sufficient generality. In these lectures I will ignore this distinction. For an introduction to the algebraic approach in the context of quantum fields in curved spacetime see [6]. For a comprehensive treatment of the algebraic approach to quantum field theory see [20].

2. See footnote 7.5.

3. Alex Maloney pointed out that it can be evaluated by expressing the binomial coefficient $2n!/(n!)^2$ as the contour integral $\oint (dz/2\pi i z)(z + 1/z)^{2n}$ and interchanging the order of the sum with the integral.

4. The two inverse factors of a coming from (78) are cancelled by the a^3 in the volume element and the a^{-1} in the relation $\partial/\partial t = a^{-1}\partial/\partial \eta$

5. The Cosmic Microwave Background observations supporting this account of primordial perturbations thus amount to quantum gravity observations of a limited kind.

6. The general solution is $c_1 a + c_2 a \int^\eta d\eta/a^2$, which is $c_3\eta^{-1} + c_4\eta^2$ in de Sitter space.

7. This can be obtained by noting from the third form of the line element (113) that along a timelike line $(1 - r_s/r)\dot{u}\dot{v} = 1$, where the dots represent the proper time derivative. As the horizon is crossed $\dot{v}$ is finite, hence $\dot{u} \sim (r - r_s)^{-1} \sim e^{-r_*/r_s} = e^{(u-v)/2r_s} \sim e^{\kappa u}$.

8. Sorkin[44] introduced the notion of black hole entanglement entropy, and with collaborators[45] computed it in the presence of a regulator. A mode-by-mode version of the thermal entropy calculation was first done by 't Hooft[46], who called it the "brick wall model" because of a Dirichlet boundary condition applied at ℓ_c.

References

[1] T. Jacobson, "Thermodynamics of space-time: The Einstein equation of state," Phys. Rev. Lett. **75**, 1260 (1995) [arXiv:gr-qc/9504004].

[2] J. Donoghue, "Perturbative dynamics of quantum general relativity," arXiv:gr-qc/9712070.

[3] N. D. Birrell and P. C. W. Davies, *Quantum Fields In Curved Space*, (Cambridge University Press, 1982).

[4] A.A. Grib, S. G. Mamaev, V. M. Mostepanenko, *Quantum Effects in Strong External Fields* (Moscow, Atomizdat, 1980) (in Russian) [English translation: *Vacuum Quantum Effects in Strong Fields*, (Friedmann Laboratory Publishing, St.Petersburg, 1994)].

[5]   S. A. Fulling, *Aspects Of Quantum Field Theory In Curved Space-Time*, London Mathematical Society Student Texts **17**, (Cambridge University Press, 1989).

[6]   R. M. Wald, *Quantum Field Theory In Curved Space-Time And Black Hole Thermodynamics*, (The University of Chicago Press, 1994).

[7]   B. S. Dewitt, "The Global Approach to Quantum Field Theory," vol. 1,2 (Oxford University Press, 2003).

[8]   B. S. Dewitt, "Quantum Field Theory In Curved Space-Time," Phys. Rept. **19**, 295 (1975).

[9]   C. J. Isham, "Quantum Field Theory In Curved Space Times: An Overview," Ann. N.Y. Acad. Sci. **302**, 114 (1977).

[10] G. W. Gibbons, "Quantum Field Theory In Curved Space-Time," in *General Relativity: An Einstein Centenary Survey*, eds. S.W. Hawking and W. Israel (Cambridge University Press, 1979).

[11] B. S. Dewitt, "Quantum Gravity: The New Synthesis," in *General Relativity: An Einstein Centenary Survey*, eds. S.W. Hawking and W. Israel (Cambridge University Press, 1979).

[12] S. A. Fulling and S. N. M. Ruijsenaars, "Temperature, periodicity, and horizons," Phys. Rept. **152**, 135-176 (1987).

[13] R. Brout, S. Massar, R. Parentani and P. Spindel, "A Primer For Black Hole Quantum Physics," Phys. Rept. **260**, 329 (1995).

[14] L. H. Ford, "Quantum Field Theory In Curved Space-Time," arXiv:gr-qc/9707062.

[15] A. Wipf, "Quantum fields near black holes," arXiv:hep-th/9801025.

[16] J. Traschen, "An introduction to black hole evaporation," arXiv:gr-qc/0010055.

[17] J. Preskill, P. Schwarz, A. D. Shapere, S. Trivedi and F. Wilczek, "Limitations on the statistical description of black holes," Mod. Phys. Lett. A **6**, 2353 (1991).

[18] A. Barrau, "Primordial black holes as a source of extremely high energy cosmic rays," Astropart. Phys. **12**, 269 (2000) [arXiv:astro-ph/9907347].

[19] W. H. Zurek, "Entropy evaporated by a black hole," Phys. Rev. Lett. **49**, 168 (1982).

[20] R. Haag, *Local quantum physics: fields, particles, algebras*, (Springer-Verlag, 1996).

[21] W. G. Unruh, "Origin Of The Particles In Black Hole Evaporation," Phys. Rev. D **15**, 365 (1977).

[22] S. Schlicht, "Considerations on the Unruh effect: Causality and regularization," arXiv:gr-qc/0306022.

[23] L. H. Ford and A. Vilenkin, "Global Symmetry Breaking In Two-Dimensional Flat Space-Time And In De Sitter Space-Time," Phys. Rev. D **33**, 2833 (1986).

[24] B. Allen, "Vacuum States In De Sitter Space," Phys. Rev. D **32**, 3136 (1985); B. Allen and A. Folacci, "The Massless Minimally Coupled Scalar Field In De Sitter Space," Phys. Rev. D **35**, 3771 (1987).

[25] G. W. Gibbons and S. W. Hawking, "Cosmological Event Horizons, Thermodynamics, And Particle Creation," Phys. Rev. D **15**, 2738 (1977).

[26] M. Spradlin, A. Strominger and A. Volovich, "Les Houches lectures on de Sitter space," arXiv:hep-th/0110007.

[27] R. H. Brandenberger, "Lectures on the theory of cosmological perturbations," arXiv:hep-th/0306071.

[28] V. F. Mukhanov, H. A. Feldman and R. H. Brandenberger, "Theory Of Cosmological Perturbations. Part 1. Classical Perturbations. Part 2. Quantum Theory Of Perturbations. Part 3. Extensions," Phys. Rept. **215**, 203 (1992).

[29] K. Fredenhagen and R. Haag, "On The Derivation Of Hawking Radiation Associated With The Formation Of A Black Hole," Commun. Math. Phys. **127**, 273 (1990).

[30] T. Jacobson, "Black hole radiation in the presence of a short distance cutoff," Phys. Rev. D **48**, 728 (1993) [arXiv:hep-th/9303103].

[31] W. G. Unruh, "Notes On Black Hole Evaporation," Phys. Rev. D **14**, 870 (1976).

[32] R. M. Wald, "Stimulated Emission Effects In Particle Creation Near Black Holes," Phys. Rev. D **13**, 3176 (1976).

[33] J. D. Bekenstein and A. Meisels, "Einstein A And B Coefficients For A Black Hole," Phys. Rev. D **15**, 2775 (1977).

[34] P. Panangaden and R. M. Wald, "Probability Distribution For Radiation From A Black Hole In The Presence Of Incoming Radiation," Phys. Rev. D **16**, 929 (1977).

[35] V. P. Frolov and K. S. Thorne, "Renormalized Stress - Energy Tensor Near The Horizon Of A Slowly Evolving, Rotating Black Hole," Phys. Rev. D **39**, 2125 (1989).

[36] W. G. Unruh and N. Weiss, "Acceleration Radiation In Interacting Field Theories," Phys. Rev. D **29**, 1656 (1984).

[37] J. J. Bisognano and E. H. Wichmann, "On The Duality Condition For A Hermitian Scalar Field," J. Math. Phys. **16**, 985 (1975); "On The Duality Condition For Quantum Fields," J. Math. Phys. **17**, 303 (1976).

[38] G. L. Sewell, "Relativity of Temperature and the Hawking Effect", Phys. Lett. A, **79**, 23, (1980); "Quantum-fields on manifolds - PCT and gravitationally induced thermal states", Ann. Phys. **141**, 201 (1982).

[39] T. Jacobson, "A Note On Hartle-Hawking Vacua," Phys. Rev. D **50**, 6031 (1994) [arXiv:gr-qc/9407022].

[40] T. A. Jacobson, "Introduction to Black Hole Microscopy," arXiv:hep-th/9510026.

[41] J. H. MacGibbon and B. R. Webber, "Quark And Gluon Jet Emission From Primordial Black Holes: The Instantaneous Spectra," Phys. Rev. D **41**, 3052 (1990).

[42] J. H. MacGibbon, "Quark And Gluon Jet Emission From Primordial Black Holes. 2. The Lifetime Emission," Phys. Rev. D **44**, 376 (1991).

[43] R. Parentani, "The Energy Momentum Tensor In Fulling-Rindler Vacuum," Class. Quant. Grav. **10**, 1409 (1993) [arXiv:hep-th/9303062].

[44] R. D. Sorkin, "On The Entropy Of The Vacuum Outside A Horizon," in *General Relativity and Gravitation*, proceedings of the GR10 Conference, Padova 1983, ed. B. Bertotti, F. de Felice, A. Pascolini (Consiglio Nazionale della Ricerche, Roma, 1983) Vol. 2.

[45] L. Bombelli, R. K. Koul, J. H. Lee and R. D. Sorkin, "A Quantum Source Of Entropy For Black Holes," Phys. Rev. D **34**, 373 (1986).

[46] G. 't Hooft, "On The Quantum Structure Of A Black Hole," Nucl. Phys. B **256**, 727 (1985).

[47] G. W. Gibbons and S. W. Hawking, "Action Integrals And Partition Functions In Quantum Gravity," Phys. Rev. D **15**, 2752 (1977).

[48] L. Susskind and J. Uglum, "Black hole entropy in canonical quantum gravity and super-string theory," *Phys. Rev. D* **50**, 2700 (1994)[arXiv:hep-th/9401070]; T. Jacobson, "Black hole entropy and induced gravity," arXiv:gr-qc/9404039; F. Larsen and F. Wilczek, "Renormalization of black hole entropy and of the gravitational coupling constant," *Nucl. Phys. B*

458, 249 (1996) [arXiv:hep-th/9506066]; V. P. Frolov and D. V. Fursaev, "Thermal fields, entropy, and black holes," Class. Quant. Grav. **15**, 2041 (1998) [arXiv:hep-th/9802010].

[49] I. Dymnikova, "Spherically symmetric space-time with the regular de Sitter center," arXiv:gr-qc/0304110.

[50] C. Barrabes and V. Frolov, "How Many New Worlds Are Inside a Black Hole?," Phys. Rev. D **53**, 3215 (1996) [arXiv:hep-th/9511136].

[51] T. Jacobson, "On the nature of black hole entropy," in *General Relativity and Relativistic Astrophysics: Eighth Canadian Conference*, AIP Conference Proceedings 493, eds. C. Burgess and R.C. Myers (AIP Press, 1999), pp. 85-97 [arXiv:gr-qc/9908031].

[52] T. Jacobson, "Black hole evaporation and ultrashort distances," Phys. Rev. D **44**, 1731 (1991).

[53] A. D. Helfer, "Do black holes radiate?," Rept. Prog. Phys. **66**, 943 (2003) [arXiv:gr-qc/0304042].

[54] G. T. Horowitz, "Quantum states of black holes," in *Black holes and relativistic stars*, ed. R. M. Wald (The University of Chicago Press, 1998) [arXiv:gr-qc/9704072].

[55] A. W. Peet, "The Bekenstein formula and string theory (N-brane theory)," Class. Quant. Grav. **15**, 3291 (1998) [arXiv:hep-th/9712253].

[56] O. Aharony, S. S. Gubser, J. M. Maldacena, H. Ooguri and Y. Oz, "Large N field theories, string theory and gravity," Phys. Rept. **323**, 183 (2000) [arXiv:hep-th/9905111].

[57] T. Jacobson, "Trans-Planckian redshifts and the substance of the space-time river," Prog. Theor. Phys. Suppl. **136**, 1 (1999) [arXiv:hep-th/0001085].

[58] T. Jacobson, "Lorentz violation and Hawking radiation," in *CPT and Lorentz Symmetry II*, ed. V.A. Kostelecky (World Scientific, Singapore, 2002) [arXiv:gr-qc/0110079].

[59] S. Corley and T. Jacobson, "Lattice black holes," Phys. Rev. D **57**, 6269 (1998) [arXiv:hep-th/9709166]; T. Jacobson and D. Mattingly, "Hawking radiation on a falling lattice," Phys. Rev. D **61**, 024017 (2000) [arXiv:hep-th/9908099].

[60] R. Parentani, "Quantum metric fluctuations and Hawking radiation," Phys. Rev. D **63**, 041503 (2001) [arXiv:gr-qc/0009011]; "Toward A Collective Treatment Of Quantum Gravitational Interactions," Int. J. Theor. Phys. **40**, 2201 (2001).

[61] J. C. Niemeyer, R. Parentani and D. Campo, "Minimal modifications of the primordial power spectrum from an adiabatic short distance cutoff," Phys. Rev. D **66**, 083510 (2002) [arXiv:hep-th/0206149]; C. Armendariz-Picon and E. A. Lim, "Vacuum choices and the predictions of inflation," arXiv:hep-th/0303103; J. Martin and R. Brandenberger, "On the dependence of the spectra of fluctuations in inflationary cosmology on trans-Planckian physics," arXiv:hep-th/0305161.

LARGE N FIELD THEORIES, STRING THEORY AND GRAVITY

Juan Maldacena

Institute for Advanced Study,
Einstein Dr. Princeton, NJ 08540, USA

Abstract   We describe the holographic correspondence between field theories and string/M theory, focusing on the relation between compactifications of string/M theory on Anti-de Sitter spaces and conformal field theories. We review the background for this correspondence and discuss its motivations and the evidence for its correctness. We describe the main results that have been derived from the correspondence in the regime that the field theory is approximated by classical or semiclassical gravity. We focus on the case of the $\mathcal{N} = 4$ supersymmetric gauge theory in four dimensions. These lecture notes are based on the Review written by O. Aharony, S. Gubser, J. Maldacena, H. Ooguri and Y. Oz, [1].

1.   General Introduction

These lecture notes are taken out of the review [1]. A more complete set of references is given there.

Even though though string theory is normally used as a theory of quantum gravity, it is not how string theory was originally discovered. String theory was discovered in an attempt to describe the large number of mesons and hadrons that were experimentally discovered in the 1960's. The idea was to view all these particles as different oscillation modes of a string. The string idea described well some features of the hadron spectrum. For example, the mass of the lightest hadron with a given spin obeys a relation like $m^2 \sim TJ^2 + const$. This is explained simply by assuming that the mass and angular momentum come from a rotating, relativistic string of tension T. It was later discovered that hadrons and mesons are actually made of quarks and that they are described by QCD.

QCD is a gauge theory based on the group $SU(3)$. This is sometimes stated by saying that quarks have three colors. QCD is asymptotically free, meaning that the effective coupling constant decreases as the energy increases. At low energies QCD becomes strongly coupled and it is not easy to perform calculations. One possible approach is to use numerical simulations on the lattice.

This is at present the best available tool to do calculations in QCD at low energies. It was suggested by 't Hooft that the theory might simplify when the number of colors N is large [7]. The hope was that one could solve exactly the theory with $N = \infty$, and then one could do an expansion in $1/N = 1/3$. Furthermore, as explained in the next section, the diagrammatic expansion of the field theory suggests that the large N theory is a free string theory and that the string coupling constant is $1/N$. If the case with $N = 3$ is similar to the case with $N = \infty$ then this explains why the string model gave the correct relation between the mass and the angular momentum. In this way the large N limit connects gauge theories with string theories. The 't Hooft argument, reviewed below, is very general, so it suggests that different kinds of gauge theories will correspond to different string theories. In this review we will study this correspondence between string theories and the large N limit of field theories. We will see that the strings arising in the large N limit of field theories are the same as the strings describing quantum gravity. Namely, string theory in some backgrounds, including quantum gravity, is equivalent (dual) to a field theory.

Strings are not consistent in four flat dimensions. Indeed, if one wants to quantize a four dimensional string theory an anomaly appears that forces the introduction of an extra field, sometimes called the "Liouville" field [8]. This field on the string worldsheet may be interpreted as an extra dimension, so that the strings effectively move in five dimensions. One might qualitatively think of this new field as the "thickness" of the string. If this is the case, why do we say that the string moves in five dimensions? The reason is that, like any string theory, this theory will contain gravity, and the gravitational theory will live in as many dimensions as the number of fields we have on the string. It is crucial then that the five dimensional geometry is curved, so that it can correspond to a four dimensional field theory, as described in detail below.

The argument that gauge theories are related to string theories in the large N limit is very general and is valid for basically any gauge theory. In particular we could consider a gauge theory where the coupling does not run (as a function of the energy scale). Then, the theory is conformally invariant. It is quite hard to find quantum field theories that are conformally invariant. In supersymmetric theories it is sometimes possible to prove exact conformal invariance. A simple example, which will be the main example in this review, is the supersymmetric $SU(N)$ (or $U(N)$) gauge theory in four dimensions with four spinor supercharges ($\mathcal{N} = 4$). Four is the maximal possible number of supercharges for a field theory in four dimensions. Besides the gauge fields (gluons) this theory contains also four fermions and six scalar fields in the adjoint representation of the gauge group. The Lagrangian of such theories is completely determined by supersymmetry. There is a global $SU(4)$ R-symmetry that rotates the six scalar fields and the four fermions. The conformal group in four dimensions is $SO(4,2)$, including the usual Poincaré transformations as well as scale transformations and special conformal transformations (which include the inversion symmetry $x^\mu \to x^\mu/x^2$). These symmetries of the field theory should be reflected in the dual string theory. The simplest way for this to happen is if the five dimensional geometry has these symmetries. Locally there is only one space with $SO(4,2)$ isometries: five dimensional Anti-de-Sitter space, or AdS_5. Anti-de Sitter space is the maximally symmetric solution of Einstein's equations with a negative cosmological constant. In this supersymmetric case we expect the strings to also be supersymmetric. We said that superstrings move in ten dimensions. Now that we have added one more dimension it is not surprising any more to add five more to get to a ten dimensional space. Since the gauge theory has an $SU(4) \simeq SO(6)$ global symmetry it is rather natural that the extra five dimensional space should be a five sphere, S^5. So, we conclude that $\mathcal{N} = 4\ U(N)$ Yang-Mills theory could be the same as ten dimensional superstring theory on $AdS_5 \times S^5$ [9]. Here we have presented a very heuristic argument for this equivalence; later we will be more precise and give more evidence for this correspondence.

The relationship we described between gauge theories and string theory on Anti-de-Sitter spaces was motivated by studies of D-branes and black holes in strings theory. D-branes are solitons in string theory [10]. They come in various dimensionalities. If they have zero spatial dimensions they are like ordinary localized, particle-type soliton solutions, analogous to the 't Hooft-Polyakov [11, 12] monopole in gauge theories. These are called D-zero-branes. If they have one extended dimension they are called D-one-branes or D-strings. They are much heavier than ordinary fundamental strings when the string coupling is small. In fact, the tension of all D-branes is proportional to $1/g_s$, where g_s is the string coupling constant. D-branes are defined in string perturbation theory in a very simple way: they are surfaces where open strings can end. These open strings have some massless modes, which describe the oscillations of the branes, a gauge field living on the brane, and their fermionic partners. If we have N coincident branes the open strings can start and end on different branes, so they carry two indices that run from one to N. This in turn implies that the low energy dynamics is described by a $U(N)$ gauge theory. D-p-branes are charged under $p + 1$-form gauge potentials, in the same way that a 0-brane (particle) can be charged under a one-form gauge potential (as in electromagnetism). These $p + 1$-form gauge potentials have $p + 2$-form field strengths, and they are part of the massless closed string modes, which belong to the supergravity (SUGRA) multiplet containing the massless fields in flat space string theory (before we put in any D-branes). If we now add D-branes they generate a flux of the corresponding field strength, and this flux in turn contributes to the stress energy tensor so the geometry becomes curved. Indeed it is possible to find solutions of the supergravity equations carrying these fluxes. Supergravity is the low-energy limit of string theory, and it is believed that these solutions may be extended to solutions of the full string theory. These solutions are very similar to extremal charged black hole solutions in general relativity, except that in this case they are black branes with p extended spatial dimensions. Like black holes they contain event horizons.

If we consider a set of N coincident D-3-branes the near horizon geometry turns out to be $AdS_5 \times S^5$. On the other hand, the low energy dynamics on their worldvolume is governed by a $U(N)$ gauge theory with $\mathcal{N} = 4$ supersymmetry [13]. These two pictures of D-branes are perturbatively valid for different regimes in the space of possible coupling constants. Perturbative field theory is valid when $g_s N$ is small, while the low-energy gravitational description is perturbatively valid when the radius of curvature is much larger than the string scale, which turns out to imply that $g_s N$ should be very large. As an object is brought closer and closer to the black brane horizon its energy measured by an outside observer is redshifted, due to the large gravitational potential, and the energy seems to be very small. On the other hand low energy excitations on the branes are governed by the Yang-Mills theory. So, it becomes natural to conjecture that Yang-Mills theory at strong coupling is describing the near horizon region of the black brane, whose geometry is $AdS_5 \times S^5$. The first indications that this is the case came from calculations of low energy graviton absorption cross sections [14, 15, 16]. It was noticed there that the calculation done using gravity and the calculation done using super Yang-Mills theory agreed. These calculations, in turn, were inspired by similar calculations for coincident D1-D5 branes. In this case the near horizon geometry involves $AdS_3 \times S^3$ and the low energy field theory living on the D-branes is a 1+1 dimensional conformal field theory. In this D1-D5 case there were numerous calculations that agreed between the field theory and gravity. First black hole entropy for extremal black holes was calculated in terms of the field theory in [17], and then agreement was shown for near extremal black holes [18, 19] and for absorption cross sections [20, 21, 22]. More generally, we will see that correlation functions in the gauge theory can be calculated using the string theory (or gravity for large $g_s N$) description, by considering the propagation of particles between different points in the boundary of AdS, the points where operators are inserted [23, 24].

Supergravities on AdS spaces were studied very extensively, see [25, 26] for reviews. See also [2, 3] for earlier hints of the correspondence.

One of the main points of these lectures will be that the strings coming from gauge theories are very much like the ordinary superstrings that have been studied during the last 20 years. The only particular feature is that they are moving on a curved geometry (anti-de Sitter space) which has a boundary at spatial infinity. The boundary is at an infinite spatial distance, but a light ray can go to the boundary and come back in finite time. Massive particles can never get to the boundary. The radius of curvature of Anti-de Sitter space depends on

N so that large N corresponds to a large radius of curvature. Thus, by taking N to be large we can make the curvature as small as we want. The theory in AdS includes gravity, since any string theory includes gravity. So in the end we claim that there is an equivalence between a gravitational theory and a field theory. However, the mapping between the gravitational and field theory degrees of freedom is quite non-trivial since the field theory lives in a lower dimension. In some sense the field theory (or at least the set of local observables in the field theory) lives on the boundary of spacetime. One could argue that in general any quantum gravity theory in AdS defines a conformal field theory (CFT) "on the boundary". In some sense the situation is similar to the correspondence between three dimensional Chern-Simons theory and a WZW model on the boundary [27]. This is a topological theory in three dimensions that induces a normal (non-topological) field theory on the boundary. A theory which includes gravity is in some sense topological since one is integrating over all metrics and therefore the theory does not depend on the metric. Similarly, in a quantum gravity theory we do not have any local observables. Notice that when we say that the theory includes "gravity on AdS" we are considering any finite energy excitation, even black holes in AdS. So this is really a sum over all spacetimes that are asymptotic to AdS at the boundary. This is analogous to the usual flat space discussion of quantum gravity, where asymptotic flatness is required, but the spacetime could have any topology as long as it is asymptotically flat. The asymptotically AdS case as well as the asymptotically flat cases are special in the sense that one can choose a natural time and an associated Hamiltonian to define the quantum theory. Since black holes might be present this time coordinate is not necessarily globally well-defined, but it is certainly well-defined at infinity. If we assume that the conjecture we made above is valid, then the $U(N)$ Yang-Mills theory gives a non-perturbative definition of string theory on AdS. And, by taking the limit $N \to \infty$, we can extract the (ten dimensional string theory) flat space physics, a procedure which is in principle (but not in detail) similar to the one used in matrix theory [28].

The fact that the field theory lives in a lower dimensional space blends in perfectly with some previous speculations about quantum gravity. It was suggested [29, 30] that quantum gravity theories should be holographic, in the sense that physics in some region can be described by a theory at the boundary with no more than one degree of freedom per Planck area. This "holographic" principle comes from thinking about the Bekenstein bound which states that the maximum amount of entropy in some region is given by the area of the region in Planck units [31]. The reason for this bound is that otherwise black hole formation could violate the second law of thermodynamics. We will see that the correspondence between field theories and string theory on AdS space (including gravity) is a concrete realization of this holographic principle.

Other reviews of this subject are [32, 33, 34, 35, 1].

2.   The Correspondence

In this section we will present an argument connecting type IIB string theory compactified on $AdS_5 \times S^5$ to $\mathcal{N} = 4$ super-Yang-Mills theory [9]. Let us start with type IIB string theory in flat, ten dimensional Minkowski space. Consider N parallel D3 branes that are sitting together or very close to each other (the precise meaning of "very close" will be defined below). The D3 branes are extended along a $(3 + 1)$ dimensional plane in $(9 + 1)$ dimensional spacetime. String theory on this background contains two kinds of perturbative excitations, closed strings and open strings. The closed strings are the excitations of empty space and the open strings end on the D-branes and describe excitations of the D-branes. If we consider the system at low energies, energies lower than the string scale $1/l_s$, then only the massless string states can be excited, and we can write an effective Lagrangian describing their interactions. The closed string massless states give a gravity supermultiplet in ten dimensions, and their low-energy effective Lagrangian is that of type IIB supergravity. The open string massless states give an $\mathcal{N} = 4$ vector supermultiplet in $(3+1)$ dimensions, and their low-energy effective Lagrangian is that of $\mathcal{N} = 4\, U(N)$ super-Yang-Mills theory [13, 36].

The complete effective action of the massless modes will have the form

$$S = S_{\text{bulk}} + S_{\text{brane}} + S_{\text{int}}. \tag{1}$$

S_{bulk} is the action of ten dimensional supergravity, plus some higher derivative corrections. Note that the Lagrangian (1) involves only the massless fields but it takes into account the effects of integrating out the massive fields. It is not renormalizable (even for the fields on the brane), and it should only be understood as an effective description in the Wilsonian sense, i.e. we integrate out all massive degrees of freedom but we do not integrate out the massless ones. The brane action S_{brane} is defined on the $(3 + 1)$ dimensional brane worldvolume, and it contains the $\mathcal{N} = 4$ super-Yang-Mills Lagrangian plus some higher derivative corrections, for example terms of the form $\alpha'^2 \text{Tr}\,(F^4)$. Finally, S_{int} describes the interactions between the brane modes and the bulk modes. The leading terms in this interaction Lagrangian can be obtained by covariantizing the brane action, introducing the background metric for the brane [37].

We can expand the bulk action as a free quadratic part describing the propagation of free massless modes (including the graviton), plus some interactions which are proportional to positive powers of the square root of the Newton constant. Schematically we have

$$S_{bulk} \sim \frac{1}{2\kappa^2} \int \sqrt{g}\mathcal{R} \sim \int (\partial h)^2 + \kappa(\partial h)^2 h + \cdots, \tag{2}$$

where we have written the metric as $g = \eta + \kappa h$. We indicate explicitly the dependence on the graviton, but the other terms in the Lagrangian, involving other fields, can be expanded in a similar way. Similarly, the interaction Lagrangian S_{int} is proportional to positive powers of κ. If we take the low energy limit, all interaction terms proportional to κ drop out. This is the well known fact that gravity becomes free at long distances (low energies).

In order to see more clearly what happens in this low energy limit it is convenient to keep the energy fixed and send $l_s \to 0$ ($\alpha' \to 0$) keeping all the dimensionless parameters fixed, including the string coupling constant and N. In this limit the coupling $\kappa \sim g_s \alpha'^2 \to 0$, so that the interaction Lagrangian relating the bulk and the brane vanishes. In addition all the higher derivative terms in the brane action vanish, leaving just the pure $\mathcal{N} = 4$ $U(N)$ gauge theory in $3 + 1$ dimensions, which is known to be a conformal field theory. And, the supergravity theory in the bulk becomes free. So, in this low energy limit we have two decoupled systems. On the one hand we have free gravity in the bulk and on the other hand we have the four dimensional gauge theory.

Next, we consider the same system from a different point of view. D-branes are massive charged objects which act as a source for the various supergravity fields. We can find a D3 brane solution [38] of supergravity, of the form

$$ds^2 = f^{-1/2}(-dt^2 + dx_1^2 + dx_2^2 + dx_3^2) + f^{1/2}(dr^2 + r^2 d\Omega_5^2) \,,$$
$$F_5 = (1 + *)dtdx_1dx_2dx_3df^{-1} \,, \tag{3}$$
$$f = 1 + \frac{R^4}{r^4} \,, \qquad R^4 \equiv 4\pi g_s \alpha'^2 N \,.$$

Note that since g_{tt} is non-constant, the energy E_p of an object as measured by an observer at a constant position r and the energy E measured by an observer at infinity are related by the redshift factor

$$E = f^{-1/4} E_p \,. \tag{4}$$

This means that the same object brought closer and closer to $r = 0$ would appear to have lower and lower energy for the observer at infinity. Now we take the low energy limit in the background described by equation (3). There are two kinds of low energy excitations (from the point of view of an observer at infinity). We can have massless particles propagating in the bulk region with wavelengths that becomes very large, or we can have any kind of excitation that we bring closer and closer to $r = 0$. In the low energy limit these two types of excitations decouple from each other. The bulk massless particles decouple from the near horizon region (around $r = 0$) because the low energy absorption cross section goes like $\sigma \sim \omega^3 R^8$ [14, 15], where ω is the energy. This can be understood from the fact that in this limit the wavelength of the particle becomes much bigger than the typical gravitational size of the brane (which is of order R). Similarly, the excitations that live very close to $r = 0$ find it harder and harder to climb the gravitational potential and escape to the asymptotic region. In conclusion, the low energy theory consists of two decoupled pieces, one is free bulk supergravity and the second is the near horizon region of the geometry. In the near horizon region, $r \ll R$, we can approximate $f \sim R^4/r^4$, and the geometry becomes

$$ds^2 = \frac{r^2}{R^2}(-dt^2 + dx_1^2 + dx_2^2 + dx_3^2) + R^2\frac{dr^2}{r^2} + R^2 d\Omega_5^2, \qquad (5)$$

which is the geometry of $AdS_5 \times S^5$.

We see that both from the point of view of a field theory of open strings living on the brane, and from the point of view of the supergravity description, we have two decoupled theories in the low-energy limit. In both cases one of the decoupled systems is supergravity in flat space. So, it is natural to identify the second system which appears in both descriptions. Thus, we are led to the conjecture that $\mathcal{N} = 4$ $U(N)$ *super-Yang-Mills theory in $3 + 1$ dimensions is the same as (or dual to) type IIB superstring theory on $AdS_5 \times S^5$* [9].

We could be a bit more precise about the near horizon limit and how it is being taken. Suppose that we take $\alpha' \to 0$, as we did when we discussed the field theory living on the brane. We want to keep fixed the energies of the objects in the throat (the near-horizon region) in string units, so that we can consider arbitrary excited string states there. This implies that $\sqrt{\alpha'} E_p \sim$ fixed. For small α' (4) reduces to $E \sim E_p r/\sqrt{\alpha'}$. Since we want to keep fixed the energy measured from infinity, which is the way energies are measured in the field theory, we need to take $r \to 0$ keeping r/α' fixed. It is then convenient to define a new variable $U \equiv r/\alpha'$, so that the metric becomes

$$ds^2 = \alpha'\left[\frac{U^2}{\sqrt{4\pi g_s N}}(-dt^2 + dx_1^2 + dx_2^2 + dx_3^2) + \sqrt{4\pi g_s N}\frac{dU^2}{U^2} + \sqrt{4\pi g_s N}d\Omega_5^2\right]. \qquad (6)$$

This can also be seen by considering a D3 brane sitting at $\vec{r}$. This corresponds to giving a vacuum expectation value to one of the scalars in the Yang-Mills theory. When we take the $\alpha' \to 0$ limit we want to keep the mass of the "W-boson" fixed. This mass, which is the mass of the string stretching between the branes sitting at $\vec{r} = 0$ and the one at $\vec{r}$, is proportional to $U = r/\alpha'$, so this quantity should remain fixed in the decoupling limit.

A $U(N)$ gauge theory is essentially equivalent to a free $U(1)$ vector multiplet times an $SU(N)$ gauge theory, up to some $\mathbb{Z}_N$ identifications (which affect only global issues). In the dual string theory all modes interact with gravity, so there are no decoupled modes. Therefore, the bulk AdS theory is describing the

$SU(N)$ part of the gauge theory. In fact we were not precise when we said that there were two sets of excitations at low energies, the excitations in the asymptotic flat space and the excitations in the near horizon region. There are also some zero modes which live in the region connecting the "throat" (the near horizon region) with the bulk, which correspond to the $U(1)$ degrees of freedom mentioned above. The $U(1)$ vector supermultiplet includes six scalars which are related to the center of mass motion of all the branes [39]. From the AdS point of view these zero modes live at the boundary, and it looks like we might or might not decide to include them in the AdS theory. Depending on this choice we could have a correspondence to an $SU(N)$ or a $U(N)$ theory. The $U(1)$ center of mass degree of freedom is related to the topological theory of B-fields on AdS [40]; if one imposes local boundary conditions for these B-fields at the boundary of AdS one finds a $U(1)$ gauge field living at the boundary [41], as is familiar in Chern-Simons theories [27, 42]. These modes living at the boundary are sometimes called singletons (or doubletons) [43, 44, 45, 46, 47, 48, 49, 50, 51].

Anti-de-Sitter space has a large group of isometries, which is $SO(4,2)$ for the case at hand. This is the same group as the conformal group in $3 + 1$ dimensions. Thus, the fact that the low-energy field theory on the brane is conformal is reflected in the fact that the near horizon geometry is Anti-de-Sitter space. We also have some supersymmetries. The number of supersymmetries is twice that of the full solution (3) containing the asymptotic region [39]. This doubling of supersymmetries is viewed in the field theory as a consequence of superconformal invariance, since the superconformal algebra has twice as many fermionic generators as the corresponding Poincare superalgebra. We also have an $SO(6)$ symmetry which rotates the S^5. This can be identified with the $SU(4)_R$ R-symmetry group of the field theory. In fact, the whole supergroup is the same for the $\mathcal{N} = 4$ field theory and the $AdS_5 \times S^5$ geometry, so both sides of the conjecture have the same spacetime symmetries. We will discuss in more detail the matching between the two sides of the correspondence in section 3.

In the above derivation the field theory is naturally defined on $\mathbb{R}^{3,1}$, but we could also think of the conformal field theory as defined on $S^3 \times \mathbb{R}$ by redefining the Hamiltonian. Since the isometries of AdS are in one to one correspondence with the generators of the conformal group of the field theory, we can conclude that this new Hamiltonian $\frac{1}{2}(P_0 + K_0)$ can be associated on AdS to the generator of translations in global time. This formulation of the conjecture is more useful since in the global coordinates there is no horizon. When we put the field theory on S^3 the Coulomb branch is lifted and there is a unique ground state. This is due to the fact that the scalars ϕ^I in the field theory are conformally coupled, so there is a term of the form $\int d^4x \mathrm{Tr}\,(\phi^2)\mathcal{R}$ in the Lagrangian, where $\mathcal{R}$ is the curvature of the four-dimensional space on which the theory is defined. Due to the positive curvature of S^3 this leads to a mass term for the scalars [24], lifting the moduli space.

The parameter N appears on the string theory side as the flux of the five-form Ramond-Ramond field strength on the S^5,

$$\int_{S^5} F_5 = N. \tag{7}$$

From the physics of D-branes we know that the Yang-Mills coupling is related to the string coupling through [10, 52]

$$\tau \equiv \frac{4\pi i}{g_{YM}^2} + \frac{\theta}{2\pi} = \frac{i}{g_s} + \frac{\chi}{2\pi}, \tag{8}$$

where we have also included the relationship of the θ angle to the expectation value of the RR scalar χ. We have written the couplings in this fashion because both the gauge theory and the string theory have an $SL(2, \mathbb{Z})$ self-duality symmetry under which $\tau \to (a\tau + b)/(c\tau + d)$ (where a, b, c, d are integers with $ad - bc = 1$). In fact, $SL(2, \mathbb{Z})$ is a conjectured strong-weak coupling duality symmetry of type IIB string theory in flat space [53], and it should also be a symmetry in the present context since all the fields that are being turned on in the $AdS_5 \times S^5$ background (the metric and the five form field strength) are invariant under this symmetry. The connection between the $SL(2, \mathbb{Z})$ duality symmetries of type IIB string theory and $\mathcal{N} = 4$ SYM was noted in [54, 55, 56]. The string theory seems to have a parameter that does not appear in the gauge theory, namely α', which sets the string tension and all other scales in the string theory. However, this is not really a parameter in the theory if we do not compare it to other scales in the theory, since only relative scales are meaningful. In fact, only the ratio of the radius of curvature to α' is a parameter, but not α' and the radius of curvature independently. Thus, α' will disappear from any final physical quantity we compute in this theory. It is sometimes convenient, especially when one is doing gravity calculations, to set the radius of curvature to one. This can be achieved by writing the metric as $ds^2 = R^2 d\tilde{s}^2$, and rewriting everything in terms of $\tilde{g}$. With these conventions $G_N \sim 1/N^2$ and $\alpha' \sim 1/\sqrt{g_s N}$. This implies that any quantity calculated purely in terms of the gravity solution, without including stringy effects, will be independent of $g_s N$ and will depend only on N. α' corrections to the gravity results give corrections which are proportional to powers of $1/\sqrt{g_s N}$.

Now, let us address the question of the validity of various approximations. The analysis of loop diagrams in the field theory shows that we can trust the perturbative analysis in the Yang-Mills theory when

$$g_{YM}^2 N \sim g_s N \sim \frac{R^4}{l_s^4} \ll 1. \tag{9}$$

Note that we need $g_{YM}^2 N$ to be small and not just g_{YM}^2. On the other hand, the classical gravity description becomes reliable when the radius of curvature R of AdS and of S^5 becomes large compared to the string length,

$$\frac{R^4}{l_s^4} \sim g_s N \sim g_{YM}^2 N \gg 1. \tag{10}$$

We see that the gravity regime (10) and the perturbative field theory regime (9) are perfectly incompatible. In this fashion we avoid any obvious contradiction due to the fact that the two theories look very different. This is the reason that this correspondence is called a "duality". The two theories are conjectured to be exactly the same, but when one side is weakly coupled the other is strongly coupled and vice versa. This makes the correspondence both hard to prove and useful, as we can solve a strongly coupled gauge theory via classical supergravity. Notice that in (9)(10) we implicitly assumed that $g_s < 1$. If $g_s > 1$ we can perform an $SL(2, \mathbb{Z})$ duality transformation and get conditions similar to (9)(10) but with $g_s \to 1/g_s$. So, we cannot get into the gravity regime (10) by taking N small ($N = 1, 2, ..$) and g_s very large, since in that case the D-string becomes light and renders the gravity approximation invalid. Another way to see this is to note that the radius of curvature in Planck units is $R^4/l_p^4 \sim N$. So, it is always necessary, but not sufficient, to have large N in order to have a weakly coupled supergravity description.

One might wonder why the above argument was not a proof rather than a conjecture. It is not a proof because we did not treat the string theory non-perturbatively (not even non-perturbatively in α'). We could also consider different forms of the conjecture. In its weakest form the gravity description would be valid for large $g_s N$, but the full string theory on AdS might not agree with the field theory. A not so weak form would say that the conjecture is valid even for finite $g_s N$, but only in the $N \to \infty$ limit (so that the α' corrections would agree with the field theory, but the g_s corrections may not). The strong form of the conjecture, which is the most interesting one and which we will assume here, is that the two theories are exactly the same for all values of g_s and N. In this conjecture the spacetime is only required to be asymptotic to $AdS_5 \times S^5$ as we approach the boundary. In the interior we can have all kinds of processes; gravitons, highly excited fundamental string states, D-branes, black holes, etc. Even the topology of spacetime can change in the interior. The Yang-Mills theory is supposed to effectively sum over all spacetimes which are asymptotic to $AdS_5 \times S^5$. This is completely analogous to the usual conditions of asymptotic flatness. We can have black holes and all kinds of topology changing processes, as long as spacetime is asymptotically flat. In this case asymptotic flatness is replaced by the asymptotic AdS behavior.

2.1   The Field $\leftrightarrow$ Operator Correspondence

A conformal field theory does not have asymptotic states or an S-matrix, so the natural objects to consider are operators. For example, in $\mathcal{N} = 4$ super-Yang-Mills we have a deformation by a marginal operator which changes the value of the coupling constant. Changing the coupling constant in the field theory is related by (8) to changing the coupling constant in the string theory, which is then related to the expectation value of the dilaton. The expectation value of the dilaton is set by the boundary condition for the dilaton at infinity. So, changing the gauge theory coupling constant corresponds to changing the boundary value of the dilaton. More precisely, let us denote by $\mathcal{O}$ the corresponding operator. We can consider adding the term $\int d^4 x \phi_0(\vec{x}) \mathcal{O}(\vec{x})$ to the Lagrangian (for simplicity we assume that such a term was not present in the original Lagrangian, otherwise we consider $\phi_0(\vec{x})$ to be the total coefficient of $\mathcal{O}(\vec{x})$ in the Lagrangian). According to the discussion above, it is natural to assume that this will change the boundary condition of the dilaton at the boundary of AdS to $\phi(\vec{x}, z)|_{z=0} = \phi_0(\vec{x})$, in the coordinate system

$$ds^2 = R_{AdS}^2 \frac{-dt^2 + dx_1^2 + \cdots + dx_3^3 + dz^2}{z^2}.$$

More precisely, as argued in [23, 24], it is natural to propose that

$$\langle e^{\int d^4 x \phi_0(\vec{x}) \mathcal{O}(\vec{x})} \rangle_{CFT} = \mathcal{Z}_{string}\left[\phi(\vec{x}, z)\Big|_{z=0} = \phi_0(\vec{x})\right], \qquad (11)$$

where the left hand side is the generating function of correlation functions in the field theory, i.e. ϕ_0 is an arbitrary function and we can calculate correlation functions of $\mathcal{O}$ by taking functional derivatives with respect to ϕ_0 and then setting $\phi_0 = 0$. The right hand side is the full partition function of string theory with the boundary condition that the field ϕ has the value ϕ_0 on the boundary of AdS. Notice that ϕ_0 is a function of the four variables parametrizing the boundary of AdS_5.

A formula like (11) is valid in general, for any field ϕ. Therefore, each field propagating on AdS space is in a one to one correspondence with an operator in the field theory. There is a relation between the mass of the field ϕ and the scaling dimension of the operator in the conformal field theory. Let us describe this more generally in AdS_{d+1}. The wave equation in Euclidean space for a field of mass m has two independent solutions, which behave like $z^{d-\Delta}$ and z^{Δ} for small z (close to the boundary of AdS), where

$$\Delta = \frac{d}{2} + \sqrt{\frac{d^2}{4} + R^2 m^2}. \qquad (12)$$

Therefore, in order to get consistent behavior for a massive field, the boundary condition on the field in the right hand side of (11) should in general be changed to

$$\phi(\vec{x}, \epsilon) = \epsilon^{d-\Delta}\phi_0(\vec{x}), \qquad (13)$$

and eventually we would take the limit where $\epsilon \to 0$. Since ϕ is dimensionless, we see that ϕ_0 has dimensions of $[\text{length}]^{\Delta-d}$ which implies, through the left hand side of (11), that the associated operator $\mathcal{O}$ has dimension Δ (12). A more detailed derivation of this relation will be given in section 4, where we will verify that the two-point correlation function of the operator $\mathcal{O}$ behaves as that of an operator of dimension Δ [23, 24]. A similar relation between fields on AdS and operators in the field theory exists also for non-scalar fields, including fermions and tensors on AdS space.

Correlation functions in the gauge theory can be computed from (11) by differentiating with respect to ϕ_0. Each differentiation brings down an insertion $\mathcal{O}$, which sends a ϕ particle (a closed string state) into the bulk. Feynman diagrams can be used to compute the interactions of particles in the bulk. In the limit where classical supergravity is applicable, the only diagrams that contribute are the tree-level diagrams of the gravity theory (see for instance figure 1).

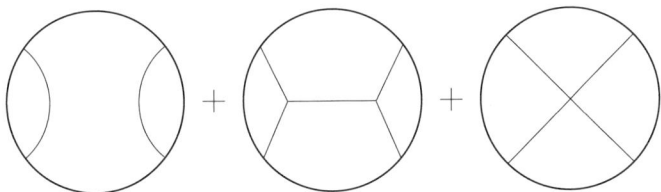

Figure 1. Correlation functions can be calculated (in the large $g_s N$ limit) in terms of super-gravity Feynman diagrams. Here we see the leading contribution coming from a disconnected diagram plus connected pieces involving interactions of the supergravity fields in the bulk of AdS. At tree level, these diagrams and those related to them by crossing are the only ones that contribute to the four-point function.

This method of defining the correlation functions of a field theory which is dual to a gravity theory in the bulk of AdS space is quite general, and it applies in principle to any theory of gravity [24]. Any local field theory contains the stress tensor as an operator. Since the correspondence described above matches the stress-energy tensor with the graviton, this implies that the AdS theory includes gravity. It should be a well defined quantum theory of gravity since we should be able to compute loop diagrams. String theory provides such a theory. But if a new way of defining quantum gravity theories comes along we could consider those gravity theories in AdS, and they should correspond to some conformal field theory "on the boundary". In particular, we could consider backgrounds of string theory of the form $AdS_5 \times M^5$ where M^5 is any Einstein manifold [63, 64, 65]. Depending on the choice of M^5 we get different dual conformal field theories. Similarly, this discussion can be extended to any AdS_{d+1} space, corresponding to a conformal field theory in d spacetime dimensions (for $d > 1$).

2.2 Holography

In this section we will describe how the AdS/CFT correspondence gives a holographic description of physics in AdS spaces.

Let us start by explaining the Bekenstein bound, which states that the maximum entropy in a region of space is $S_{max} = \text{Area}/4G_N$ [31], where the area is that of the boundary of the region. Suppose that we had a state with more entropy than S_{max}, then we show that we could violate the second law of thermodynamics. We can throw in some extra matter such that we form a black hole. The entropy should not decrease. But if a black hole forms inside the region its entropy is just the area of its horizon, which is smaller than the area of the boundary of the region (which by our assumption is smaller than the initial entropy). So, the second law has been violated.

Note that this bound implies that the number of degrees of freedom inside some region grows as the area of the boundary of a region and not like the volume of the region. In standard quantum field theories this is certainly not possible. Attempting to understand this behavior leads to the "holographic principle", which states that in a quantum gravity theory all physics within some volume can be described in terms of some theory on the boundary which has less than one degree of freedom per Planck area [29, 30] (so that its entropy satisfies the Bekenstein bound).

In the AdS/CFT correspondence we are describing physics in the bulk of AdS space by a field theory of one less dimension (which can be thought of as living on the boundary), so it looks like holography. However, it is hard to check what the number of degrees of freedom per Planck area is, since the theory, being conformal, has an infinite number of degrees of freedom, and the area of the boundary of AdS space is also infinite. Thus, in order to compare things properly we should introduce a cutoff on the number of degrees of freedom in the field theory and see what it corresponds to in the gravity theory. For this purpose let us write the metric of AdS as

$$ds^2 = R^2 \left[-\left(\frac{1+r^2}{1-r^2} \right)^2 dt^2 + \frac{4}{(1-r^2)^2}(dr^2 + r^2 d\Omega^2) \right]. \quad (14)$$

In these coordinates the boundary of AdS is at $r = 1$. We saw above that when we calculate correlation functions we have to specify boundary conditions at $r = 1 - \delta$ and then take the limit of $\delta \to 0$. It is clear by studying the action of the conformal group on Poincaré coordinates that the radial position plays the role of some energy scale, since we approach the boundary when we do a conformal transformation that localizes objects in the CFT. So, the limit $\delta \to 0$ corresponds to going to the UV of the field theory. When we are close to the boundary we could also use the Poincaré coordinates

$$ds^2 = R^2 \frac{-dt^2 + d\vec{x}^2 + dz^2}{z^2}, \tag{15}$$

in which the boundary is at $z = 0$. If we consider a particle or wave propagating in (15) or (14) we see that its motion is independent of R in the supergravity approximation. Furthermore, if we are in Euclidean space and we have a wave that has some spatial extent λ in the $\vec{x}$ directions, it will also have an extent λ in the z direction. This can be seen from (15) by eliminating λ through the change of variables $x \to \lambda x$, $z \to \lambda z$. This implies that a cutoff at

$$z \sim \delta \tag{16}$$

corresponds to a UV cutoff in the field theory at distances δ, with no factors of R (δ here is dimensionless, in the field theory it is measured in terms of the radius of the S^4 or S^3 that the theory lives on). Equation (16) is called the UV-IR relation [66].

Consider the case of $\mathcal{N} = 4$ SYM on a three-sphere of radius one. We can estimate the number of degrees of freedom in the field theory with a UV cutoff δ. We get

$$S \sim N^2 \delta^{-3}, \tag{17}$$

since the number of cells into which we divide the three-sphere is of order $1/\delta^3$. In the gravity solution (14) the area in Planck units of the surface at $r = 1 - \delta$, for $\delta \ll 1$, is

$$\frac{\text{Area}}{4G_N} = \frac{V_{S^5} R^3 \delta^{-3}}{4G_N} \sim N^2 \delta^{-3}. \tag{18}$$

Thus, we see that the AdS/CFT correspondence saturates the holographic bound [66].

One could be a little suspicious of the statement that gravity in AdS is holographic, since it does not seem to be saying much because in AdS space the volume and the boundary area of a given region scale in the same fashion as we increase the size of the region. In fact, *any* field theory in AdS would be holographic in the sense that the number of degrees of freedom within some (large enough) volume is proportional to the area (and also to the volume). What makes this case different is that we have the additional parameter R, and then we can take AdS spaces of different radii (corresponding to different values of N in the SYM theory), and then we can ask whether the number of degrees of freedom goes like the volume or the area, since these have a different dependence on R.

One might get confused by the fact that the surface $r = 1 - \delta$ is really nine dimensional as opposed to four dimensional. From the form of the full metric on $AdS_5 \times S^5$ we see that as we take $\delta \to 0$ the physical size of four of the dimensions of this nine dimensional space grow, while the other five, the S^5, remain constant. So, we see that the theory on this nine dimensional surface becomes effectively four dimensional, since we need to multiply the metric by a factor that goes to zero as we approach the boundary in order to define a finite metric for the four dimensional gauge theory.

3.   Tests of the AdS/CFT Correspondence

In this section we review the direct tests of the AdS/CFT correspondence. In section 2 we saw how string theory on AdS defines a partition function which can be used to define a field theory. Here we will review the evidence showing that this field theory is indeed the same as the conjectured dual field theory. We will focus here only on tests of the correspondence between the $\mathcal{N} = 4 \, SU(N)$ SYM theory and the type IIB string theory compactified on $AdS_5 \times S^5$; most of the tests described here can be generalized also to cases in other dimensions and/or with less supersymmetry, which will be described below.

As described in section 2, the AdS/CFT correspondence is a strong/weak coupling duality. In the 't Hooft large N limit, it relates the region of weak field theory coupling $\lambda = g_{YM}^2 N$ in the SYM theory to the region of high curvature (in string units) in the string theory, and vice versa. Thus, a direct comparison of correlation functions is generally not possible, since (with our current knowledge) we can only compute most of them perturbatively in λ on the field theory side and perturbatively in $1/\sqrt{\lambda}$ on the string theory side. For example, as described below, we can compute the equation of state of the SYM theory and also the quark-anti-quark potential both for small λ and for large λ, and we obtain different answers, which we do not know how to compare since we can only compute them perturbatively on both sides. A similar situation arises also in many field theory dualities that were analyzed in the last few years (such as the electric/magnetic $SL(2, \mathbb{Z})$ duality of the $\mathcal{N} = 4$ SYM theory itself), and it was realized that there are several properties of these theories which do not depend on the coupling, so they can be compared to test the duality. These are:

- The global symmetries of the theory, which cannot change as we change the coupling (except for extreme values of the coupling). As discussed in section 2, in the case of the AdS/CFT correspondence we have the same supergroup $SU(2, 2|4)$ (whose bosonic subgroup is $SO(4, 2) \times SU(4)$) as the global symmetry of both theories. Also, both theories are believed to have a non-perturbative $SL(2, \mathbb{Z})$ duality symmetry acting on their coupling constant τ. These are the only symmetries of the theory on $\mathbb{R}^4$. Additional $\mathbb{Z}_N$ symmetries arise when the theories are compactified on non-simply-connected manifolds, and these were also successfully matched in [67, 40][1].

- Some correlation functions, which are usually related to anomalies, are protected from any quantum corrections and do not depend on λ. The matching of these correlation functions will be described in section 3.2 below.

- The spectrum of chiral operators does not change as the coupling varies, and it will be compared in section 3.1 below.

- The moduli space of the theory also does not depend on the coupling. In the $SU(N)$ field theory the moduli space is $\mathbb{R}^{6(N-1)}/S_N$, parametrized by the eigenvalues of six commuting traceless $N \times N$ matrices. On the AdS side it is not clear exactly how to define the moduli space. As described in section **??**, there is a background of string theory corresponding to any point in the field theory moduli space, but it is not clear how to see that this is the exact moduli space on the string theory side (especially since high curvatures arise for generic points in the moduli space).

- The qualitative behavior of the theory upon deformations by relevant or marginal operators also does not depend on the coupling (at least for chiral operators whose dimension does not depend on the coupling, and in the absence of phase transitions).

  There are many more qualitative tests of the correspondence, such as the existence of confinement for the finite temperature theory [68], which we will not discuss in this section. We will also not discuss here tests involving the behavior of the theory on its moduli space [60, 69, 61].

3.1 The Spectrum of Chiral Primary Operators

3.1.1 The Field Theory Spectrum. The $\mathcal{N} = 4$ supersymmetry algebra in $d = 4$ has four generators Q_α^A (and their complex conjugates $\bar{Q}_{\dot{\alpha}A}$), where α is a Weyl-spinor index (in the **2** of the $SO(3,1)$ Lorentz group) and A is an index in the **4** of the $SU(4)_R$ R-symmetry group (lower indices A will be taken to transform in the $\bar{\mathbf{4}}$ representation). They obey the algebra

$$\{Q_\alpha^A, \bar{Q}_{\dot{\alpha}B}\} = 2(\sigma^\mu)_{\alpha\dot{\alpha}} P_\mu \delta_B^A,$$
$$\{Q_\alpha^A, Q_\beta^B\} = \{\bar{Q}_{\dot{\alpha}A}, \bar{Q}_{\dot{\beta}B}\} = 0, \tag{19}$$

where σ^i ($i = 1, 2, 3$) are the Pauli matrices and $(\sigma^0)_{\alpha\dot{\alpha}} = -\delta_{\alpha\dot{\alpha}}$ (we use the conventions of Wess and Bagger [70]).

---

[1]Unlike most of the other tests described here, this test actually tests the finite N duality and not just the large N limit.

$\mathcal{N} = 4$ supersymmetry in four dimensions has a unique multiplet which does not include spins greater than one, which is the vector multiplet. It includes a vector field A_μ (μ is a vector index of the $SO(3, 1)$ Lorentz group), four complex Weyl fermions $\lambda_{\alpha A}$ (in the $\bar{\mathbf{4}}$ of $SU(4)_R$), and six real scalars ϕ^I (where I is an index in the $\mathbf{6}$ of $SU(4)_R$). The classical action of the supersymmetry generators on these fields is schematically given (for on-shell fields) by

$$[Q_\alpha^A, \phi^I] \sim \lambda_{\alpha B},$$
$$\{Q_\alpha^A, \lambda_{\beta B}\} \sim (\sigma^{\mu\nu})_{\alpha\beta} F_{\mu\nu} + \epsilon_{\alpha\beta}[\phi^I, \phi^J],$$
$$\{Q_\alpha^A, \bar{\lambda}_{\dot{\beta}}^B\} \sim (\sigma^\mu)_{\alpha\dot{\beta}} \mathcal{D}_\mu \phi^I, \tag{20}$$
$$[Q_\alpha^A, A_\mu] \sim (\sigma_\mu)_{\alpha\dot{\alpha}} \bar{\lambda}_{\dot{\beta}}^A \epsilon^{\dot{\alpha}\dot{\beta}},$$

with similar expressions for the action of the $\bar{Q}$'s, where $\sigma^{\mu\nu}$ are the generators of the Lorentz group in the spinor representation, $\mathcal{D}_\mu$ is the covariant derivative, the field strength $F_{\mu\nu} \equiv [\mathcal{D}_\mu, \mathcal{D}_\nu]$, and we have suppressed the $SU(4)$ Clebsch-Gordan coefficients corresponding to the products $\mathbf{4} \times \mathbf{6} \to \bar{\mathbf{4}}$, $\mathbf{4} \times \bar{\mathbf{4}} \to \mathbf{1} + \mathbf{15}$ and $\mathbf{4} \times \mathbf{4} \to \mathbf{6}$ in the first three lines of (20).

An $\mathcal{N} = 4$ supersymmetric field theory is uniquely determined by specifying the gauge group, and its field content is a vector multiplet in the adjoint of the gauge group. Such a field theory is equivalent to an $\mathcal{N} = 2$ theory with one hypermultiplet in the adjoint representation, or to an $\mathcal{N} = 1$ theory with three chiral multiplets Φ^i in the adjoint representation (in the $\mathbf{3}_{2/3}$ of the $SU(3) \times U(1)_R \subset SU(4)_R$ which is left unbroken by the choice of a single $\mathcal{N} = 1$ SUSY generator) and a superpotential of the form $W \propto \epsilon_{ijk} \text{Tr}\,(\Phi^i \Phi^j \Phi^k)$. The interactions of the theory include a scalar potential proportional to $\sum_{I,J} \text{Tr}\,([\phi^I, \phi^J]^2)$, such that the moduli space of the theory is the space of commuting matrices ϕ^I ($I = 1, \cdots, 6$).

The spectrum of operators in this theory includes all the gauge invariant quantities that can be formed from the fields described above. In this section we will focus on local operators which involve fields taken at the same point in space-time. For the $SU(N)$ theory described above, properties of the adjoint representation of $SU(N)$ determine that such operators necessarily involve a product of traces of products of fields (or the sum of such products). It is natural to divide the operators into single-trace operators and multiple-trace operators. In the 't Hooft large N limit correlation functions involving multiple-trace operators are suppressed by powers of N compared to those of single-trace operators involving the same fields. We will discuss here in detail only the single-trace operators; the multiple-trace operators appear in operator product expansions of products of single-trace operators.

It is natural to classify the operators in a conformal theory into primary operators and their descendants. In a superconformal theory it is also natural to distinguish between chiral primary operators, which are in short representations of the superconformal algebra and are annihilated by some of the supercharges, and non-chiral primary operators. Representations of the superconformal algebra are formed by starting with some state of lowest dimension, which is annihilated by the operators S and K_μ, and acting on it with the operators Q and P_μ. The $\mathcal{N} = 4$ supersymmetry algebra involves 16 real supercharges. A generic primary representation of the superconformal algebra will thus include 2^{16} primaries of the conformal algebra, generated by acting on the lowest state with products of different supercharges; acting with additional supercharges always leads to descendants of the conformal algebra (i.e. derivatives). Since the supercharges have helicities $\pm 1/2$, the primary fields in such representations will have a range of helicities between $\lambda - 4$ (if the lowest dimension operator ψ has helicity λ) and $\lambda + 4$ (acting with more than 8 supercharges of the same helicity either annihilates the state or leads to a conformal descendant). In non-generic representations of the superconformal algebra a product of less than 16 different Q's annihilates the lowest dimension operator, and the range of helicities appearing is smaller. In particular, in the small representations of the $\mathcal{N} = 4$ superconformal algebra only up to 4 Q's of the same helicity acting on the lowest dimension operator give a non-zero result, and the range of helicities is between $\lambda - 2$ and $\lambda + 2$. For the $\mathcal{N} = 4$ supersymmetry algebra (not including the conformal algebra) it is known that medium representations, whose range of helicities is 6, can also exist (they arise, for instance, on the moduli space of the $SU(N)$ $\mathcal{N} = 4$ SYM theory [71, 72, 73, 74, 75, 76, 77, 78]); it is not clear if such medium representations of the superconformal algebra [79] can appear in physical theories or not (there are no known examples). More details on the structure of representations of the $\mathcal{N} = 4$ superconformal algebra may be found in [80, 81, 82, 83, 84, 85, 79] and references therein.

In the $U(1)$ $\mathcal{N} = 4$ SYM theory (which is a free theory), the only gauge-invariant "single trace" operators are the fields of the vector multiplet itself (which are $\phi^I, \lambda_A, \bar\lambda^A$ and $F_{\mu\nu} = \partial_{[\mu} A_{\nu]}$). These operators form an ultra-short representation of the $\mathcal{N} = 4$ algebra whose range of helicities is from (-1) to 1 (acting with more than two supercharges of the same helicity on any of these states gives either zero or derivatives, which are descendants of the conformal algebra). All other local gauge invariant operators in the theory involve derivatives or products of these operators. This representation is usually called the doubleton representation, and it does not appear in the $SU(N)$ SYM theory (though the representations which do appear can all be formed by tensor products of the doubleton representation). In the context of AdS space one can think of this multiplet as living purely on the boundary of the space [86, 87, 88, 89, 90, 46, 91, 92, 93, 94, 95], as expected for the $U(1)$ part of the original $U(N)$ gauge group of the D3-branes (see the discussion in section 2).

There is no known simple systematic way to compute the full spectrum of chiral primary operators of the $\mathcal{N} = 4$ $SU(N)$ SYM theory, so we will settle for presenting the known chiral primary operators. The lowest component of a superconformal-primary multiplet is characterized by the fact that it cannot be written as a supercharge Q acting on any other operator. Looking at the action of the supersymmetry charges (20) suggests that generally operators built from the fermions and the gauge fields will be descendants (given by Q acting on some other fields), so one would expect the lowest components of the chiral primary representations to be built only from the scalar fields, and this turns out to be correct.

Let us analyze the behavior of operators of the form $\mathcal{O}^{I_1 I_2 \cdots I_n} \equiv \operatorname{Tr} (\phi^{I_1} \phi^{I_2} \cdots \phi^{I_n})$. First we can ask if this operator can be written as $\{Q, \psi\}$ for any field ψ. In the SUSY algebra (20) only commutators of ϕ^I's appear on the right hand side, so we see that if some of the indices are antisymmetric the field will be a descendant. Thus, only symmetric combinations of the indices will be lowest components of primary multiplets. Next, we should ask if the multiplet built on such an operator is a (short) chiral primary multiplet or not. There are several different ways to answer this question. One possibility is to use the relation between the dimension of chiral primary operators and their R-symmetry representation [96, 97, 98, 99, 100], and to check if this relation is obeyed in the free field theory, where $[\mathcal{O}^{I_1 I_2 \cdots I_n}] = n$. In this way we find that the representation is chiral primary if and only if the indices form a symmetric traceless product of n **6**'s (traceless representations are defined as those who give zero when any two indices are contracted). This is a representation of weight $(0, n, 0)$ of $SU(4)_R$; in this section we will refer to $SU(4)_R$ representations either by their dimensions in boldface or by their weights.

Another way to check this is to see if by acting with Q's on these operators we get the most general possible states or not, namely if the representation contains "null vectors" or not (it turns out that in all the relevant cases "null vectors" appear already at the first level by acting with a single Q, though in principle there could be representations where "null vectors" appear only at higher levels). Using the SUSY algebra (20) it is easy to see that for symmetric traceless representations we get "null vectors" while for other representations we do not. For instance, let us analyze in detail the case $n = 2$. The symmetric product of two **6**'s is given by $\mathbf{6} \times \mathbf{6} \to \mathbf{1} + \mathbf{20'}$. The field in the **1** representation is $\operatorname{Tr} (\phi^I \phi^I)$, for which $[Q_\alpha^A, \operatorname{Tr} (\phi^I \phi^I)] \sim C^{AJB} \operatorname{Tr} (\lambda_{\alpha B} \phi^J)$ where C^{AIB} is a Clebsch-Gordan coefficient for $\bar{\mathbf{4}} \times \mathbf{6} \to \mathbf{4}$. The right-hand side is in the **4** representation, which is the most general representation that can appear in the product $\mathbf{4} \times \mathbf{1}$, so we find no null vectors at this level. On the other hand, if we look at the symmetric traceless product $\operatorname{Tr} (\phi^{\{I} \phi^{J\}}) \equiv \operatorname{Tr} (\phi^I \phi^J) - \frac{1}{6} \delta^{IJ} \operatorname{Tr} (\phi^K \phi^K)$ in the **20'** representation, we find that $\{Q_\alpha^A, \operatorname{Tr} (\phi^{\{I} \phi^{J\}})\} \sim \operatorname{Tr} (\lambda_{\alpha B} \phi^K)$ with the right-hand side being in the **20** representation (appearing in $\bar{\mathbf{4}} \times \mathbf{6} \to \mathbf{4} + \mathbf{20}$), while the left-hand side could in principle be in the $4 \times \mathbf{20'} \to \mathbf{20} + \mathbf{60}$. Since the $\mathbf{60}$ does not appear on the right-hand side (it is a "null vector") we identify that the representation built on the $\mathbf{20'}$ is a short representation of the SUSY algebra. By similar manipulations (see [24, 101, 81, 84] for more details) one can verify that chiral primary representations correspond exactly to symmetric traceless products of $\mathbf{6}$'s.

It is possible to analyze the chiral primary spectrum also by using $\mathcal{N} = 1$ subalgebras of the $\mathcal{N} = 4$ algebra. If we use an $\mathcal{N} = 1$ subalgebra of the $\mathcal{N} = 4$ algebra, as described above, the operators $\mathcal{O}_n$ include the chiral operators of the form $\mathrm{Tr}\,(\Phi^{i_1}\Phi^{i_2}\cdots\Phi^{i_n})$ (in a representation of $SU(3)$ which is a symmetric product of $\mathbf{3}$'s), but for a particular choice of the $\mathcal{N} = 1$ subalgebra not all the operators $\mathcal{O}_n$ appear to be chiral (a short multiplet of the $\mathcal{N} = 4$ algebra includes both short and long multiplets of the $\mathcal{N} = 1$ subalgebra).

The last issue we should discuss is what is the range of values of n. The product of more than N commuting[2] $N \times N$ matrices can always be written as a sum of products of traces of less than N of the matrices, so it does not form an independent operator. This means that for $n > N$ we can express the operator $\mathcal{O}^{I_1 I_2 \cdots I_n}$ in terms of other operators, up to operators including commutators which (as explained above) are descendants of the SUSY algebra. Thus, we find that the short chiral primary representations are built on the operators $\mathcal{O}_n = \mathcal{O}^{\{I_1 I_2 \cdots I_n\}}$ with $n = 2, 3, \cdots, N$, for which the indices are in the symmetric traceless product of n $\mathbf{6}$'s (in a $U(N)$ theory we would find the same spectrum with the additional representation corresponding to $n = 1$). The superconformal algebra determines the dimension of these fields to be $[\mathcal{O}_n] = n$, which is the same as their value in the free field theory. We argued above that these are the only short chiral primary representations in the $SU(N)$ gauge theory, but we will not attempt to rigorously prove this here.

The full chiral primary representations are obtained by acting on the fields $\mathcal{O}_n$ by the generators Q and P of the supersymmetry algebra. The representation built on $\mathcal{O}_n$ contains a total of $256 \times \frac{1}{12}n^2(n^2 - 1)$ primary states, of which half are bosonic and half are fermionic. Since these multiplets are built on a field of helicity zero, they will contain primary fields of helicities between (-2) and 2. The highest dimension primary field in the multiplet is (generically) of the form $Q^4 \bar{Q}^4 \mathcal{O}_n$, and its dimension is $n + 4$. There is an elegant way to write these multiplets as traces of products of "twisted chiral $\mathcal{N} = 4$ superfields" [101, 81]; see also [102] which checks some components of these superfields against the couplings to supergravity modes predicted on the basis of the DBI action for D3-branes in anti-de Sitter space [4].

---

[2] We can limit the discussion to commuting matrices since, as discussed above, commutators always lead to descendants, and we can write any product of matrices as a product of commuting matrices plus terms with commutators.

It is easy to find the form of all the fields in such a multiplet by using the algebra (20). For example, let us analyze here in detail the bosonic primary fields of dimension $n+1$ in the multiplet. To get a field of dimension $n+1$ we need to act on $\mathcal{O}_n$ with two supercharges (recall that $[Q] = \frac{1}{2}$). If we act with two supercharges Q_α^A of the same chirality, their Lorentz indices can be either antisymmetrized or symmetrized. In the first case we get a Lorentz scalar field in the $(2, n-2, 0)$ representation of $SU(4)_R$, which is of the schematic form

$$\epsilon^{\alpha\beta}\{Q_\alpha, [Q_\beta, \mathcal{O}_n]\} \sim \epsilon^{\alpha\beta}\mathrm{Tr}\,(\lambda_{\alpha A}\lambda_{\beta B}\phi^{J_1}\cdots\phi^{J_{n-2}}) + \mathrm{Tr}\,([\phi^{K_1}, \phi^{K_2}]\phi^{L_1}\cdots\phi^{L_{n-1}}) \tag{21}$$

Using an $\mathcal{N}=1$ subalgebra some of these operators may be written as the lowest components of the chiral superfields $\mathrm{Tr}\,(W_\alpha^2 \Phi^{j_1}\cdots\Phi^{j_{n-2}})$. In the second case we get an anti-symmetric 2-form of the Lorentz group, in the $(0, n-1, 0)$ representation of $SU(4)_R$, of the form

$$\{Q_{\{\alpha}, [Q_{\beta\}}, \mathcal{O}_n]\} \sim \mathrm{Tr}\,((\sigma^{\mu\nu})_{\alpha\beta}F_{\mu\nu}\phi^{J_1}\cdots\phi^{J_{n-1}}) + \mathrm{Tr}\,(\lambda_{\alpha A}\lambda_{\beta B}\phi^{K_1}\cdots\phi^{K_{n-2}}). \tag{22}$$

Both of these fields are complex, with the complex conjugate fields given by the action of two $\bar{Q}$'s. Acting with one Q and one $\bar{Q}$ on the state $\mathcal{O}_n$ gives a (real) Lorentz-vector field in the $(1, n-2, 1)$ representation of $SU(4)_R$, of the form

$$\{Q_\alpha, [\bar{Q}_{\dot{\alpha}}, \mathcal{O}_n]\} \sim \mathrm{Tr}\,(\lambda_{\alpha A}\bar{\lambda}_{\dot{\alpha}}^B \phi^{J_1}\cdots\phi^{J_{n-2}}) + (\sigma^\mu)_{\alpha\dot{\alpha}}\mathrm{Tr}\,((\mathcal{D}_\mu\phi^J)\phi^{K_1}\cdots\phi^{K_{n-1}}). \tag{23}$$

At dimension $n+2$ (acting with four supercharges) we find :

- A complex scalar field in the $(0, n-2, 0)$ representation, given by $Q^4\mathcal{O}_n$, of the form $\mathrm{Tr}\,(F_{\mu\nu}^2\phi^{I_1}\cdots\phi^{I_{n-2}}) + \cdots$.

- A real scalar field in the $(2, n-4, 2)$ representation, given by $Q^2\bar{Q}^2\mathcal{O}_n$, of the form $\epsilon^{\alpha\beta}\epsilon^{\dot{\alpha}\dot{\beta}}\mathrm{Tr}\,(\lambda_{\alpha A_1}\lambda_{\beta A_2}\bar{\lambda}_{\dot{\alpha}}^{B_1}\bar{\lambda}_{\dot{\beta}}^{B_2}\phi^{I_1}\cdots\phi^{I_{n-4}}) + \cdots$.

- A complex vector field in the $(1, n-4, 1)$ representation, given by $Q^3\bar{Q}\mathcal{O}_n$, of the form $\mathrm{Tr}\,(F_{\mu\nu}\mathcal{D}^\nu\phi^J\phi^{I_1}\cdots\phi^{I_{n-2}}) + \cdots$.

- An complex anti-symmetric 2-form field in the $(2, n-3, 0)$ representation, given by $Q^2\bar{Q}^2\mathcal{O}_n$, of the form $\mathrm{Tr}\,(F_{\mu\nu}[\phi^{J_1}, \phi^{J_2}]\phi^{I_1}\cdots\phi^{I_{n-2}}) + \cdots$.

- A symmetric tensor field in the $(0, n-2, 0)$ representation, given by $Q^2\bar{Q}^2\mathcal{O}_n$, of the form $\mathrm{Tr}\,(\mathcal{D}_{\{\mu}\phi^J\mathcal{D}_{\nu\}}\phi^K\phi^{I_1}\cdots\phi^{I_{n-2}}) + \cdots$.

The spectrum of primary fields at dimension $n+3$ is similar to that of dimension $n+1$ (the same fields appear but in smaller $SU(4)_R$ representations), and at dimension $n+4$ there is a single primary field, which is a real scalar in the $(0, n - 4, 0)$ representation, given by $Q^4 \bar{Q}^4 \mathcal{O}_n$, of the form $\text{Tr}\,(F^4_{\mu\nu} \phi^{I_1} \cdots \phi^{I_{n-4}}) + \cdots$. Note that fields with more than four $F_{\mu\nu}$'s or more than eight λ's are always descendants or non-chiral primaries.

For $n = 2, 3$ the short multiplets are even shorter since some of the representations appearing above vanish. In particular, for $n = 2$ the highest-dimension primaries in the chiral primary multiplet have dimension $n + 2 = 4$. The $n = 2$ representation includes the currents of the superconformal algebra. It includes a vector of dimension 3 in the **15** representation which is the $SU(4)_R$ R-symmetry current, and a symmetric tensor field of dimension 4 which is the energy-momentum tensor (the other currents of the superconformal algebra are descendants of these). The $n = 2$ multiplet also includes a complex scalar field which is an $SU(4)_R$-singlet, whose real part is the Lagrangian density coupling to $\frac{1}{4g^2_{YM}}$ (of the form $\text{Tr}\,(F^2_{\mu\nu}) + \cdots$) and whose imaginary part is the Lagrangian density coupling to θ (of the form $\text{Tr}\,(F \wedge F)$). For later use we note that the chiral primary multiplets which contain scalars of dimension $\Delta \leq 4$ are the $n = 2$ multiplet (which has a scalar in the **20′** of dimension 2, a complex scalar in the **10** of dimension 3, and a complex scalar in the **1** of dimension 4), the $n = 3$ multiplet (which contains a scalar in the **50** of dimension 3 and a complex scalar in the **45** of dimension 4), and the $n = 4$ multiplet which contains a scalar in the **105** of dimension 4.

3.1.2   The String Theory Spectrum and the Matching.   As discussed in section 2.1, fields on AdS_5 are in a one-to-one correspondence with operators in the dual conformal field theory. Thus, the spectrum of operators described in section 3.1.1 should agree with the spectrum of fields of type IIB string theory on $AdS_5 \times S^5$. Fields on AdS naturally lie in the same multiplets of the conformal group as primary operators; the second Casimir of these representations is $C_2 = \Delta(\Delta - 4)$ for a primary scalar field of dimension Δ in the field theory, and $C_2 = m^2 R^2$ for a field of mass m on an AdS_5 space with a radius of curvature R. Single-trace operators in the field theory may be identified with single-particle states in AdS_5, while multiple-trace operators correspond to multi-particle states.

Unfortunately, it is not known how to compute the full spectrum of type IIB string theory on $AdS_5 \times S^5$. In fact, the only known states are the states which arise from the dimensional reduction of the ten-dimensional type IIB supergravity multiplet. These fields all have helicities between (-2) and 2, so it is clear that they all lie in small multiplets of the superconformal algebra, and we will describe below how they match with the small multiplets of the field theory described above. String theory on $AdS_5 \times S^5$ is expected to have many additional states, with masses of the order of the string scale $1/l_s$ or of the Planck scale $1/l_p$. Such states would correspond (using the mass/dimension relation described above) to operators in the field theory with dimensions of order $\Delta \sim (g_s N)^{1/4}$ or $\Delta \sim N^{1/4}$ for large $N, g_s N$. Presumably none of these states are in small multiplets of the superconformal algebra (at least, this would be the prediction of the AdS/CFT correspondence).

The spectrum of type IIB supergravity compactified on $AdS_5 \times S^5$ was computed in [103]. The computation involves expanding the ten dimensional fields in appropriate spherical harmonics on S^5, plugging them into the supergravity equations of motion, linearized around the $AdS_5 \times S^5$ background, and diagonalizing the equations to give equations of motion for free (massless or massive) fields[3]. For example, the ten dimensional dilaton field τ may be expanded as $\tau(x, y) = \sum_{k=0}^{\infty} \tau^k(x) Y^k(y)$ where x is a coordinate on AdS_5, y is a coordinate on S^5, and the Y^k are the scalar spherical harmonics on S^5. These spherical harmonics are in representations corresponding to symmetric traceless products of **6**'s of $SU(4)_R$; they may be written as $Y^k(y) \sim y^{I_1} y^{I_2} \cdots y^{I_k}$ where the y^I, for $I = 1, 2, \cdots, 6$ and with $\sum_{I=1}^{6} (y^I)^2 = 1$, are coordinates on S^5. Thus, we find a field $\tau^k(x)$ on AdS_5 in each such $(0, k, 0)$ representation of $SU(4)_R$, and the equations of motion determine the mass of this field to be $m_k^2 = k(k+4)/R^2$. A similar expansion may be performed for all other fields.

If we organize the results of [103] into representations of the superconformal algebra [80], we find representations of the form described in the previous section, which are built on a lowest dimension field which is a scalar in the $(0, n, 0)$ representation of $SU(4)_R$ for $n = 2, 3, \cdots, \infty$. The lowest dimension scalar field in each representation turns out to arise from a linear combination of spherical harmonic modes of the S^5 components of the graviton h_a^a (expanded around the $AdS_5 \times S^5$ vacuum) and the 4-form field D_{abcd}, where a, b, c, d are indices on S^5. The scalar fields of dimension $n + 1$ correspond to 2-form fields B_{ab} with indices in the S^5. The symmetric tensor fields arise from the expansion of the AdS_5-components of the graviton. The dilaton fields described above are the complex scalar fields arising with dimension $n + 2$ in the multiplet (as described in the previous subsection).

In particular, the $n = 2$ representation is called the supergraviton representation, and it includes the field content of $d = 5, \mathcal{N} = 8$ gauged supergravity. The field/operator correspondence matches this representation to the representation including the superconformal currents in the field theory. It includes a massless graviton field, which (as expected) corresponds to the energy-momentum tensor in the field theory, and massless $SU(4)_R$ gauge fields which correspond to (or couple to) the global $SU(4)_R$ currents in the field theory.

In the naive dimensional reduction of the type IIB supergravity fields, the $n = 1$ doubleton representation, corresponding to a free $U(1)$ vector multiplet in the dual theory, also appears. However, the modes of this multiplet are all

---

[3]The fields arising from different spherical harmonics are related by a "spectrum generating algebra", see [104].

pure gauge modes in the bulk of AdS_5, and they may be set to zero there. This is one of the reasons why it seems more natural to view the corresponding gauge theory as an $SU(N)$ gauge theory and not a $U(N)$ theory. It may be possible (and perhaps even natural) to add the doubleton representation to the theory (even though it does not include modes which propagate in the bulk of AdS_5, but instead it is equivalent to a topological theory in the bulk) to obtain a theory which is dual to the $U(N)$ gauge theory, but this will not affect most of our discussion in this review so we will ignore this possibility here.

Comparing the results described above with the results of section 3.1.1, we see that we find the same spectrum of chiral primary operators for $n = 2, 3, \cdots, N$. The supergravity results cannot be trusted for masses above the order of the string scale (which corresponds to $n \sim (g_s N)^{1/4}$) or the Planck scale (which corresponds to $n \sim N^{1/4}$), so the results agree within their range of validity. The field theory results suggest that the exact spectrum of chiral representations in type IIB string theory on $AdS_5 \times S^5$ actually matches the naive supergravity spectrum up to a mass scale $m^2 \sim N^2/R^2 \sim N^{3/2} M_p^2$ which is much higher than the string scale and the Planck scale, and that there are no chiral fields above this scale. It is not known how to check this prediction; tree-level string theory is certainly not enough for this since when $g_s = 0$ we must take $N = \infty$ to obtain a finite value of $g_s N$. Thus, with our current knowledge the matching of chiral primaries of the $\mathcal{N} = 4$ SYM theory with those of string theory on $AdS_5 \times S^5$ tests the duality only in the large N limit. In some generalizations of the AdS/CFT correspondence the string coupling goes to zero at the boundary even for finite N, and then classical string theory should lead to exactly the same spectrum of chiral operators as the field theory. This happens in particular for the near-horizon limit of NS5-branes, in which case the exact spectrum was successfully compared in [105]. In other instances of the AdS/CFT correspondence (such as the ones discussed in [106, 107, 108]) there exist also additional chiral primary multiplets with n of order N, and these have been successfully matched with wrapped branes on the string theory side.

The fact that there seem to be no non-chiral fields on AdS_5 with a mass below the string scale suggests that for large N and large $g_s N$, the dimension of all non-chiral operators in the field theory, such as $\text{Tr}\,(\phi^I \phi^I)$, grows at least as $(g_s N)^{1/4} \sim (g_{YM}^2 N)^{1/4}$. The reason for this behavior on the field theory side is not clear; it is a prediction of the AdS/CFT correspondence.

3.2 Matching of Correlation Functions and Anomalies

The classical $\mathcal{N} = 4$ theory has a scale invariance symmetry and an $SU(4)_R$ R-symmetry, and (unlike many other theories) these symmetries are exact also in the full quantum theory. However, when the theory is coupled to external gravitational or $SU(4)_R$ gauge fields, these symmetries are broken by quantum effects. In field theory this breaking comes from one-loop diagrams and does not receive any further corrections; thus it can be computed also in the strong coupling regime and compared with the results from string theory on AdS space.

We will begin by discussing the anomaly associated with the $SU(4)_R$ global currents. These currents are chiral since the fermions $\lambda_{\alpha A}$ are in the $\bar{\mathbf{4}}$ representation while the fermions of the opposite chirality $\bar{\lambda}^A_{\dot{\alpha}}$ are in the $\mathbf{4}$ representation. Thus, if we gauge the $SU(4)_R$ global symmetry, we will find an Adler-Bell-Jackiw anomaly from the triangle diagram of three $SU(4)_R$ currents, which is proportional to the number of charged fermions. In the $SU(N)$ gauge theory this number is $N^2 - 1$. The anomaly can be expressed either in terms of the 3-point function of the $SU(4)_R$ global currents,

$$\left\langle J_\mu^a(x) J_\nu^b(y) J_\rho^c(z) \right\rangle_{-} = -\frac{N^2 - 1}{32\pi^6} i d^{abc} \frac{\mathrm{Tr}\left[\gamma_5 \gamma_\mu (\not{x} - \not{y}) \gamma_\nu (\not{y} - \not{z}) \gamma_\rho (\not{z} - \not{x}) \right]}{(x - y)^4 (y - z)^4 (z - x)^4},$$
$$(24)$$

where $d^{abc} = 2\mathrm{Tr}\left(T^a \{ T^b, T^c \} \right)$ and we take only the negative parity component of the correlator, or in terms of the non-conservation of the $SU(4)_R$ current when the theory is coupled to external $SU(4)_R$ gauge fields $F_{\mu\nu}^a$,

$$(\mathcal{D}^\mu J_\mu)^a = \frac{N^2 - 1}{384\pi^2} i d^{abc} \epsilon^{\mu\nu\rho\sigma} F_{\mu\nu}^b F_{\rho\sigma}^c. \qquad (25)$$

How can we see this effect in string theory on $AdS_5 \times S^5$? One way to see it is, of course, to use the general prescription of section 4 to compute the 3-point function (24), and indeed one finds [109, 110] the correct answer to leading order in the large N limit (namely, one recovers the term proportional to N^2). It is more illuminating, however, to consider directly the meaning of the anomaly (25) from the point of view of the AdS theory [24]. In the AdS theory we have gauge fields A_μ^a which couple, as explained above, to the $SU(4)_R$ global currents J_μ^a of the gauge theory, but the anomaly means that when we turn on non-zero field strengths for these fields the theory should no longer be gauge invariant. This effect is precisely reproduced by a Chern-Simons term which exists in the low-energy supergravity theory arising from the compactification of type IIB supergravity on $AdS_5 \times S^5$, which is of the form

$$\frac{iN^2}{96\pi^2} \int_{AdS_5} d^5 x (d^{abc} \epsilon^{\mu\nu\lambda\rho\sigma} A_\mu^a \partial_\nu A_\lambda^b \partial_\rho A_\sigma^c + \cdots). \qquad (26)$$

This term is gauge invariant up to total derivatives, which means that if we take a gauge transformation $A_\mu^a \to A_\mu^a + (\mathcal{D}_\mu \Lambda)^a$ for which Λ does not vanish on the boundary of AdS_5, the action will change by a boundary term of the form

$$-\frac{iN^2}{384\pi^2} \int_{\partial AdS_5} d^4 x \epsilon^{\mu\nu\rho\sigma} d^{abc} \Lambda^a F_{\mu\nu}^b F_{\rho\sigma}^c. \qquad (27)$$

From this we can read off the anomaly in $(\mathcal{D}^\mu J_\mu)$ since, when we have a coupling of the form $\int d^4x A_a^\mu J_\mu^a$, the change in the action under a gauge transformation is given by $\int d^4x (\mathcal{D}^\mu \Lambda)_a J_\mu^a = -\int d^4x \Lambda_a (\mathcal{D}^\mu J_\mu^a)$, and we find exact agreement with (25) for large N.

The other anomaly in the $\mathcal{N} = 4$ SYM theory is the conformal (or Weyl) anomaly (see [111, 112] and references therein), indicating the breakdown of conformal invariance when the theory is coupled to a curved external metric (there is a similar breakdown of conformal invariance when the theory is coupled to external $SU(4)_R$ gauge fields, which we will not discuss here). The conformal anomaly is related to the 2-point and 3-point functions of the energy-momentum tensor [113, 114, 115, 116]. In four dimensions, the general form of the conformal anomaly is

$$\langle g^{\mu\nu} T_{\mu\nu} \rangle = -a E_4 - c I_4, \tag{28}$$

where

$$E_4 = \frac{1}{16\pi^2}(\mathcal{R}_{\mu\nu\rho\sigma}^2 - 4\mathcal{R}_{\mu\nu}^2 + \mathcal{R}^2),$$
$$I_4 = -\frac{1}{16\pi^2}(\mathcal{R}_{\mu\nu\rho\sigma}^2 - 2\mathcal{R}_{\mu\nu}^2 + \frac{1}{3}\mathcal{R}^2), \tag{29}$$

where $\mathcal{R}_{\mu\nu\rho\sigma}$ is the curvature tensor, $\mathcal{R}_{\mu\nu} \equiv \mathcal{R}_{\mu\rho\nu}^\rho$ is the Riemann tensor, and $\mathcal{R} \equiv \mathcal{R}_\mu^\mu$ is the scalar curvature. A free field computation in the $SU(N)$ $\mathcal{N} = 4$ SYM theory leads to $a = c = (N^2 - 1)/4$. In supersymmetric theories the supersymmetry algebra relates $g^{\mu\nu} T_{\mu\nu}$ to derivatives of the R-symmetry current, so it is protected from any quantum corrections. Thus, the same result should be obtained in type IIB string theory on $AdS_5 \times S^5$, and to leading order in the large N limit it should be obtained from type IIB supergravity on $AdS_5 \times S^5$. This was indeed found to be true in [117, 118, 119, 120][4], where the conformal anomaly was shown to arise from subtleties in the regularization of the (divergent) supergravity action on AdS space. The result of [117, 118, 119, 120] implies that a computation using gravity on AdS_5 always gives rise to theories with $a = c$, so generalizations of the AdS/CFT correspondence which have (for large N) a supergravity approximation are limited to conformal theories which have $a = c$ in the large N limit. Of course, if we do not require the string theory to have a supergravity approximation then there is no such restriction.

For both of the anomalies we described the field theory and string theory computations agree for the leading terms, which are of order N^2. Thus, they are successful tests of the duality in the large N limit. For other instances of the AdS/CFT correspondence there are corrections to anomalies at order

---

[4]A generalization with more varying fields may be found in [121].

$1/N \sim g_s(\alpha'/R^2)^2$; such corrections were discussed in [122] and success-fully compared in [123, 124, 125][5]. It would be interesting to compare other corrections to the large N result.

4.   Correlation Functions

A useful statement of the AdS/CFT correspondence is that the partition function of string theory on $AdS_5 \times S^5$ should coincide with the partition function of $\mathcal{N} = 4$ super-Yang-Mills theory "on the boundary" of AdS_5 [23, 24]. The basic idea was explained in section 2.1, but before summarizing the actual calcula-tions of Green's functions, it seems worthwhile to motivate the methodology from a somewhat different perspective.

Throughout this section, we approximate the string theory partition function by $e^{-I_{SUGRA}}$, where I_{SUGRA} is the supergravity action evaluated on $AdS_5 \times S^5$ (or on small deformations of this space). This approximation amounts to ignor-ing all the stringy α' corrections that cure the divergences of supergravity, and also all the loop corrections, which are controlled essentially by the gravitational coupling $\kappa \sim g_{st}\alpha'^2$. On the gauge theory side, as explained in section 2.1, this approximation amounts to taking both N and $g_{YM}^2 N$ large, and the basic relation becomes

$$e^{-I_{SUGRA}} \simeq Z_{\text{string}} = Z_{\text{gauge}} = e^{-W} , \qquad (30)$$

where W is the generating functional for connected Green's functions in the gauge theory. At finite temperature, $W = \beta F$ where β is the inverse tempera-ture and F is the free energy of the gauge theory. When we apply this relation to a Schwarzschild black hole in AdS_5, which is thought to be reflected in the gauge theory by a thermal state at the Hawking temperature of the black hole, we arrive at the relation $I_{SUGRA} \simeq \beta F$. Calculating the free energy of a black hole from the Euclidean supergravity action has a long tradition in the super-gravity literature [126], so the main claim that is being made here is that the dual gauge theory provides a description of the state of the black hole which is physically equivalent to the one in string theory. We will discuss the finite temperature case further in section 6, and devote the rest of this section to the partition function of the field theory on $\mathbb{R}^4$.

The main technical idea behind the bulk-boundary correspondence is that the boundary values of string theory fields (in particular, supergravity fields) act as sources for gauge-invariant operators in the field theory. From a D-brane perspective, we think of closed string states in the bulk as sourcing gauge singlet

---

[5]Computing such corrections tests the conjecture that the correspondence holds order by order in $1/N$; however, this is weaker than the statement that the correspondence holds for finite N, since the $1/N$ expansion is not expected to converge.

operators on the brane which originate as composite operators built from open strings. We will write the bulk fields generically as $\phi(\vec{x}, z)$ (in the coordinate system (15)), with value $\phi_0(\vec{x})$ for $z = \epsilon$. The true boundary of anti-de Sitter space is $z = 0$, and $\epsilon \neq 0$ serves as a cutoff which will eventually be removed. In the supergravity approximation, we think of choosing the values ϕ_0 arbitrarily and then extremizing the action $I_{SUGRA}[\phi]$ in the region $z > \epsilon$ subject to these boundary conditions. In short, we solve the equations of motion in the bulk subject to Dirichlet boundary conditions on the boundary, and evaluate the action on the solution. If there is more than one solution, then we have more than one saddle point contributing to the string theory partition function, and we must determine which is most important. In this section, multiple saddle points will not be a problem. So, we can write

$$W_{\text{gauge}}[\phi_0] = -\log \left\langle e^{\int d^4 x \, \phi_0(x)\mathcal{O}(x)} \right\rangle_{CFT} \simeq \underset{\phi|_{z=\epsilon}=\phi_0}{\text{extremum}} I_{SUGRA}[\phi] \,. \quad (31)$$

That is, the generator of connected Green's functions in the gauge theory, in the large N, $g_{YM}^2 N$ limit, is the on-shell supergravity action.

Note that in (31) we have not attempted to be prescient about inserting factors of ϵ. Instead our strategy will be to use (31) without modification to compute two-point functions of $\mathcal{O}$, and then perform a wave-function renormalization on either $\mathcal{O}$ or ϕ so that the final answer is independent of the cutoff. This approach should be workable even in a space (with boundary) which is not asymptotically anti-de Sitter, corresponding to a field theory which does not have a conformal fixed point in the ultraviolet.

A remark is in order regarding the relation of (31) to the old approach of extracting Green's functions from an absorption cross-section [16]. In absorption calculations one is keeping the whole D3-brane geometry, not just the near-horizon $AdS_5 \times S^5$ throat. The usual treatment is to split the space into a near region (the throat) and a far region. The incoming wave from asymptotically flat infinity can be regarded as fixing the value of a supergravity field at the outer boundary of the near region. As usual, the supergravity description is valid at large N and large 't Hooft coupling. At small 't Hooft coupling, there is a different description of the process: a cluster of D3-branes sits at some location in flat ten-dimensional space, and the incoming wave impinges upon it. In the low-energy limit, the value of the supergravity field which the D3-branes feel is the same as the value in the curved space description at the boundary of the near horizon region. Equation (31) is just a mathematical expression of the fact that the throat geometry should respond identically to the perturbed supergravity fields as the low-energy theory on the D3-branes.

Following [23, 24], a number of papers—notably [127, 128, 109, 129, 110, 130, 131, 132, 133, 134, 135, 136, 137, 138, 139, 140, 141]—have undertaken the program of extracting explicit n-point correlation functions of gauge singlet operators by developing both sides of (31) in a power series in ϕ_0. Because the right hand side is the extremization of a classical action, the power series has a graphical representation in terms of tree-level Feynman graphs for fields in the supergravity. There is one difference: in ordinary Feynman graphs one assigns the wavefunctions of asymptotic states to the external legs of the graph, but in the present case the external leg factors reflect the boundary values ϕ_0. They are special limits of the usual gravity propagators in the bulk, and are called bulk-to-boundary propagators. We will encounter their explicit form in the next two sections.

4.1   Two-point Functions

For two-point functions, only the part of the action which is quadratic in the relevant field perturbation is needed. For massive scalar fields in AdS_5, this has the generic form

$$S = \eta \int d^5x \, \sqrt{g} \left[\tfrac{1}{2}(\partial\phi)^2 + \tfrac{1}{2}m^2\phi^2 \right] , \tag{32}$$

where η is some normalization which in principle follows from the ten-dimensional origin of the action. The bulk-to-boundary propagator is a particular solution of the equation of motion, $(\Box - m^2)\phi = 0$, which has special asymptotic properties. We will start by considering the momentum space propagator, which is useful for computing the two-point function and also in situations where the bulk geometry loses conformal invariance; then, we will discuss the position space propagator, which has proven more convenient for the study of higher point correlators in the conformal case. We will always work in Euclidean space[6]. A coordinate system in the bulk of AdS_5 such that

$$ds^2 = \frac{R^2}{z^2} \left(d\vec{x}^2 + dz^2 \right) \tag{33}$$

provides manifest Euclidean symmetry on the directions parametrized by $\vec{x}$. To avoid divergences associated with the small z region of integration in (32), we will employ an explicit cutoff, $z \geq \epsilon$.

A complete set of solutions for the linearized equation of motion, $(\Box - m^2)\phi = 0$, is given by $\phi = e^{i\vec{p}\cdot\vec{x}} Z(pz)$, where the function $Z(u)$ satisfies the radial equation

$$\left[u^5 \partial_u \frac{1}{u^3} \partial_u - u^2 - m^2 R^2 \right] Z(u) = 0 . \tag{34}$$

---

[6]The results may be analytically continued to give the correlation functions of the field theory on Minkowskian $\mathbb{R}^4$, which corresponds to the Poincaré coordinates of AdS space.

There are two independent solutions to (34), namely $Z(u) = u^2 I_{\Delta-2}(u)$ and $Z(u) = u^2 K_{\Delta-2}(u)$, where I_ν and K_ν are Bessel functions and

$$\Delta = 2 + \sqrt{4 + m^2 R^2} \ . \tag{35}$$

The second solution is selected by the requirement of regularity in the interior: $I_{\Delta-2}(u)$ increases exponentially as $u \to \infty$ and does not lead to a finite action configuration. Imposing the boundary condition $\phi(\vec{x}, z) = \phi_0(\vec{x}) = e^{i\vec{p}\cdot\vec{x}}$ at $z = \epsilon$, we find the bulk-to-boundary propagator

$$\phi(\vec{x}, z) = K_{\vec{p}}(\vec{x}, z) = \frac{(pz)^2 K_{\Delta-2}(pz)}{(p\epsilon)^2 K_{\Delta-2}(p\epsilon)} e^{i\vec{p}\cdot\vec{x}} \ . \tag{36}$$

To compute a two-point function of the operator $\mathcal{O}$ for which ϕ_0 is a source, we write

$$\langle \mathcal{O}(\vec{p})\mathcal{O}(\vec{q})\rangle = \frac{\partial^2 W\left[\phi_0 = \lambda_1 e^{i\vec{p}\cdot x} + \lambda_2 e^{i\vec{q}\cdot x}\right]}{\partial\lambda_1 \partial\lambda_2}\bigg|_{\lambda_1=\lambda_2=0}$$

$$= \text{(leading analytic terms in } (\epsilon p)^2)$$

$$- \eta \epsilon^{2\Delta-8}(2\Delta - 4)\frac{\Gamma(3-\Delta)}{\Gamma(\Delta-1)}\delta^4(\vec{p}+\vec{q})\left(\frac{\vec{p}}{2}\right)^{2\Delta-4} \tag{37}$$

$$+ \text{(higher order terms in } (\epsilon p)^2),$$

$$\langle \mathcal{O}(\vec{x})\mathcal{O}(\vec{y})\rangle = \eta \epsilon^{2\Delta-8}\frac{2\Delta - 4}{\Delta}\frac{\Gamma(\Delta+1)}{\pi^2\Gamma(\Delta-2)}\frac{1}{|\vec{x}-\vec{y}|^{2\Delta}} \ .$$

Several explanatory remarks are in order:

- To establish the second equality in (37) we have used (36), substituted in (32), performed the integral and expanded in ϵ. The leading analytic terms give rise to contact terms in position space, and the higher order terms are unimportant in the limit where we remove the cutoff. Only the leading nonanalytic term is essential. We have given the expression for generic real values of Δ. Expanding around integer $\Delta \geq 2$ one obtains finite expressions involving $\log \epsilon p$.

- The Fourier transforms used to obtain the last line are singular, but they can be defined for generic complex Δ by analytic continuation and for positive integer Δ by expanding around a pole and dropping divergent terms, in the spirit of differential regularization [142]. The result is a pure power law dependence on the separation $|\vec{x} - \vec{y}|$, as required by conformal invariance.

- We have assumed a coupling $\int d^4x\, \phi(\vec{x}, z = \epsilon)\mathcal{O}(\vec{x})$ to compute the Green's functions. The explicit powers of the cutoff in the final position space answer can be eliminated by absorbing a factor of $\epsilon^{\Delta-4}$ into the definition of $\mathcal{O}$. From here on we will take that convention, which amounts to inserting a factor of $\epsilon^{4-\Delta}$ on the right hand side of (36). In fact, precise matchings between the normalizations in field theory and in string theory for all the chiral primary operators have not been worked out. In part this is due to the difficulty of determining the coupling of bulk fields to field theory operators (or in stringy terms, the coupling of closed string states to composite open string operators on the brane). See [15] for an early approach to this problem. For the dilaton, the graviton, and their superpartners (including gauge fields in AdS_5), the couplings can be worked out explicitly. In some of these cases all normalizations have been worked out unambiguously and checked against field theory predictions (see for example [23, 109, 134]).

- The mass-dimension relation (35) holds even for string states that are not included in the Kaluza-Klein supergravity reduction: the mass and the dimension are just different expressions of the second Casimir of $SO(4,2)$. For instance, excited string states, with $m \sim 1/\sqrt{\alpha'}$, are expected to correspond to operators with dimension $\Delta \sim (g_{YM}^2 N)^{1/4}$. The remarkable fact is that all the string theory modes with $m \sim 1/R$ (which is to say, all closed string states which arise from massless ten dimensional fields) fall in short multiplets of the supergroup $SU(2,2|4)$. All other states have a much larger mass. The operators in short multiplets have algebraically protected dimensions. The obvious conclusion is that all operators whose dimensions are not algebraically protected have large dimension in the strong 't Hooft coupling, large N limit to which supergravity applies. This is no longer true for theories of reduced supersymmetry: the supergroup gets smaller, but the Kaluza-Klein states are roughly as numerous as before, and some of them escape the short multiplets and live in long multiplets of the smaller supergroups. They still have a mass on the order of $1/R$, and typically correspond to dimensions which are finite (in the large $g_{YM}^2 N$ limit) but irrational.

Correlation functions of non-scalar operators have been widely studied following [24]; the literature includes [143, 144, 145, 146, 147, 148, 149, 150, 151, 152, 153]. For $\mathcal{N} = 4$ super-Yang-Mills theory, all correlation functions of fields in chiral multiplets should follow by application of supersymmetries once those of the chiral primary fields are known, so in this case it should be enough to study the scalars. It is worthwhile to note however that the mass-dimension formula changes for particles with spin. In fact the definition of mass has some convention-dependence. Conventions seem fairly uniform in the literature, and a table of mass-dimension relations in AdS_{d+1} with unit radius was made in [154] from the various sources cited above (see also [101]):

- scalars: $\Delta_\pm = \frac{1}{2}(d \pm \sqrt{d^2 + 4m^2})$,

- spinors: $\Delta = \frac{1}{2}(d + 2|m|)$,

- vectors: $\Delta_{\pm} = \frac{1}{2}(d \pm \sqrt{(d-2)^2 + 4m^2})$,

- p-forms: $\Delta = \frac{1}{2}(d \pm \sqrt{(d-2p)^2 + 4m^2})$,

- first-order $(d/2)$-forms (d even): $\Delta = \frac{1}{2}(d + 2|m|)$,

- spin-3/2: $\Delta = \frac{1}{2}(d + 2|m|)$,

- massless spin-2: $\Delta = d$.

In the case of fields with second order lagrangians, we have not attempted to pick which of $\Delta_{\pm}$ is the physical dimension. Usually the choice $\Delta = \Delta_+$ is clear from the unitarity bound, but in some cases (notably $m^2 = 15/4$ in AdS_5) there is a genuine ambiguity. In practice this ambiguity is usually resolved by appealing to some special algebraic property of the relevant fields, such as transformation under supersymmetry or a global bosonic symmetry.

For brevity we will omit a further discussion of higher spins, and instead refer the reader to the (extensive) literature.

4.2   Three-point Functions

Working with bulk-to-boundary propagators in the momentum representation is convenient for two-point functions, but for higher point functions position space is preferred because the full conformal invariance is more obvious. (However, for non-conformal examples of the bulk-boundary correspondence, the momentum representation seems uniformly more convenient). The boundary behavior of position space bulk-to-boundary propagators is specified in a slightly more subtle way: following [109] we require

$$K_\Delta(\vec{x}, z; \vec{y}) \to z^{4-\Delta} \delta^4(\vec{x} - \vec{y}) \quad \text{as} \quad z \to 0. \tag{38}$$

Here $\vec{y}$ is the point on the boundary where we insert the operator, and $(\vec{x}, z)$ is a point in the bulk. The unique regular K_Δ solving the equation of motion and satisfying (38) is

$$K_\Delta(\vec{x}, z; \vec{y}) = \frac{\Gamma(\Delta)}{\pi^2 \Gamma(\Delta - 2)} \left(\frac{z}{z^2 + (\vec{x} - \vec{y})^2} \right)^\Delta. \tag{39}$$

At a fixed cutoff, $z = \epsilon$, the bulk-to-boundary propagator $K_\Delta(\vec{x}, \epsilon; \vec{y})$ is a continuous function which approximates $\epsilon^{4-\Delta} \delta^4(\vec{x} - \vec{y})$ better and better as $\epsilon \to 0$. Thus at any finite ϵ, the Fourier transform of (39) only approximately coincides with (36) (modified by the factor of $\epsilon^{4-\Delta}$ as explained after (37)). This apparently innocuous subtlety turned out to be important for two-point functions, as

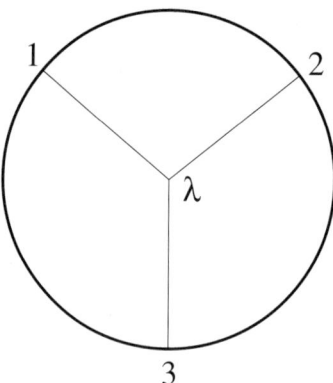

Figure 2.   The Feynman graph for the three-point function as computed in supergravity. The legs correspond to factors of K_{Δ_i}, and the cubic vertex to a factor of λ. The position of the vertex is integrated over AdS_5.

discovered in [109]. A correct prescription is to specify boundary conditions at finite $z = \epsilon$, cut off all bulk integrals at that boundary, and only afterwards take $\epsilon \to 0$. That is what we have done in (37). Calculating two-point functions directly using the position-space propagators (38), but cutting the bulk integrals off again at ϵ, and finally taking the same $\epsilon \to 0$ answer, one arrives at a different answer. This is not surprising since the $z = \epsilon$ boundary conditions were not used consistently. The authors of [109] checked that using the cutoff consistently (i.e. with the momentum space propagators) gave two-point functions $\langle \mathcal{O}(\vec{x}_1)\mathcal{O}(\vec{x}_2)\rangle$ a normalization such that Ward identities involving the three-point function $\langle \mathcal{O}(\vec{x}_1)\mathcal{O}(\vec{x}_2)J_\mu(\vec{x}_3)\rangle$, where J_μ is a conserved current, were obeyed. Two-point functions are uniquely difficult because of the poor convergence properties of the integrals over z. The integrals involved in three-point functions are sufficiently benign that one can ignore the issue of how to impose the cutoff.

If one has a Euclidean bulk action for three scalar fields ϕ_1, ϕ_2, and ϕ_3, of the form

$$S = \int d^5x \sqrt{g} \left[\sum_i \tfrac{1}{2}(\partial \phi_i)^2 + \tfrac{1}{2}m_i^2\phi_i^2 + \lambda\phi_1\phi_2\phi_3 \right], \qquad (40)$$

and if the ϕ_i couple to operators in the field theory by interaction terms $\int d^4x\, \phi_i \mathcal{O}_i$, then the calculation of $\langle \mathcal{O}_1\mathcal{O}_2\mathcal{O}_3\rangle$ reduces, via (31), to the evaluation of the graph shown in figure 2. That is,

$$\langle \mathcal{O}_1(\vec{x}_1)\mathcal{O}_2(\vec{x}_2)\mathcal{O}_3(\vec{x}_3)\rangle = -\lambda \int d^5x \sqrt{g} K_{\Delta_1}(x;\vec{x}_1) K_{\Delta_2}(x;\vec{x}_2) K_{\Delta_3}(x;\vec{x}_3)$$

$$= \frac{\lambda a_1}{|\vec{x}_1-\vec{x}_2|^{\Delta_1+\Delta_2-\Delta_3}|\vec{x}_1-\vec{x}_3|^{\Delta_1+\Delta_3-\Delta_2}|\vec{x}_2-\vec{x}_3|^{\Delta_2+\Delta_3-\Delta_1}},$$

$$(41)$$

for some constant a_1. The dependence on the $\vec{x}_i$ is dictated by the conformal invariance, but the only way to compute a_1 is by performing the integral over x. The result [109] is

$$a_1 = -\frac{\Gamma\left[\frac{1}{2}(\Delta_1+\Delta_2-\Delta_3)\right]\Gamma\left[\frac{1}{2}(\Delta_1+\Delta_3-\Delta_2)\right]\Gamma\left[\frac{1}{2}(\Delta_2+\Delta_3-\Delta_1)\right]}{2\pi^4\Gamma(\Delta_1-2)\Gamma(\Delta_2-2)\Gamma(\Delta_3-2)} \cdot$$

$$\Gamma\left[\frac{1}{2}(\Delta_1+\Delta_2+\Delta_3)-2\right].$$

$$(42)$$

In principle one could also have couplings of the form $\phi_1\partial\phi_2\partial\phi_3$. This leads only to a modification of the constant a_1.

The main technical difficulty with three-point functions is that one must figure out the cubic couplings of supergravity fields. Because of the difficulties in writing down a covariant action for type IIB supergravity in ten dimensions (see however [155, 156, 157]), it is most straightforward to read off these "cubic couplings" from quadratic terms in the equations of motion. In flat ten-dimensional space these terms can be read off directly from the original type IIB supergravity papers [158, 159]. For $AdS_5 \times S^5$, one must instead expand in fluctuations around the background metric and five-form field strength. The old literature [103] only dealt with the linearized equations of motion; for 3-point functions it is necessary to go to one higher order of perturbation theory. This was done for a restricted set of fields in [132]. The fields considered were those dual to operators of the form $\mathrm{Tr}\,\phi^{(J_1}\phi^{J_2}\ldots\phi^{J_\ell)}$ in field theory, where the parentheses indicate a symmetrized traceless product. These operators are the chiral primaries of the gauge theory: all other single trace operators of protected dimension descend from these by commuting with supersymmetry generators. Only the metric and the five-form are involved in the dual supergravity fields, and we are interested only in modes which are scalars in AdS_5. The result of [132] is that the equations of motion for the scalar modes $\tilde{s}_I$ dual to

$$\mathcal{O}^I = \mathcal{C}^I_{J_1\ldots J_\ell}\mathrm{Tr}\,\phi^{(J_1}\ldots\phi^{J_\ell)} \tag{43}$$

follow from an action of the form

$$S = \frac{4N^2}{(2\pi)^5}\int d^5x \sqrt{g}\left\{\sum_I \frac{A_I(w^I)^2}{2}\left[-(\nabla\tilde{s}_I)^2 - l(l-4)\tilde{s}_I^2\right]\right.$$

$$\left. + \sum_{I_1,I_2,I_3} \frac{\mathcal{G}_{I_1I_2I_3}w^{I_1}w^{I_2}w^{I_3}}{3}\tilde{s}_{I_1}\tilde{s}_{I_2}\tilde{s}_{I_3}\right\}.$$

$$(44)$$

Derivative couplings of the form $\tilde{s}\partial\tilde{s}\partial\tilde{s}$ are expected *a priori* to enter into (44), but an appropriate field redefinition eliminates them. The notation in (43) and (44) requires some explanation. I is an index which runs over the weight vectors of all possible representations constructed as symmetric traceless products of the **6** of $SU(4)_R$. These are the representations whose Young diagrams are $\Box$, $\Box\Box$, $\Box\Box\Box$, $\cdots$. $\mathcal{C}^I_{J_1\dots J_\ell}$ is a basis transformation matrix, chosen so that $\mathcal{C}^I_{J_1\dots J_\ell}\mathcal{C}^J_{J_1\dots J_\ell} = \delta^{IJ}$. As commented in the previous section, there is generally a normalization ambiguity on how supergravity fields couple to operators in the gauge theory. We have taken the coupling to be $\int d^4x\, \tilde{s}_I \mathcal{O}^I$, and the normalization ambiguity is represented by the "leg factors" w^I. It is the combination $s^I = w^I \tilde{s}^I$ rather than $\tilde{s}^I$ itself which has a definite relation to supergravity fields. We refer the reader to [132] for explicit expressions for A_I and the symmetric tensor $\mathcal{G}_{I_1 I_2 I_3}$. To get rid of factors of w^I, we introduce operators $\mathcal{O}^I = \tilde{w}^I \mathcal{O}^I$. One can choose $\tilde{w}^I$ so that a two-point function computation along the lines of section 4.1 leads to

$$\langle \mathcal{O}^{I_1}(\vec{x}) \mathcal{O}^{I_2}(0) \rangle = \frac{\delta^{I_1 I_2}}{x^{2\Delta_1}} \, . \tag{45}$$

With this choice, the three-point function, as calculated using (41), is

$$\langle \mathcal{O}^{I_1}(\vec{x_1}) \mathcal{O}^{I_2}(\vec{x_2}) \mathcal{O}^{I_3}(\vec{x_3}) \rangle = \tag{46}$$
$$\frac{1}{N} \frac{\sqrt{\Delta_1 \Delta_2 \Delta_3} \langle \mathcal{C}^{I_1} \mathcal{C}^{I_2} \mathcal{C}^{I_3} \rangle}{|\vec{x_1} - \vec{x_2}|^{\Delta_1 + \Delta_2 - \Delta_3} |\vec{x_1} - \vec{x_3}|^{\Delta_1 + \Delta_3 - \Delta_2} |\vec{x_2} - \vec{x_3}|^{\Delta_2 + \Delta_3 - \Delta_1}} \, ,$$

where we have defined

$$\langle \mathcal{C}^{I_1} \mathcal{C}^{I_2} \mathcal{C}^{I_3} \rangle = \mathcal{C}^{I_1}_{J_1 \dots J_i K_1 \dots K_j} \mathcal{C}^{I_2}_{J_1 \dots J_i L_1 \dots L_k} \mathcal{C}^{I_3}_{K_1 \dots K_j L_1 \dots L_k} \, . \tag{47}$$

Remarkably, (47) is the same result one obtains from free field theory by Wick contracting all the ϕ^J fields in the three operators. This suggests that there is a non-renormalization theorem for this correlation function, but such a theorem has not yet been proven (see however comments at the end of section 3.2). It is worth emphasizing that the normalization ambiguity in the bulk-boundary coupling is circumvented essentially by considering invariant ratios of three-point functions and two-point functions, into which the "leg factors" w^I do not enter. This is the same strategy as was pursued in comparing matrix models of quantum gravity to Liouville theory.

4.3   Four-point Functions

The calculation of four-point functions is difficult because there are several graphs which contribute, and some of them inevitably involve bulk-to-bulk propagators of fields with spin. The computation of four-point functions of the operators $\mathcal{O}_\phi$ and $\mathcal{O}_C$ dual to the dilaton and the axion was completed in

[160]. See also [128, 133, 135, 136, 161, 162, 139, 137, 163, 5] for earlier contributions. One of the main technical results, further developed in [164], is that diagrams involving an internal propagator can be reduced by integration over one of the bulk vertices to a sum of quartic graphs expressible in terms of the functions

$$D_{\Delta_1\Delta_2\Delta_3\Delta_4}(\vec{x}_1, \vec{x}_2, \vec{x}_3, \vec{x}_4) = \int d^5x \sqrt{g} \prod_{i=1}^{4} \tilde{K}_{\Delta_i}(\vec{x}, z; \vec{x}_i),$$

$$\tilde{K}_\Delta(\vec{x}, z; \vec{y}) = \left(\frac{z}{z^2 + (\vec{x} - \vec{y})^2} \right)^\Delta . \tag{48}$$

The integration is over the bulk point $(\vec{x}, z)$. There are two independent conformally invariant combinations of the $\vec{x}_i$:

$$s = \frac{1}{2} \frac{\vec{x}_{13}^2 \vec{x}_{24}^2}{\vec{x}_{12}^2 \vec{x}_{34}^2 + \vec{x}_{14}^2 \vec{x}_{23}^2} \qquad t = \frac{\vec{x}_{12}^2 \vec{x}_{34}^2 - \vec{x}_{14}^2 \vec{x}_{23}^2}{\vec{x}_{12}^2 \vec{x}_{34}^2 + \vec{x}_{14}^2 \vec{x}_{23}^2} . \tag{49}$$

One can write the connected four-point function as

$$\langle \mathcal{O}_\phi(\vec{x}_1) \mathcal{O}_C(\vec{x}_2) \mathcal{O}_\phi(\vec{x}_3) \mathcal{O}_C(\vec{x}_4) \rangle = \left(\frac{6}{\pi^2} \right)^4 \left[16\vec{x}_{24}^2 \left(\frac{1}{2s} - 1 \right) D_{4455} + \frac{64}{9} \frac{\vec{x}_{24}^2}{\vec{x}_{13}^2} \frac{1}{s} D_{3355} \right.$$

$$\left. + \frac{16}{3} \frac{\vec{x}_{24}^2}{\vec{x}_{13}^2} \frac{1}{s} D_{2255} - 14 D_{4444} - \frac{46}{9\vec{x}_{13}^2} D_{3344} - \frac{40}{9\vec{x}_{13}^2} D_{2244} - \frac{8}{3\vec{x}_{13}^6} D_{1144} + 64\vec{x}_{24}^2 D_{4455} \right] . \tag{50}$$

An interesting limit of (50) is to take two pairs of points close together. Following [160], let us take the pairs $(\vec{x}_1, \vec{x}_3)$ and $(\vec{x}_2, \vec{x}_4)$ close together while holding $\vec{x}_1$ and $\vec{x}_2$ a fixed distance apart. Then the existence of an OPE expansion implies that

$$\langle \mathcal{O}_{\Delta_1}(\vec{x}_1) \mathcal{O}_{\Delta_2}(\vec{x}_2) \mathcal{O}_{\Delta_3}(\vec{x}_3) \mathcal{O}_{\Delta_4}(\vec{x}_4) \rangle = \sum_{n,m} \frac{\alpha_n \langle \mathcal{O}_n(\vec{x}_1) \mathcal{O}_m(\vec{x}_2) \rangle \beta_m}{\vec{x}_{13}^{\Delta_1 + \Delta_3 - \Delta_m} \vec{x}_{24}^{\Delta_2 + \Delta_4 - \Delta_n}}, \tag{51}$$

at least as an asymptotic series, and hopefully even with a finite radius of convergence for $\vec{x}_{13}$ and $\vec{x}_{24}$. The operators $\mathcal{O}_n$ are the ones that appear in the OPE of $\mathcal{O}_1$ with $\mathcal{O}_3$, and the operators $\mathcal{O}_m$ are the ones that appear in the OPE of $\mathcal{O}_2$ with $\mathcal{O}_4$. $\mathcal{O}_\phi$ and $\mathcal{O}_C$ are descendants of chiral primaries, and so have protected dimensions. The product of descendants of chiral fields is not itself necessarily the descendent of a chiral field: an appropriately normal ordered product : $\mathcal{O}_\phi \mathcal{O}_\phi$: is expected to have an unprotected dimension of the form $8 + O(1/N^2)$. This is the natural result from the field theory point of view because there are $O(N^2)$ degrees of freedom contributing to each factor, and the

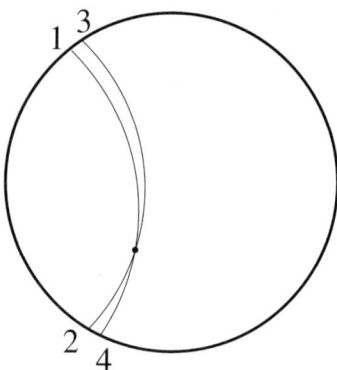

Figure 3.   A nearly degenerate quartic graph contributing to the four-point function in the limit $|\vec{x}_{13}|, |\vec{x}_{24}| \ll |\vec{x}_{12}|$.

commutation relations between them are non-trivial only a fraction $1/N^2$ of the time. From the supergravity point of view, a composite operator like : $\mathcal{O}_\phi \mathcal{O}_\phi$: corresponds to a two-particle bulk state, and the $O(1/N^2) = O(\kappa^2/R^8)$ correction to the mass is interpreted as the correction to the mass of the two-particle state from gravitational binding energy. Roughly one is thinking of graviton exchange between the legs of figure 3 that are nearly coincident.

If (51) is expanded in inverse powers of N, then the $O(1/N^2)$ correction to Δ_n and Δ_m shows up to leading order as a term proportional to a logarithm of some combination of the separations $\vec{x}_{ij}$. Logarithms also appear in the expansion of (50) in the $|\vec{x}_{13}|, |\vec{x}_{24}| \ll |\vec{x}_{12}|$ limit in which (51) applies: the leading log in this limit is $\frac{1}{(\vec{x}_{12})^{16}} \log \left(\frac{\vec{x}_{13}\vec{x}_{24}}{\vec{x}_{12}^2} \right)$. This is the correct form to be interpreted in terms of the propagation of a two-particle state dual to an operator whose dimension is slightly different from 8.

5.   Wilson Loops

In this section we consider Wilson loop operators in the gauge theory. The Wilson loop operator

$$W(\mathcal{C}) = \text{Tr} \left[P \exp \left(i \oint_{\mathcal{C}} A \right) \right] \tag{52}$$

depends on a loop $\mathcal{C}$ embedded in four dimensional space, and it involves the path-ordered integral of the gauge connection along the contour. The trace is taken over some representation of the gauge group; we will discuss here only the case of the fundamental representation (see [165] for a discussion of other representations). From the expectation value of the Wilson loop operator $\langle W(\mathcal{C}) \rangle$ we can calculate the quark-antiquark potential. For this purpose we consider a rectangular loop with sides of length T and L in Euclidean space. Then, viewing T as the time direction, it is clear that for large T the expectation value will behave as e^{-TE} where E is the lowest possible energy of the quark-anti-quark configuration. Thus, we have

$$\langle W \rangle \sim e^{-TV(L)} , \tag{53}$$

where $V(L)$ is the quark anti-quark potential. For large N and large $g_{YM}^2 N$, the AdS/CFT correspondence maps the computation of $\langle W \rangle$ in the CFT into a problem of finding a minimum surface in AdS [166, 167].

5.1 Wilson Loops and Minimum Surfaces

In QCD, we expect the Wilson loop to be related to the string running from the quark to the antiquark. We expect this string to be analogous to the string in our configuration, which is a superstring which lives in ten dimensions, and which can stretch between two points on the boundary of AdS. In order to motivate this prescription let us consider the following situation. We start with the gauge group $U(N + 1)$, and we break it to $U(N) \times U(1)$ by giving an expectation value to one of the scalars. This corresponds, as discussed in section 2, to having a D3 brane sitting at some radial position U in AdS, and at a point on S^5. The off-diagonal states, transforming in the $\mathbf{N}$ of $U(N)$, get a mass proportional to U, $m = U/2\pi$. So, from the point of view of the $U(N)$ gauge theory, we can view these states as massive quarks, which act as a source for the various $U(N)$ fields. Since they are charged they will act as a source for the vector fields. In order to get a non-dynamical source (an "external quark" with no fluctuations of its own, which will correspond precisely to the Wilson loop operator) we need to take $m \rightarrow \infty$, which means U should also go to infinity. Thus, the string should end on the boundary of AdS space.

These stretched strings will also act as a source for the scalar fields. The coupling to the scalar fields can be seen qualitatively by viewing the quarks as strings stretching between the N branes and the single separated brane. These strings will pull the N branes and will cause a deformation of the branes, which is described by the scalar fields. A more formal argument for this coupling is that these states are BPS, and the coupling to the scalar (Higgs) fields is determined by supersymmetry. Finally, one can see this coupling explicitly by writing the full $U(N + 1)$ Lagrangian, putting in the Higgs expectation value and calculating the equation of motion for the massive fields [166]. The precise definition of the Wilson loop operator corresponding to the superstring will actually include also the field theory fermions, which will imply some particular boundary conditions for the worldsheet fermions at the boundary of AdS. However, this will not affect the leading order computations we describe here.

So, the final conclusion is that the stretched strings couple to the operator

$$W(\mathcal{C}) = \mathrm{Tr} \left[P \exp \left(\oint (iA_\mu \dot{x}^\mu + \theta^I \phi^I \sqrt{\dot{x}^2}) d\tau \right) \right], \qquad (54)$$

where $x^\mu(\tau)$ is any parametrization of the loop and θ^I ($I = 1, \cdots, 6$) is a unit vector in $\mathbb{R}^6$ (the point on S^5 where the string is sitting). This is the expression when the signature of $\mathbb{R}^4$ is Euclidean. In the Minkowski signature case, the phase factor associated to the trajectory of the quark has an extra factor "i" in front of θ^I [7].

Generalizing the prescription of section 4 for computing correlation functions, the discussion above implies that in order to compute the expectation value of the operator (54) in $\mathcal{N} = 4$ SYM we should consider the string theory partition function on $AdS_5 \times S^5$, with the condition that we have a string worldsheet ending on the loop $\mathcal{C}$, as in figure 4 [167, 166]. In the supergravity regime, when $g_s N$ is large, the leading contribution to this partition function will come from the area of the string worldsheet. This area is measured with the AdS metric, and it is generally not the same as the area enclosed by the loop $\mathcal{C}$ in four dimensions.

Figure 4.   The Wilson loop operator creates a string worldsheet ending on the corresponding loop on the boundary of AdS.

The area as defined above is divergent. The divergence arises from the fact that the string worldsheet is going all the way to the boundary of AdS. If we evaluate the area up to some radial distance $U = r$, we see that for large r it diverges as $r|\mathcal{C}|$, where $|\mathcal{C}|$ is the length of the loop in the field theory [166, 167]. On the other hand, the perturbative computation in the field theory shows that $\langle W \rangle$, for W given by (54), is finite, as it should be since a divergence in the Wilson loop would have implied a mass renormalization

---

[7] The difference in the factor of i between the Euclidean and the Minkowski cases can be traced to the analytic continuation of $\sqrt{\dot{x}^2}$. A detailed derivation of (54) can be found in [168].

of the BPS particle. The apparent discrepancy between the divergence of the area of the minimum surface in AdS and the finiteness of the field theory computation can be reconciled by noting that the appropriate action for the string worldsheet is not the area itself but its Legendre transform with respect to the string coordinates corresponding to θ^I and the radial coordinate u [168]. This is because these string coordinates obey the Neumann boundary conditions rather than the Dirichlet conditions. When the loop is smooth, the Legendre transformation simply subtracts the divergent term $r|\mathcal{C}|$, leaving the resulting action finite.

As an example let us consider a circular Wilson loop. Take $\mathcal{C}$ to be a circle of radius a on the boundary, and let us work in the Poincaré coordinates. We could find the surface that minimizes the area by solving the Euler-Lagrange equations. However, in this case it is easier to use conformal invariance. Note that there is a conformal transformation in the field theory that maps a line to a circle. In the case of the line, the minimum area surface is clearly a plane that intersects the boundary and goes all the way to the horizon (which is just a point on the boundary in the Euclidean case). Using the conformal transformation to map the line to a circle we obtain the minimal surface we want. It is, using the coordinates (15) for AdS_5,

$$\vec{x} = \sqrt{a^2 - z^2}(\vec{e}_1 \cos\theta + \vec{e}_2 \sin\theta), \tag{55}$$

where $\vec{e}_1$, $\vec{e}_2$ are two orthonormal vectors in four dimensions (which define the orientation of the circle) and $0 \le z \le a$. We can calculate the area of this surface in AdS, and we get a contribution to the action

$$S \sim \frac{1}{2\pi\alpha'}\mathcal{A} = \frac{R^2}{2\pi\alpha'}\int d\theta \int_\epsilon^a \frac{dza}{z^2} = \frac{R^2}{\alpha'}(\frac{a}{\epsilon} - 1), \tag{56}$$

where we have regularized the area by putting a an IR cutoff at $z = \epsilon$ in AdS, which is equivalent to a UV cutoff in the field theory [66]. Subtracting the divergent term we get

$$\langle W \rangle \sim e^{-S} \sim e^{R^2/\alpha'} = e^{\sqrt{4\pi g_s N}}. \tag{57}$$

This is independent of a as required by conformal invariance.

We could similarly consider a "magnetic" Wilson loop, which is also called a 't Hooft loop [169]. This case is related by electric-magnetic duality to the previous case. Since we identify the electric-magnetic duality with the $SL(2,\mathbb{Z})$ duality of type IIB string theory, we should consider in this case a D-string worldsheet instead of a fundamental string worldsheet. We get the same result as in (57) but with $g_s \to 1/g_s$.

Using (53) it is possible to compute the quark-antiquark potential in the supergravity approximation [167, 166]. In this case we consider a configuration which is invariant under (Euclidean) time translations. We take both particles to have the same scalar charge, which means that the two ends of the string are at the same point in S^5 (one could consider also the more general case with a string ending at different points on S^5 [166]). We put the quark at $x = -L/2$ and the anti-quark at $x = L/2$. Here "quark" means an infinitely massive W-boson connecting the N branes with one brane which is (infinitely) far away. The classical action for a string worldsheet is

$$S = \frac{1}{2\pi\alpha'} \int d\tau d\sigma \sqrt{\det(G_{MN}\partial_\alpha X^M \partial_\beta X^N)}, \tag{58}$$

where G_{MN} is the Euclidean $AdS_5 \times S^5$ metric. Note that the factors of α' cancel out in (58), as they should. Since we are interested in a static configuration we take $\tau = t$, $\sigma = x$, and then the action becomes

$$S = \frac{TR^2}{2\pi} \int_{-L/2}^{L/2} dx \frac{\sqrt{(\partial_x z)^2 + 1}}{z^2}. \tag{59}$$

We need to solve the Euler-Lagrange equations for this action. Since the action does not depend on x explicitly the solution satisfies

$$\frac{1}{z^2\sqrt{(\partial_x z)^2 + 1}} = \text{constant}. \tag{60}$$

Defining z_0 to be the maximum value of $z(x)$, which by symmetry occurs at $x = 0$, we find that the solution is[8]

$$x = z_0 \int_{z/z_0}^1 \frac{dy y^2}{\sqrt{1 - y^4}}, \tag{61}$$

where z_0 is determined by the condition

$$\frac{L}{2} = z_0 \int_0^1 \frac{dy y^2}{\sqrt{1 - y^4}} = z_0 \frac{\sqrt{2}\pi^{3/2}}{\Gamma(1/4)^2}. \tag{62}$$

The qualitative form of the solution is shown in figure 5(b). Notice that the string quickly approaches $x = L/2$ for small z (close to the boundary),

$$\frac{L}{2} - x \sim z^3, \quad z \to 0. \tag{63}$$

Now we compute the total energy of the configuration. We just plug in the solution (61) in (59), subtract the infinity as explained above (which can be

---

[8] All integrals in this section can be calculated in terms of elliptic or Beta functions.

interpreted as the energy of two separated massive quarks, as in figure 5(a)), and we find

$$E = V(L) = -\frac{4\pi^2 (2g_{YM}^2 N)^{1/2}}{\Gamma(\frac{1}{4})^4 L}. \qquad (64)$$

We see that the energy goes as $1/L$, a fact which is determined by conformal invariance. Note that the energy is proportional to $(g_{YM}^2 N)^{1/2}$, as opposed to $g_{YM}^2 N$ which is the perturbative result. This indicates some screening of the charges at strong coupling. The above calculation makes sense for all distances L when $g_s N$ is large, independently of the value of g_s. Some subleading corrections coming from quantum fluctuations of the worldsheet were calculated in [170, 171, 172].

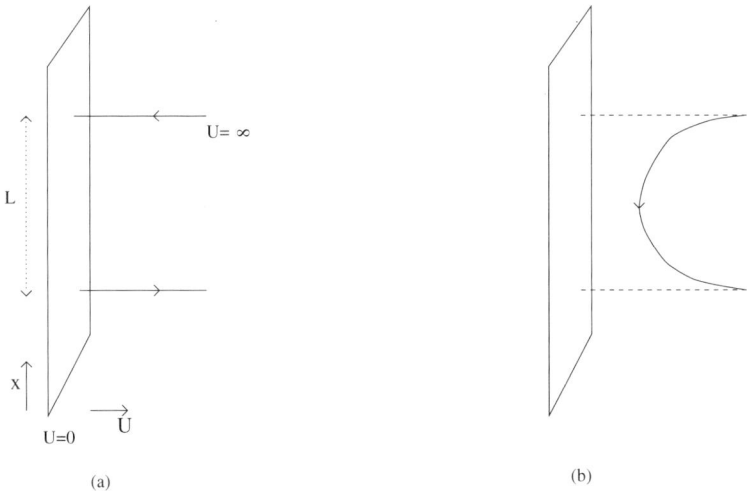

(a)   (b)

Figure 5.   (a) Initial configuration corresponding to two massive quarks before we turn on their coupling to the $U(N)$ gauge theory. (b) Configuration after we consider the coupling to the $U(N)$ gauge theory. This configuration minimizes the action. The quark-antiquark energy is given by the difference of the total length of the strings in (a) and (b).

In a similar fashion we could compute the potential between two magnetic monopoles in terms of a D-string worldsheet, and the result will be the same as (64) but with $g_{YM} \to 4\pi/g_{YM}$. One can also calculate the interaction between a magnetic monopole and a quark. In this case the fundamental string (ending on the quark) will attach to the D-string (ending on the monopole), and they will connect to form a $(1, 1)$ string which will go into the horizon. The resulting potential is a complicated function of g_{YM} [173], but in the limit that g_{YM} is small (but still with $g_{YM}^2 N$ large) we get that the monopole-quark potential is just $1/4$ of the quark-quark potential. This can be understood from the fact that when g is small the D-string is very rigid and the fundamental string will end almost perpendicularly on the D-string. Therefore, the solution for the fundamental string will be half of the solution we had above, leading to a factor of $1/4$ in the potential. Calculations of Wilson loops in the Higgs phase were done in [174].

Another interesting case one can study analytically is a surface near a cusp on $\mathbb{R}^4$. In this case, the perturbative computation in the gauge theory shows a logarithmic divergence with a coefficient depending on the angle at the cusp. The area of the minimum surface also contains a logarithmic divergence depending on the angle [168]. Other aspects of the gravity calculation of Wilson loops were discussed in [175, 176, 177, 178, 179].

5.2   Other Branes Ending on the Boundary

We could also consider other branes that are ending at the boundary [180]. The simplest example would be a zero-brane (i.e. a particle) of mass m. In Euclidean space a zero-brane describes a one dimensional trajectory in anti-de-Sitter space which ends at two points on the boundary. Therefore, it is associated with the insertion of two local operators at the two points where the trajectory ends. In the supergravity approximation the zero-brane follows a geodesic. Geodesics in the hyperbolic plane (Euclidean AdS) are semicircles. If we compute the action we get

$$S = m \int ds = -2mR \int_{\epsilon}^{a} \frac{a\,dz}{z\sqrt{a^2 - z^2}}, \tag{65}$$

where we took the distance between the two points at the boundary to be $L = 2a$ and regulated the result. We find a logarithmic divergence when $\epsilon \to 0$, proportional to $\log(\epsilon/a)$. If we subtract the logarithmic divergence we get a residual dependence on a. Naively we might have thought that (as in the previous subsection) the answer had to be independent of a due to conformal invariance. In fact, the dependence on a is very important, since it leads to a result of the form

$$e^{-S} \sim e^{-2mR\log a} \sim \frac{1}{a^{2mR}}, \tag{66}$$

which is precisely the result we expect for the two-point function of an operator of dimension $\Delta = mR$. This is precisely the large mR limit of the formula (12), so we reproduce in the supergravity limit the 2-point function described in section 4. In general, this sort of logarithmic divergence arises when the brane worldvolume is odd dimensional [180], and it implies that the expectation value of the corresponding operator depends on the overall scale. In particular one could consider the "Wilson surfaces" that arise in the six dimensional $\mathcal{N} = (2, 0)$ theory. In that case one has to consider a two-brane, with a three dimensional worldvolume, ending on a two dimensional surface on the boundary of AdS_7. Again, one gets a logarithmic term, which is proportional to the rigid string action of the two dimensional surface living on the string in the $\mathcal{N} = (2,0)$ field theory [181, 180].

One can also compute correlation functions involving more than one Wilson loop. To leading order in N this will be just the product of the expectation values of each Wilson loop. On general grounds one expects that the subleading corrections are given by surfaces that end on more than one loop. One limiting case is when the surfaces look similar to the zeroth order surfaces but with additional thin tubes connecting them. These thin tubes are nothing else than massless particles being exchanged between the two string worldsheets [165, 181].

6. Theories at Finite Temperature

As discussed in section 3, the quantities that can be most successfully compared between gauge theory and string theory are those with some protection from supersymmetry and/or conformal invariance — for instance, dimensions of chiral primary operators. Finite temperature breaks both supersymmetry and conformal invariance, and the insights we gain from examining the $T > 0$ physics will be of a more qualitative nature. They are no less interesting for that: we shall see in section 6.1 how the entropy of near-extremal D3-branes comes out identical to the free field theory prediction up to a factor of a power of $4/3$; then in section 6.2 we explain how a phase transition studied by Hawking and Page in the context of quantum gravity is mapped into a confinement-deconfinement transition in the gauge theory.

6.1 Construction

The gravity solution describing the gauge theory at finite temperature can be obtained by starting from the general black three-brane solution and taking the decoupling limit of section 2 keeping the energy density above extremality finite. The resulting metric can be written as

$$ds^2 = R^2 \left[u^2(-h\,dt^2 + dx_1^2 + dx_2^2 + dx_3^2) + \frac{du^2}{hu^2} + d\Omega_5^2 \right] \tag{67}$$

$$h = 1 - \frac{u_0^4}{u^4}, \qquad u_0 = \pi T.$$

It will often be useful to Wick rotate by setting $t_E = it$, and use the relation between the finite temperature theory and the Euclidean theory with a compact time direction.

The first computation which indicated that finite-temperature $U(N)$ Yang-Mills theory might be a good description of the microstates of N coincident D3-branes was the calculation of the entropy [182, 183]. On the supergravity side, the entropy of near-extremal D3-branes is just the usual Bekenstein-Hawking result, $S = A/4G_N$, and it is expected to be a reliable guide to the entropy of the gauge theory at large N and large $g^2_{YM}N$. There is no problem on the gauge theory side in working at large N, but large $g^2_{YM}N$ at finite temperature is difficult indeed. The analysis of [182] was limited to a free field computation in the field theory, but nevertheless the two results for the entropy agreed up to a factor of a power of $4/3$. In the canonical ensemble, where temperature and volume are the independent variables, one identifies the field theory volume with the world-volume of the D3-branes, and one sets the field theory temperature equal to the Hawking temperature in supergravity. The result is

$$F_{SUGRA} = -\frac{\pi^2}{8}N^2VT^4,$$

$$F_{SYM} = \frac{4}{3}F_{SUGRA}. \tag{68}$$

The supergravity result is at leading order in l_s/R, and it would acquire corrections suppressed by powers of TR if we had considered the full D3-brane metric rather than the near-horizon limit, (67). These corrections do not have an interpretation in the context of CFT because they involve R as an intrinsic scale. Two equivalent methods to evaluate F_{SUGRA} are a) to use $F = E - TS$ together with standard expressions for the Bekenstein-Hawking entropy, the Hawking temperature, and the ADM mass; and b) to consider the gravitational action of the Euclidean solution, with a periodicity in the Euclidean time direction (related to the temperature) which eliminates a conical deficit angle at the horizon.[9]

The $4/3$ factor is a long-standing puzzle into which we still have only qualitative insight. The gauge theory computation was performed at zero 't Hooft coupling, whereas the supergravity is supposed to be valid at strong 't Hooft coupling, and unlike in the 1+1-dimensional case where the entropy is essentially fixed by the central charge, there is no non-renormalization theorem for the coefficient of T^4 in the free energy. Indeed, it was suggested in [184] that the leading term in the $1/N$ expansion of F has the form

$$F = -f(g^2_{YM}N)\frac{\pi^2}{6}N^2VT^4, \tag{69}$$

where $f(g^2_{YM}N)$ is a function which smoothly interpolates between a weak coupling limit of 1 and a strong coupling limit of $3/4$. It was pointed out early [185] that the quartic potential $g^2_{YM}\text{Tr}\,[\phi^I, \phi^J]^2$ in the $\mathcal{N} = 4$ Yang-Mills action might be expected to freeze out more and more degrees of freedom as the

---

[9]The result of [182], $S_{SYM} = (4/3)^{1/4}S_{SUGRA}$, differs superficially from (68), but it is only because the authors worked in the microcanonical ensemble: rather than identifying the Hawking temperature with the field theory temperature, the ADM mass above extremality was identified with the field theory energy.

coupling was increased, which would suggest that $f(g_{YM}^2 N)$ is monotone decreasing. An argument has been given [186], based on the non-renormalization of the two-point function of the stress tensor, that $f(g_{YM}^2 N)$ should remain finite at strong coupling.

The leading corrections to the limiting value of $f(g_{YM}^2 N)$ at strong and weak coupling were computed in [184] and [187], respectively. The results are

$$\begin{aligned} f(g_{YM}^2 N) &= 1 - \frac{3}{2\pi^2} g_{YM}^2 N + \dots & \text{for small } g_{YM}^2 N, \\ f(g_{YM}^2 N) &= \frac{3}{4} + \frac{45}{32} \frac{\zeta(3)}{(g_{YM}^2 N)^{3/2}} + \dots & \text{for large } g_{YM}^2 N. \end{aligned} \tag{70}$$

The weak coupling result is a straightforward although somewhat tedious application of the diagrammatic methods of perturbative finite-temperature field theory. The constant term is from one loop, and the leading correction is from two loops. The strong coupling result follows from considering the leading α' corrections to the supergravity action. The relevant one involves a particular contraction of four powers of the Weyl tensor. It is important now to work with the Euclidean solution, and one restricts attention further to the near-horizon limit. The Weyl curvature comes from the non-compact part of the metric, which is no longer AdS_5 but rather the AdS-Schwarzschild solution which we will discuss in more detail in section 6.2. The action including the α' corrections no longer has the Einstein-Hilbert form, and correspondingly the Bekenstein-Hawking prescription no longer agrees with the free energy computed as βI where I is the Euclidean action. In keeping with the basic prescription for computing Green's functions, where a free energy in field theory is equated (in the appropriate limit) with a supergravity action, the relation $I = \beta F$ is regarded as the correct one. (See [188].) It has been conjectured that the interpolating function $f(g_{YM}^2 N)$ is not smooth, but exhibits some phase transition at a finite value of the 't Hooft coupling. We regard this as an unsettled question. The arguments in [189, 190] seem as yet incomplete. In particular, they rely on analyticity properties of the perturbation expansion which do not seem to be proven for finite temperature field theories.

6.2   Thermal Phase Transition

The holographic prescription of [23, 24], applied at large N and $g_{YM}^2 N$ where loop and stringy corrections are negligible, involves extremizing the supergravity action subject to particular asymptotic boundary conditions. We can think of this as the saddle point approximation to the path integral over supergravity fields. That path integral is ill-defined because of the non-renormalizable nature of supergravity. String amplitudes (when we can calculate them) render on-shell quantities well-defined. Despite the conceptual difficulties we can use some simple intuition about path integrals to illustrate an important point about the AdS/CFT correspondence: namely, there can be more than one saddle point in the range of integration, and when there is we should sum $e^{-I_{SUGRA}}$ over the classical configurations to obtain the saddle-point approximation to the gauge theory partition function. Multiple classical configurations are possible because of the general feature of boundary value problems in differential equations: there can be multiple solutions to the classical equations satisfying the same asymptotic boundary conditions. The solution which globally minimizes I_{SUGRA} is the one that dominates the path integral.

When there are two or more solutions competing to minimize I_{SUGRA}, there can be a phase transition between them. An example of this was studied in [191] long before the AdS/CFT correspondence, and subsequently resurrected, generalized, and reinterpreted in [24, 68] as a confinement-deconfinement transition in the gauge theory. Since the qualitative features are independent of the dimension, we will restrict our attention to AdS_5. It is worth noting however that if the AdS_5 geometry is part of a string compactification, it doesn't matter what the internal manifold is except insofar as it fixes the cosmological constant, or equivalently the radius R of anti-de Sitter space.

There is an embedding of the Schwarzschild black hole solution into anti-de Sitter space which extremizes the action

$$I = -\frac{1}{16\pi G_5} \int d^5x \sqrt{g} \left(\mathcal{R} + \frac{12}{R^2}\right) . \tag{71}$$

Explicitly, the metric is

$$ds^2 = f dt^2 + \frac{1}{f} dr^2 + r^2 d\Omega_3^2,$$

$$f = 1 + \frac{r^2}{R^2} - \frac{\mu}{r^2} . \tag{72}$$

The radial variable r is restricted to $r \geq r_+$, where r_+ is the largest root of $f = 0$. The Euclidean time is periodically identified, $t \sim t + \beta$, in order to eliminate the conical singularity at $r = r_+$. This requires

$$\beta = \frac{2\pi R^2 r_+}{2r_+^2 + R^2} . \tag{73}$$

Topologically, this space is $S^3 \times B^2$, and the boundary is $S^3 \times S^1$ (which is the relevant space for the field theory on S^3 with finite temperature). We will call this space X_2. Another space with the same boundary which is also a local extremum of (71) is given by the metric in (72) with $\mu = 0$ and again with periodic time. This space, which we will call X_1, is not only metrically distinct from the first (being locally conformally flat), but also topologically $B^4 \times S^1$ rather than $S^3 \times B^2$. Because the S^1 factor is not simply connected, there are two possible spin structures on X_1, corresponding to thermal (anti-periodic) or supersymmetric (periodic) boundary conditions on fermions. In contrast, X_2 is simply connected and hence admits a unique spin structure, corresponding to thermal boundary conditions. For the purpose of computing the twisted partition function, $\operatorname{Tr}(-1)^F e^{-\beta H}$, in a saddle-point approximation, only X_1 contributes. But, X_1 and X_2 make separate saddle-point contributions to the usual thermal partition function, $\operatorname{Tr} e^{-\beta H}$, and the more important one is the one with the smaller Euclidean action.

Actually, both $I(X_1)$ and $I(X_2)$ are infinite, so to compute $I(X_2) - I(X_1)$ a regulation scheme must be adopted. The one used in [68, 184] is to cut off both X_1 and X_2 at a definite coordinate radius $r = R_0$. For X_2, the elimination of the conical deficit angle at the horizon fixes the period of Euclidean time; but for X_1, the period is arbitrary. In order to make the comparison of $I(X_1)$ and $I(X_2)$ meaningful, we fix the period of Euclidean time on X_1 so that the proper circumference of the S_1 at $r = R_0$ is the same as the proper length on X_2 of an orbit of the Killing vector $\partial/\partial t$, also at $r = R_0$. In the limit $R_0 \to \infty$, one finds

$$I(X_2) - I(X_1) = \frac{\pi^2 r_+^3 (R^2 - r_+^2)}{4G_5 (2r_+^2 + R^2)} \,, \tag{74}$$

where again r_+ is the largest root of $f = 0$. The fact that (74) (or more precisely its AdS_4 analog) can change its sign was interpreted in [191] as indicating a phase transition between a black hole in AdS and a thermal gas of particles in AdS (which is the natural interpretation of the space X_1). The black hole is the thermodynamically favored state when the horizon radius r_+ exceeds the radius of curvature R of AdS. In the gauge theory we interpret this transition as a confinement-deconfinement transition. Since the theory is conformally invariant, the transition temperature must be proportional to the inverse radius of the space S^3 which the field theory lives on. Similar transitions, and also local thermodynamic instability due to negative specific heats, have been studied in the context of spinning branes and charged black holes in [192, 193, 194, 195, 196, 197, 198]. Most of these works are best understood on the CFT side as explorations of exotic thermal phenomena in finite-temperature gauge theories. Connections with Higgsed states in gauge theory are clearer in [199, 200]. The relevance to confinement is explored in [197]. See also [201, 202, 203, 204] for other interesting contributions to the finite temperature literature.

Deconfinement at high temperature can be characterized by a spontaneous breaking of the center of the gauge group. In our case the gauge group is $SU(N)$ and its center is $\mathbb{Z}_N$. The order parameter for the breaking of the center is the expectation value of the Polyakov (temporal) loop $\langle W(C) \rangle$. The boundary of the spaces X_1, X_2 is $S^3 \times S^1$, and the path C wraps around the circle. An element of the center $g \in \mathbb{Z}_N$ acts on the Polyakov loop by

$\langle W(C) \rangle \to g \langle W(C) \rangle$. The expectation value of the Polyakov loop measures the change of the free energy of the system $F_q(T)$ induced by the presence of the external charge q, $\langle W(C) \rangle \sim exp\left(-F_q(T)/T\right)$. In a confining phase $F_q(T)$ is infinite and therefore $\langle W(C) \rangle = 0$. In the deconfined phase $F_q(T)$ is finite and therefore $\langle W(C) \rangle \neq 0$.

As discussed in section 5, in order to compute $\langle W(C) \rangle$ we have to evaluate the partition function of strings with a worldsheet D that is bounded by the loop C. Consider first the low temperature phase. The relevant space is X_1 which, as discussed above, has the topology $B^4 \times S^1$. The contour C wraps the circle and is not homotopic to zero in X_1. Therefore C is not a boundary of any D, which immediately implies that $\langle W(C) \rangle = 0$. This is the expected behavior at low temperatures (compared to the inverse radius of the S^3), where the center of the gauge group is not broken.

For the high temperature phase the relevant space is X_2, which has the topology $S^3 \times B^2$. The contour C is now a boundary of a string worldsheet $D = B^2$ (times a point in S^3). This seems to be in agreement with the fact that in the high temperature phase $\langle W(C) \rangle \neq 0$ and the center of the gauge group is broken. It was pointed out in [68] that there is a subtlety with this argument, since the center should not be broken in finite volume (S^3), but only in the infinite volume limit ($\mathbb{R}^3$). Indeed, the solution X_2 is not unique and we can add to it an expectation value for the integral of the NS-NS 2-form field B on B^2, with vanishing field strength. This is an angular parameter ψ with period 2π, which contributes $i\psi$ to the string worldsheet action. The string theory partition function includes now an integral over all values of ψ, making $\langle W(C) \rangle = 0$ on S^3. In contrast, on $\mathbb{R}^3$ one integrates over the local fluctuations of ψ but not over its vacuum expectation value. Now $\langle W(C) \rangle \neq 0$ and depends on the value of $\psi \in U(1)$, which may be understood as the dependence on the center $\mathbb{Z}_N$ in the large N limit. Explicit computations of Polyakov loops at finite temperature were done in [205, 6].

In [68] the Euclidean black hole solution (72) was suggested to be holographically dual to a theory related to pure QCD in three dimensions. In the large volume limit the solution corresponds to the $\mathcal{N} = 4$ gauge theory on $\mathbb{R}^3 \times S^1$ with thermal boundary conditions, and when the S^1 is made small (corresponding to high temperature T) the theory at distances larger than $1/T$ effectively reduces to pure Yang-Mills on $\mathbb{R}^3$. Some of the non-trivial successes of this approach to QCD are summarized in [1].

Acknowledgments

These notes were made by selected pieces of [1]. I thank O. Aharony, S. Gubser, H. Ooguri and Y. Oz for collaborating on writing that review. I also thank the organizers for a very nice school. This work was supported in part by by the Packard foundation and the DOE grant DE-FGO2-91ER40654.

References

[1] O. Aharony, S. S. Gubser, J. Maldacena, H. Ooguri and Y. Oz, Phys. Rept. **323**, 183 (2000) [hep-th/9905111].

[2] H. J. Boonstra, B. Peeters and K. Skenderis, Phys. Lett. **B411**, 59 (1997) [hep-th/9706192].

[3] K. Sfetsos and K. Skenderis, Nucl. Phys. **B517**, 179 (1998) [hep-th/9711138].

[4] S. R. Das and S. P. Trivedi, Phys. Lett. **B445**, 142 (1998) [hep-th/9804149].

[5] F. Gonzalez-Rey, I. Park and K. Schalm, Phys. Lett. **B448**, 37 (1999) [hep-th/9811155].

[6] A. Brandhuber, N. Itzhaki, J. Sonnenschein and S. Yankielowicz, Phys. Lett. **B434**, 36 (1998) [hep-th/9803137].

[7] G. 't Hooft, "A Planar Diagram Theory for Strong Interactions," *Nucl. Phys.* **B72** (1974) 461.

[8] A. M. Polyakov, "Quantum Geometry of Bosonic Strings," *Phys. Lett.* **103B** (1981) 207.

[9] J. Maldacena, "The Large N limit of superconformal field theories and supergravity," *Adv. Theor. Math. Phys.* **2** (1998) 231, hep-th/9711200.

[10] J. Polchinski, "Dirichlet Branes and Ramond-Ramond charges," *Phys. Rev. Lett.* **75** (1995) 4724–4727, hep-th/9510017.

[11] G. 't Hooft, "Magnetic monopoles in unified gauge theories," *Nucl. Phys.* **B79** (1974) 276–284.

[12] A. M. Polyakov, "Particle spectrum in the quantum field theory," *JETP Lett.* **20** (1974) 194–195.

[13] E. Witten, "Bound States of Strings and p-Branes," *Nucl. Phys.* **B460** (1996) 335–350, hep-th/9510135.

[14] I. R. Klebanov, "World volume approach to absorption by nondilatonic branes," *Nucl. Phys.* **B496** (1997) 231, hep-th/9702076.

[15] S. S. Gubser, I. R. Klebanov, and A. A. Tseytlin, "String theory and classical absorption by three-branes," *Nucl. Phys.* **B499** (1997) 217, hep-th/9703040.

[16] S. S. Gubser and I. R. Klebanov, "Absorption by branes and Schwinger terms in the world volume theory," *Phys. Lett.* **B413** (1997) 41–48, hep-th/9708005.

[17] A. Strominger and C. Vafa, "Microscopic origin of the Bekenstein-Hawking entropy," *Phys. Lett.* **B379** (1996) 99–104, hep-th/9601029.

[18] C. G. Callan and J. M. Maldacena, "D-Brane Approach to Black Hole Quantum Mechanics," *Nucl. Phys.* **B472** (1996) 591–610, hep-th/9602043.

[19] G. T. Horowitz and A. Strominger, "Counting States of Near Extremal Black Holes," *Phys. Rev. Lett.* **77** (1996) 2368–2371, hep-th/9602051.

[20] S. R. Das and S. D. Mathur, "Comparing decay rates for black holes and D-branes," *Nucl. Phys.* **B478** (1996) 561–576, hep-th/9606185.

[21] A. Dhar, G. Mandal, and S. R. Wadia, "Absorption vs. decay of black holes in string theory and T symmetry," *Phys. Lett.* **B388** (1996) 51–59, hep-th/9605234.

[22] J. Maldacena and A. Strominger, "Black hole grey body factors and d-brane spectroscopy," *Phys. Rev.* **D55** (1997) 861–870, hep-th/9609026.

[23] S. S. Gubser, I. R. Klebanov, and A. M. Polyakov, "Gauge theory correlators from non-critical string theory," *Phys. Lett.* **B428** (1998) 105, hep-th/9802109.

[24] E. Witten, "Anti-de Sitter space and holography," *Adv. Theor. Math. Phys.* **2** (1998) 253, hep-th/9802150.

[25] A. Salam and E. Sezgin, eds., *Supergravities in Diverse Dimensions*, vol. 1 and 2. North-Holland, Amsterdam, Netherlands, 1989.

[26] M. J. Duff, B. E. W. Nilsson, and C. N. Pope, "Kaluza-Klein Supergravity," *Phys. Rept.* **130** (1986) 1.

[27] E. Witten, "Quantum Field Theory and the Jones Polynomial," *Commun. Math. Phys.* **121** (1989) 351.

[28] T. Banks, W. Fischler, S. H. Shenker, and L. Susskind, "M Theory as a Matrix Model: A Conjecture," *Phys. Rev.* **D55** (1997) 5112–5128, hep-th/9610043.

[29] G. 't Hooft, "Dimensional Reduction in Quantum Gravity," gr-qc/9310026.

[30] L. Susskind, "The World as a hologram," *J. Math. Phys.* **36** (1995) 6377–6396, hep-th/9409089.

[31] J. D. Bekenstein, "Entropy Bounds and Black Hole Remnants," *Phys. Rev.* **D49** (1994) 1912–1921, gr-qc/9307035.

[32] P. D. Vecchia, "An Introduction to AdS/CFT correspondence," hep-th/9903007.

[33] M. R. Douglas and S. Randjbar-Daemi, "Two lectures on the AdS / CFT correspondence," hep-th/9902022.

[34] J. L. Petersen, "Introduction to the Maldacena conjecture on AdS / CFT," hep-th/9902131.

[35] I. R. Klebanov, "From three-branes to large N gauge theories," hep-th/9901018.

[36] J. Polchinski, "String Theory,". Cambridge University Press (1998).

[37] R. G. Leigh, "Dirac-Born-Infeld Action from Dirichlet Sigma-Model," *Mod. Phys. Lett.* **A4** (1989) 2767.

[38] G. T. Horowitz and A. Strominger, "Black strings and P-branes," *Nucl. Phys.* **B360** (1991) 197–209.

[39] G. W. Gibbons and P. K. Townsend, "Vacuum interpolation in supergravity via super p-branes," *Phys. Rev. Lett.* **71** (1993) 3754–3757, hep-th/9307049.

[40] E. Witten, "AdS / CFT correspondence and topological field theory," *JHEP* **12** (1998) 012, hep-th/9812012.

[41] N. Seiberg, private communication, February 1999.

[42] S. Elitzur, G. Moore, A. Schwimmer, and N. Seiberg, "Remarks on the Canonical Quantization of the Chern-Simons-Witten Theory," *Nucl. Phys.* **B326** (1989) 108.

[43] C. Fronsdal, "The Dirac Supermultiplet," *Phys. Rev.* **D26** (1982) 1988.

[44] D. Z. Freedman and H. Nicolai, "Multiplet Shortening in $OSp(N, 4)$," *Nucl. Phys.* **B237** (1984) 342.

[45] K. Pilch, P. van Nieuwenhuizen, and P. K. Townsend, "Compactification of $d = 11$ Supergravity on S^4 (Or $11 = 7 + 4$, Too)," *Nucl. Phys.* **B242** (1984) 377.

[46] M. Gunaydin, P. van Nieuwenhuizen, and N. P. Warner, "General construction of the unitary representations of anti-de Sitter superalgebras and the spectrum of the S^4 compactification of eleven dimensional supergravity," *Nucl. Phys.* **B255** (1985) 63.

[47] M. Gunaydin and N. P. Warner, "Unitary supermultiplets of $OSp(8|4, \mathbb{R})$ and the spectrum of the S^7 compactification of eleven dimensional supergravity," *Nucl. Phys.* **B272** (1986) 99.

[48] M. Gunaydin, B. E. W. Nilsson, G. Sierra, and P. K. Townsend, "Singletons and Superstrings," *Phys. Lett.* **176B** (1986) 45.

[49] E. Bergshoeff, A. Salam, E. Sezgin, and Y. Tanii, "Singletons, Higher Spin Massless States and the Supermembrane," *Phys. Lett.* **205B** (1988) 237.

[50] E. Bergshoeff, A. Salam, E. Sezgin, and Y. Tanii, "$N = 8$ Supersingleton Quantum Field Theory," *Nucl. Phys.* **B305** (1988) 497.

[51] E. Bergshoeff, M. J. Duff, C. N. Pope, and E. Sezgin, "Compactifications of the Eleven-Dimensional Supermembrane," *Phys. Lett.* **B224** (1989) 71.

[52] M. R. Douglas, "Branes within branes," hep-th/9512077.

[53] C. M. Hull and P. K. Townsend, "Unity of superstring dualities," *Nucl. Phys.* **B438** (1995) 109–137, hep-th/9410167.

[54] A. A. Tseytlin, "Selfduality of Born-Infeld action and Dirichlet three-brane of type IIB superstring theory," *Nucl. Phys.* **B469** (1996) 51–67, hep-th/9602064.

[55] M. B. Green and M. Gutperle, "Comments on three-branes," *Phys. Lett.* **B377** (1996) 28–35, hep-th/9602077.

[56] M. R. Douglas and M. Li, "D-brane realization of N=2 superYang-Mills theory in four-dimensions," hep-th/9604041.

[57] N. Seiberg and E. Witten, "Monopoles, duality and chiral symmetry breaking in N=2 supersymmetric QCD," *Nucl. Phys.* **B431** (1994) 484–550, hep-th/9408099.

[58] M. Dine and N. Seiberg, "Comments on higher derivative operators in some SUSY field theories," *Phys. Lett.* **B409** (1997) 239–244, hep-th/9705057.

[59] M. R. Douglas, D. Kabat, P. Pouliot, and S. H. Shenker, "D-branes and short distances in string theory," *Nucl. Phys.* **B485** (1997) 85–127, hep-th/9608024.

[60] M. R. Douglas and W. Taylor, "Branes in the bulk of Anti-de Sitter space," hep-th/9807225.

[61] S. R. Das, "Holograms of branes in the bulk and acceleration terms in SYM effective action," hep-th/9905037.

[62] S. R. Das, "Brane waves, Yang-Mills theories and causality," *JHEP* **02** (1999) 012, hep-th/9901004.

[63] A. Kehagias, "New type IIB vacua and their F theory interpretation," *Phys. Lett.* **B435** (1998) 337, hep-th/9805131.

[64] S. S. Gubser, "Einstein manifolds and conformal field theories," *Phys. Rev.* **D59** (1999) 025006, hep-th/9807164.

[65] L. J. Romans, "New Compactifications of Chiral $N = 2, d = 10$ Supergravity," *Phys. Lett.* **153B** (1985) 392.

[66] L. Susskind and E. Witten, "The Holographic bound in anti-de Sitter space," hep-th/9805114.

[67] O. Aharony and E. Witten, "Anti-de Sitter Space and the Center of the Gauge Group," *JHEP* **11** (1998) 018, hep-th/9807205.

[68] E. Witten, "Anti-de Sitter space, thermal phase transition, and confinement in gauge theories," *Adv. Theor. Math. Phys.* **2** (1998) 505, hep-th/9803131.

[69] F. Gonzalez-Rey, B. Kulik, I. Y. Park, and M. Rocek, "Selfdual effective action of N=4 superYang-Mills," *Nucl. Phys.* **B544** (1999) 218, hep-th/9810152.

[70] J. Bagger and J. Wess, *Supersymmetry and Supergravity*. Princeton Series in Physics. Princeton University Press, Princeton, 1992.

[71] O. Bergman, "Three Pronged Strings and 1/4 BPS States in $N = 4$ Super Yang-Mills Theory," *Nucl. Phys.* **B525** (1998) 104, hep-th/9712211.

[72] K. Hashimoto, H. Hata, and N. Sasakura, "Three - string junction and BPS saturated solutions in SU(3) supersymmetric Yang-Mills theory," *Phys. Lett.* **B431** (1998) 303–310, hep-th/9803127.

[73] T. Kawano and K. Okuyama, "String network and 1/4 BPS states in N=4 SU(N) supersymmetric Yang-Mills theory," *Phys. Lett.* **B432** (1998) 338–342, hep-th/9804139.

[74] O. Bergman and B. Kol, "String Webs and 1/4 BPS Monopoles," *Nucl. Phys.* **B536** (1998) 149, hep-th/9804160.

[75] K. Hashimoto, H. Hata, and N. Sasakura, "Multipronged strings and BPS saturated solutions in SU(N) supersymmetric Yang-Mills theory," *Nucl. Phys.* **B535** (1998) 83, hep-th/9804164.

[76] K. Lee and P. Yi, "Dyons in N=4 supersymmetric theories and three pronged strings," *Phys. Rev.* **D58** (1998) 066005, hep-th/9804174.

[77] N. Sasakura and S. Sugimoto, "M theory description of 1/4 BPS states in N=4 supersymmetric Yang-Mills theory," hep-th/9811087.

[78] D. Tong, "A Note on 1/4 BPS states," hep-th/9902005.

[79] M. Gunaydin, D. Minic, and M. Zagermann, "Novel supermultiplets of $SU(2,2|4)$ and the AdS$_5$ / CFT$_4$ duality," hep-th/9810226.

[80] M. Gunaydin and N. Marcus, "The Spectrum of the S^5 Compactification of the Chiral $N = 2, D = 10$ Supergravity and the Unitary Supermultiplets of $U(2,2/4)$.," *Class. Quant. Grav.* **2** (1985) L11.

[81] L. Andrianopoli and S. Ferrara, "K-K Excitations on $AdS_5 \times S^5$ as $N = 4$ 'Primary' Superfields," *Phys. Lett.* **B430** (1998) 248–253, hep-th/9803171.

[82] L. Andrianopoli and S. Ferrara, "'Nonchiral' Primary Superfields in the AdS$_{d+1}$ / CFT$_d$ Correspondence," *Lett. Math. Phys.* **46** (1998) 265, hep-th/9807150.

[83] S. Ferrara and A. Zaffaroni, "Bulk gauge fields in AdS supergravity and supersingletons," hep-th/9807090.

[84] M. Gunaydin, D. Minic, and M. Zagermann, "4-D doubleton conformal theories, CPT and II B string on AdS$_5 \times S^5$," *Nucl. Phys.* **B534** (1998) 96–120, hep-th/9806042.

[85] L. Andrianopoli and S. Ferrara, "On Short and Long $SU(2,2/4)$ Multiplets in the AdS/CFT Correspondence," hep-th/9812067.

[86] M. Flato and C. Fronsdal, "One massless particle equals two Dirac singletons: elementary particles in a curved space. 6," *Lett. Math. Phys.* **2** (1978) 421.

[87] C. Fronsdal and M. Flato, "On Dis and Racs," *Phys. Lett.* **97B** (1980) 236.

[88] M. Flato and C. Fronsdal, "Quantum field theory of singletons: the Rac," *J. Math. Phys.* **22** (1981) 1100.

[89] E. Angelopoulos, M. Flato, C. Fronsdal, and D. Sternheimer, "Massless Particles, Conformal Group and de Sitter Universe," *Phys. Rev.* **D23** (1981) 1278.

[90] H. Nicolai and E. Sezgin, "Singleton representations of $OSp(N,4)$," *Phys. Lett.* **143B** (1984) 389.

[91] M. Gunaydin and N. Marcus, "The unitary supermultiplet of $N = 8$ conformal superalgebra involving fields of spin ≤ 2," *Class. Quant. Grav.* **2** (1985) L19.

[92] M. Flato and C. Fronsdal, "Quarks of singletons?," *Phys. Lett.* **B172** (1986) 412.

[93] S. Ferrara and C. Fronsdal, "Conformal Maxwell theory as a singleton field theory on AdS$_5$, IIB three-branes and duality," *Class. Quant. Grav.* **15** (1998) 2153, hep-th/9712239.

[94] S. Ferrara and C. Fronsdal, "Gauge fields and singletons of AdS$_{2p+1}$," *Lett. Math. Phys.* **46** (1998) 157, hep-th/9806072.

[95] S. Ferrara and C. Fronsdal, "Gauge fields as composite boundary excitations," *Phys. Lett.* **B433** (1998) 19–28, hep-th/9802126.

[96] V. G. Kac, "Representations of classical Lie superalgebras,". Proceedings, Differential Geometrical Methods In Mathematical Physics.Ii., Berlin 1977, 597-626.

[97] V. K. Dobrev and V. B. Petkova, "Group theoretical approach to extended conformal supersymmetry: function space realizations and invariant differential operators," *Fortschr. Phys.* **35** (1987) 537.

[98] V. K. Dobrev and V. B. Petkova, "All positive energy unitary irreducible representations of extended conformal supersymmetry," *Phys. Lett.* **162B** (1985) 127.

[99] N. Seiberg, "Notes on theories with 16 supercharges," *Nucl. Phys. Proc. Suppl.* **67** (1998) 158, hep-th/9705117.

[100] S. Minwalla, "Restrictions imposed by superconformal invariance on quantum field theories," *Adv. Theor. Math. Phys.* **2** (1998) 781–846, hep-th/9712074.

[101] S. Ferrara, C. Fronsdal, and A. Zaffaroni, "On N=8 supergravity on AdS$_5$ and N=4 superconformal Yang- Mills theory," *Nucl. Phys.* **B532** (1998) 153, hep-th/9802203.

[102] S. Ferrara, M. A. Lledo, and A. Zaffaroni, "Born-Infeld corrections to D3-brane action in AdS(5) x S(5) and N=4, d = 4 primary superfields," *Phys. Rev.* **D58** (1998) 105029, hep-th/9805082.

[103] H. J. Kim, L. J. Romans, and P. van Nieuwenhuizen, "The mass spectrum of chiral $\mathcal{N} = 2$ $d = 10$ supergravity on S^5," *Phys. Rev.* **D32** (1985) 389.

[104] P. Berglund, E. G. Gimon, and D. Minic, "The AdS/CFT correspondence and spectrum generating algebras," hep-th/9905097.

[105] O. Aharony, M. Berkooz, D. Kutasov, and N. Seiberg, "Linear Dilatons, NS Five-Branes and Holography," *JHEP* **10** (1998) 004, hep-th/9808149.

[106] E. Witten, "Baryons and branes in anti-de Sitter space," *JHEP* **07** (1998) 006, hep-th/9805112.

[107] I. R. Klebanov and E. Witten, "Superconformal field theory on three-branes at a Calabi-Yau singularity," *Nucl. Phys.* **B536** (1998) 199, hep-th/9807080.

[108] S. S. Gubser and I. R. Klebanov, "Baryons and domain walls in an N=1 superconformal gauge theory," *Phys. Rev.* **D58** (1998) 125025, hep-th/9808075.

[109] D. Z. Freedman, S. D. Mathur, A. Matusis, and L. Rastelli, "Correlation functions in the CFT$_d$ / AdS$_{d+1}$ correspondence," hep-th/9804058.

[110] G. Chalmers, H. Nastase, K. Schalm, and R. Siebelink, "R Current Correlators in $N = 4$ Super Yang-Mills Theory from Anti-de Sitter Supergravity," *Nucl. Phys.* **B540** (1999) 247, hep-th/9805105.

[111] S. Deser and A. Schwimmer, "Geometric classification of conformal anomalies in arbitrary dimensions," *Phys. Lett.* **B309** (1993) 279–284, hep-th/9302047.

[112] M. J. Duff, "Twenty years of the Weyl anomaly," *Class. Quant. Grav.* **11** (1994) 1387–1404, hep-th/9308075.

[113] H. Osborn and A. Petkos, "Implications of conformal invariance in field theories for general dimensions," *Ann. Phys.* **231** (1994) 311–362, hep-th/9307010.

[114] D. Anselmi, M. Grisaru, and A. Johansen, "A Critical Behavior of Anomalous Currents, Electric-Magnetic Universality and CFT in Four-Dimensions," *Nucl. Phys.* **B491** (1997) 221–248, hep-th/9601023.

[115] J. Erdmenger and H. Osborn, "Conserved currents and the energy momentum tensor in conformally invariant theories for general dimensions," *Nucl. Phys.* **B483** (1997) 431–474, hep-th/9605009.

[116] D. Anselmi, D. Z. Freedman, M. T. Grisaru, and A. A. Johansen, "Nonperturbative Formulas for Central Functions of Supersymmetric Gauge Theories," *Nucl. Phys.* **B526** (1998) 543, hep-th/9708042.

[117] M. Henningson and K. Skenderis, "The Holographic Weyl anomaly," *JHEP* **07** (1998) 023, hep-th/9806087.

[118] M. Henningson and K. Skenderis, "Holography and the Weyl anomaly," hep-th/9812032.

[119] V. Balasubramanian and P. Kraus, "A Stress Tensor for Anti-de Sitter Gravity," hep-th/9902121.

[120] W. Mueck and K. S. Viswanathan, "Counterterms for the Dirichlet prescription of the AdS/CFT correspondence," hep-th/9905046.

[121] S. Nojiri and S. D. Odintsov, "Conformal anomaly for dilaton coupled theories from AdS / CFT correspondence," *Phys. Lett.* **B444** (1998) 92, hep-th/9810008.

[122] D. Anselmi and A. Kehagias, "Subleading corrections and central charges in the AdS / CFT correspondence," hep-th/9812092.

[123] O. Aharony, J. Pawelczyk, S. Theisen, and S. Yankielowicz, "A Note on Anomalies in the AdS / CFT Correspondence," hep-th/9901134.

[124] M. Blau, K. S. Narain, and E. Gava, "On subleading contributions to the AdS / CFT trace anomaly," hep-th/9904179.

[125] S. Nojiri and S. D. Odintsov, "On the conformal anomaly from higher derivative gravity in AdS / CFT correspondence," hep-th/9903033.

[126] G. W. Gibbons and S. W. Hawking, "Action integrals and partition functions in quantum gravity," *Phys. Rev.* **D15** (1977) 2752–2756.

[127] I. Y. Aref'eva and I. V. Volovich, "On large N conformal theories, field theories in anti-de Sitter space and singletons," hep-th/9803028.

[128] W. Muck and K. S. Viswanathan, "Conformal field theory correlators from classical scalar field theory on AdS$_{d+1}$," *Phys. Rev.* **D58** (1998) 041901, hep-th/9804035.

[129] H. Liu and A. A. Tseytlin, "D = 4 superYang-Mills, D = 5 gauged supergravity, and D = 4 conformal supergravity," *Nucl. Phys.* **B533** (1998) 88, hep-th/9804083.

[130] W. Muck and K. S. Viswanathan, "Conformal field theory correlators from classical field theory on anti-de Sitter space. 2. Vector and spinor fields," *Phys. Rev.* **D58** (1998) 106006, hep-th/9805145.

[131] S. N. Solodukhin, "Correlation functions of boundary field theory from bulk Green's functions and phases in the boundary theory," *Nucl. Phys.* **B539** (1999) 403, hep-th/9806004.

[132] S. Lee, S. Minwalla, M. Rangamani, and N. Seiberg, "Three point functions of chiral operators in D = 4, N=4 SYM at large N," *Adv. Theor. Math. Phys.* **2** (1999) 697, hep-th/9806074.

[133] H. Liu and A. A. Tseytlin, "On four point functions in the CFT / AdS correspondence," *Phys. Rev.* **D59** (1999) 086002, hep-th/9807097.

[134] E. D'Hoker, D. Z. Freedman, and W. Skiba, "Field theory tests for correlators in the AdS / CFT correspondence," *Phys. Rev.* **D59** (1999) 045008, hep-th/9807098.

[135] D. Z. Freedman, S. D. Mathur, A. Matusis, and L. Rastelli, "Comments on 4 point functions in the CFT / AdS correspondence," hep-th/9808006.

[136] E. D'Hoker and D. Z. Freedman, "Gauge boson exchange in AdS_{d+1}," hep-th/9809179.

[137] G. Chalmers and K. Schalm, "The Large N_c Limit of Four-Point Functions in $N = 4$ Super Yang-Mills Theory from Anti-de Sitter Supergravity," hep-th/9810051.

[138] W. Muck and K. S. Viswanathan, "The Graviton in the AdS-CFT correspondence: Solution via the Dirichlet boundary value problem," hep-th/9810151.

[139] E. D'Hoker and D. Z. Freedman, "General scalar exchange in AdS_{d+1}," hep-th/9811257.

[140] P. Minces and V. O. Rivelles, "Chern-Simons theories in the AdS / CFT correspondence," hep-th/9902123.

[141] G. Arutyunov and S. Frolov, "Three-Point Green Function of the Stress-Energy Tensor in the AdS / CFT Correspondence," hep-th/9901121.

[142] D. Z. Freedman, K. Johnson, and J. I. Latorre, "Differential regularization and renormalization: A New method of calculation in quantum field theory," *Nucl. Phys.* **B371** (1992) 353–414.

[143] M. Henningson and K. Sfetsos, "Spinors and the AdS / CFT correspondence," *Phys. Lett.* **B431** (1998) 63–68, hep-th/9803251.

[144] A. M. Ghezelbash, K. Kaviani, I. Shahrokh AF Tehran Parvizi, and A. H. Fatollahi, "Interacting spinors - scalars and AdS / CFT correspondence," *Phys. Lett.* **B435** (1998) 291, hep-th/9805162.

[145] G. E. Arutyunov and S. A. Frolov, "On the origin of supergravity boundary terms in the AdS / CFT correspondence," *Nucl. Phys.* **B544** (1999) 576, hep-th/9806216.

[146] G. E. Arutyunov and S. A. Frolov, "Antisymmetric tensor field on AdS(5)," *Phys. Lett.* **B441** (1998) 173, hep-th/9807046.

[147] W. S. l'Yi, "Holographic projection of massive vector fields in AdS / CFT correspondence," hep-th/9808051.

[148] A. Volovich, "Rarita-Schwinger field in the AdS / CFT correspondence," *JHEP* **09** (1998) 022, hep-th/9809009.

[149] W. S. l'Yi, "Generating functionals of correlation functions of p form currents in AdS / CFT correspondence," hep-th/9809132.

[150] W. S. l'Yi, "Correlators of currents corresponding to the massive p form fields in AdS / CFT correspondence," *Phys. Lett.* **B448** (1999) 218, hep-th/9811097.

[151] A. S. Koshelev and O. A. Rytchkov, "Note on the massive Rarita-Schwinger field in the AdS / CFT correspondence," *Phys. Lett.* **B450** (1999) 368, hep-th/9812238.

[152] R. C. Rashkov, "Note on the boundary terms in AdS / CFT correspondence for Rarita-Schwinger field," hep-th/9904098.

[153] A. Polishchuk, "Massive symmetric tensor field on AdS," `hep-th/9905048`.

[154] D. Z. Freedman, S. S. Gubser, K. Pilch, and N. P. Warner, "Renormalization group flows from holography supersymmetry and a c theorem," `hep-th/9904017`.

[155] G. Dall'Agata, K. Lechner, and D. Sorokin, "Covariant actions for the bosonic sector of d = 10 IIB supergravity," *Class. Quant. Grav.* **14** (1997) L195–L198, `hep-th/9707044`.

[156] G. Dall'Agata, K. Lechner, and M. Tonin, "D = 10, N = IIB supergravity: Lorentz invariant actions and duality," *JHEP* **07** (1998) 017, `hep-th/9806140`.

[157] G. E. Arutyunov and S. A. Frolov, "Quadratic action for Type IIB supergravity on $AdS_5 \times S^5$," `hep-th/9811106`.

[158] J. H. Schwarz, "Covariant Field Equations of Chiral $N = 2, D = 10$ Supergravity," *Nucl. Phys.* **B226** (1983) 269.

[159] P. S. Howe and P. C. West, "The Complete $N = 2, d = 10$ Supergravity," *Nucl. Phys.* **B238** (1984) 181.

[160] E. D'Hoker, D. Z. Freedman, S. D. Mathur, A. Matusis, and L. Rastelli, "Graviton exchange and complete four point functions in the AdS / CFT correspondence," `hep-th/9903196`.

[161] E. D'Hoker, D. Z. Freedman, S. D. Mathur, A. Matusis, and L. Rastelli, "Graviton and gauge boson propagators in AdS(d+1)," `hep-th/9902042`.

[162] H. Liu, "Scattering in anti-de Sitter space and operator product expansion," `hep-th/9811152`.

[163] G. Chalmers and K. Schalm, "Holographic normal ordering and multiparticle states in the AdS / CFT correspondence," `hep-th/9901144`.

[164] E. D'Hoker, D. Z. Freedman, and L. Rastelli, "AdS/CFT four point functions: How to succeed at Z integrals without really trying," `hep-th/9905049`.

[165] D. J. Gross and H. Ooguri, "Aspects of Large N Gauge Theory Dynamics as seen by String Theory," *Phys. Rev.* **D58** (1998) 106002, `hep-th/9805129`.

[166] J. Maldacena, "Wilson loops in large N field theories," *Phys. Rev. Lett.* **80** (1998) 4859, `hep-th/9803002`.

[167] S.-J. Rey and J. Yee, "Macroscopic strings as heavy quarks in large N gauge theory and anti-de Sitter supergravity," `hep-th/9803001`.

[168] N. Drukker, D. J. Gross, and H. Ooguri, "Wilson Loops and Minimal Surfaces," `hep-th/9904191`.

[169] G. 't Hooft, "A Property of Electric and Magnetic Flux in Nonabelian Gauge Theories," *Nucl. Phys.* **B153** (1979) 141.

[170] S. Forste, D. Ghoshal, and S. Theisen, "Stringy corrections to the Wilson loop in N=4 superYang- Mills theory," `hep-th/9903042`.

[171] S. Naik, "Improved heavy quark potential at finite temperature from Anti-de Sitter supergravity," `hep-th/9904147`.

[172] J. Greensite and P. Olesen, "Worldsheet Fluctuations and the Heavy Quark Potential in the AdS/CFT Approach," `hep-th/9901057`.

[173] J. A. Minahan, "Quark - monopole potentials in large N superYang-Mills," *Adv. Theor. Math. Phys.* **2** (1998) 559, `hep-th/9803111`.

[174] J. A. Minahan and N. P. Warner, "Quark potentials in the Higgs phase of large N supersymmetric Yang-Mills theories," *JHEP* **06** (1998) 005, `hep-th/9805104`.

[175] I. I. Kogan and O. A. Solovev, "Gravitationally dressed RG flows and zigzag invariant strings," *Phys. Lett.* **B442** (1998) 136, hep-th/9807223.

[176] I. I. Kogan and O. A. Solovev, "On zigzag invariant strings," hep-th/9901131.

[177] S. Nojiri and S. D. Odintsov, "Running gauge coupling and quark - anti-quark potential from dilatonic gravity," hep-th/9904036.

[178] K. Zarembo, "Wilson loop correlator in the AdS / CFT correspondence," hep-th/9904149.

[179] E. Alvarez, C. Gomez, and T. Ortin, "String representation of Wilson loops," *Nucl. Phys.* **B545** (1999) 217, hep-th/9806075.

[180] C. R. Graham and E. Witten, "Conformal anomaly of submanifold observables in AdS / CFT correspondence," hep-th/9901021.

[181] D. Berenstein, R. Corrado, W. Fischler, and J. Maldacena, "The Operator Product Expansion for Wilson Loops and Surfaces in the Large N Limit," hep-th/9809188.

[182] S. S. Gubser, I. R. Klebanov, and A. W. Peet, "Entropy and temperature of black 3-branes," *Phys. Rev.* **D54** (1996) 3915–3919, hep-th/9602135.

[183] A. Strominger, unpublished notes, November, 1997.

[184] S. S. Gubser, I. R. Klebanov, and A. A. Tseytlin, "Coupling constant dependence in the thermodynamics of N=4 supersymmetric Yang-Mills theory," *Nucl. Phys.* **B534** (1998) 202, hep-th/9805156.

[185] G. T. Horowitz and J. Polchinski, "A Correspondence principle for black holes and strings," *Phys. Rev.* **D55** (1997) 6189–6197, hep-th/9612146.

[186] N. Itzhaki, "A Comment on the entropy of strongly coupled N=4," hep-th/9904035.

[187] A. Fotopoulos and T. R. Taylor, "Comment on two loop free energy in N=4 supersymmetric Yang- Mills theory at finite temperature," hep-th/9811224.

[188] R. M. Wald, "Black hole entropy in Noether charge," *Phys. Rev.* **D48** (1993) 3427–3431, gr-qc/9307038.

[189] M. Li, "Evidence for large N phase transition in N=4 superYang- Mills theory at finite temperature," *JHEP* **03** (1999) 004, hep-th/9807196.

[190] Y. hong Gao and M. Li, "Large N strong / weak coupling phase transition and the correspondence principle," hep-th/9810053.

[191] S. W. Hawking and D. N. Page, "Thermodynamics of black holes in anti-de Sitter space," *Commun. Math. Phys.* **87** (1983) 577.

[192] S. S. Gubser, "Thermodynamics of spinning D3-branes," hep-th/9810225.

[193] K. Landsteiner, "String corrections to the Hawking-Page phase transition," *Mod. Phys. Lett.* **A14** (1999) 379, hep-th/9901143.

[194] R.-G. Cai and K.-S. Soh, "Critical behavior in the rotating D-branes," hep-th/9812121.

[195] M. Cvetic and S. S. Gubser, "Phases of R charged black holes, spinning branes and strongly coupled gauge theories," hep-th/9902195.

[196] A. Chamblin, R. Emparan, C. V. Johnson, and R. C. Myers, "Charged AdS Black Holes and Catastrophic Holography," hep-th/9902170.

[197] M. Cvetic and S. S. Gubser, "Thermodynamic stability and phases of general spinning branes," hep-th/9903132.

[198] M. M. Caldarelli and D. Klemm, "M theory and stringy corrections to Anti-de Sitter black holes and conformal field theories," hep-th/9903078.

[199] P. Kraus, F. Larsen, and S. P. Trivedi, "The Coulomb branch of gauge theory from rotating branes," *JHEP* **03** (1999) 003, hep-th/9811120.

[200] A. A. Tseytlin and S. Yankielowicz, "Free energy of N=4 superYang-Mills in Higgs phase and nonextremal D3-brane interactions," *Nucl. Phys.* **B541** (1999) 145, hep-th/9809032.

[201] D. Birmingham, "Topological black holes in Anti-de Sitter space," *Class. Quant. Grav.* **16** (1999) 1197, hep-th/9808032.

[202] J. Louko and D. Marolf, "Single exterior black holes and the AdS / CFT conjecture," *Phys. Rev.* **D59** (1999) 066002, hep-th/9808081.

[203] S. W. Hawking, C. J. Hunter, and M. M. Taylor-Robinson, "Rotation and the AdS / CFT correspondence," *Phys. Rev.* **D59** (1999) 064005, hep-th/9811056.

[204] A. W. Peet and S. F. Ross, "Microcanonical phases of string theory on $AdS_m \times S^n$," *JHEP* **12** (1998) 020, hep-th/9810200.

[205] S.-J. Rey, S. Theisen, and J. Yee, "Wilson-Polyakov Loop at Finite Temperature in Large N Gauge Theory and Anti-de Sitter Supergravity," *Nucl.Phys.* **B527** (1998) 171–186, hep-th/9803135.

LECTURES ON D-BRANES, TACHYON CONDENSATION, AND STRING FIELD THEORY

Washington Taylor
Center for Theoretical Physics
MIT, Bldg. 6-308
Cambridge, MA 02139, U.S.A.
wati@mit.edu

Abstract   These lectures provide an introduction to the subject of tachyon condensation in the open bosonic string. The problem of tachyon condensation is first described in the context of the low-energy Yang-Mills description of a system of multiple D-branes, and then using the language of string field theory. An introduction is given to Witten's cubic open bosonic string field theory. The Sen conjectures on tachyon condensation in open bosonic string field theory are introduced, and evidence confirming these conjectures is reviewed.

1.   Introduction

The last seven years have been a very exciting time for string theory. A new understanding of nonperturbative features of string theory, such as D-branes, has led to exciting new developments relating string theory to physically interesting systems such as black holes and supersymmetric gauge field theories, as well as to a new understanding of the relationship between Yang-Mills theories and quantum theories of gravity.

Despite remarkable progress in these directions, however, a consistent non-perturbative background-independent formulation of string theory is still lacking. This situation makes it impossible at this point, even in principle, to directly address cosmological questions using string theory. String field theory is a nonperturbative approach to string theory which holds some promise towards providing a background-independent definition of the theory. These lecture notes give an introduction to string field theory and review some recent work which incorporates D-branes into the framework of string field theory. This work shows that string field theory is a sufficiently robust framework that distinct string backgrounds can arise as disconnected solutions of the theory,

151

at least for open strings. It remains to be seen whether this success can be replicated in the closed string sector.

In this section we review briefly the situation in string theory as a whole, and summarize the goals of this set of lectures. In Section 2 we review some basic aspects of D-branes. In Section 3, we describe a particular D-brane configuration which exhibits a tachyonic instability. This tachyon can be seen in the low-energy super Yang-Mills description of the D-brane geometry. This field theory tachyon provides a simple model which embodies much of the physics of the more complicated string field theory tachyon discussed in the later lectures. In Section 4 we give an introduction to Witten's cubic bosonic open string field theory and summarize the conjectures made by Sen in 1999, which suggested that the tachyonic instability of the open bosonic string can be interpreted in terms of an unstable space-filling D-brane, and that this system can be analytically described through open string field theory. Section 5 gives a more detailed analytic description of Witten's cubic string field theory. In Section 6 we summarize evidence from string field theory for Sen's conjectures. Section 7 contains a brief review of some more recent developments. Section 8 contains concluding remarks and lists some open problems.

Much new work has been done in this area since these lectures were presented at Valdivia in January 2002. Except for a few references to more recent developments in footnotes and in the last two sections, these lecture notes primarily cover work done before January 2002. Previous articles reviewing related work include those of Ohmori [1], de Smet [2], and Aref'eva *et al.* [3]. An expanded set of lecture notes, based on lectures given by the author and Barton Zwiebach at TASI '01, will appear in [4]; the larger set of notes will include further details on a number of topics.

1.1   The status of string theory: a brief review

To understand the significance of developments over the last seven years, it is useful to recall the situation of string theory as it was in early 1995. At that time it was clearly understood that there were 5 distinct ways in which a supersymmetric closed string could be quantized to give a microscopic definition of a theory of quantum gravity in ten dimensions. Each of these approaches to quantizing the string gives a set of rules for calculating scattering amplitudes between on-shell string states which describe gravitational quanta as well as an infinite family of massive particles in a ten-dimensional spacetime. These five string theories are known as the type IIA, IIB, I, heterotic $SO(32)$, and heterotic $E_8 \times E_8$ superstring theories. While these string theories give perturbative descriptions of quantum gravity, in 1995 there was little understanding of nonperturbative aspects of these theories.

In the years between 1995 and 2000, several new ideas dramatically transformed our understanding of string theory. We now briefly summarize these ideas and mention some aspects of these developments relevant to the main topic of these lectures.

Dualities: The five different perturbative formulations of superstring theory are all related to one another through duality symmetries [5, 6], whereby the degrees of freedom in one theory can be described through a duality transformation in terms of the degrees of freedom of another theory. Some of these duality symmetries are nonperturbative, in the sense that the string coupling g in one theory is related to the inverse string coupling $1/g$ in a dual theory. The web of dualities relating the different theories gives a picture in which, rather than describing five distinct possibilities for a fundamental theory, each of the perturbative superstring theories appears to be a particular perturbative limit of some other, as yet unknown, underlying theoretical structure.

M-theory: In addition to the five perturbative string theories, the web of dualities also seems to include a limit which describes a quantum theory of gravity in eleven dimensions. This new theory has been dubbed "M-theory". Although no covariant definition for M-theory has been given, this theory can be related to type IIA and heterotic $E_8 \times E_8$ string theories through compactification on a circle S^1 and the space S^1/Z_2 respectively [7, 6, 8]. For example, in relating to the type IIA theory, the compactification radius R_{11} of M-theory becomes the product gl_s of the string coupling and string length in the 10D IIA theory. Thus, M-theory in flat space, which arises in the limit $R_{11} \to \infty$, can be thought of as the strong coupling limit of type IIA string theory. It is also suspected that M-theory may be describable as a quantum theory of membranes in 11 dimensions [7], although a covariant formulation of such a theory is still lacking.

Branes: In addition to strings, all five superstring theories, as well as M-theory, contain extended objects of higher dimensionality known as "branes". M-theory has M2-branes and M5-branes, which have two and five dimensions of spatial extent (whereas a string has one). The different superstring theories each have different complements of D-branes as well as the fundamental string and Neveu-Schwarz 5-brane; in particular, the IIA/IIB superstring theories contain D-branes of all even/odd dimensions. The branes of one theory can be related to the branes of another through the duality transformations mentioned above. Through an appropriate sequence of dualities, any brane can be mapped to any other brane, including the string itself. This suggests that none of these objects are really any more fundamental than any others; this idea is known as "brane democracy".

M(atrix) theory and AdS/CFT: One of the most remarkable results of the developments just mentioned is the realization that in certain space-time backgrounds, M-theory and string theory can be completely described through simple supersymmetric quantum mechanics and field theory models related to the low-energy description of systems of branes. The M(atrix) model of M-theory is a simple supersymmetric matrix quantum mechanics which is believed to capture all of the physics of M-theory in asymptotically flat spacetime (in light-cone coordinates). A closely related set of higher-dimensional supersymmetric Yang-Mills theories are related to string theory in backgrounds described by the product of anti-de Sitter space and a sphere through the AdS/CFT correspondence. It is believed that these models of M-theory and string theory give true nonperturbative descriptions of quantum gravity in space-time backgrounds which have the asymptotic geometry relevant to each model. For reviews of M(atrix) theory and AdS/CFT, see [9, 10].

The set of ideas just summarized have greatly increased our understanding of nonperturbative aspects of string theory. In particular, through M(atrix) theory and the AdS/CFT correspondences we now have nonperturbative definitions of M-theory and string theory in certain asymptotic space-time backgrounds which could, in principle, be used to calculate any local result in quantum gravity. While these new insights are very powerful, however, we are still lacking a truly background-independent formulation of string theory.

1.2   The goal of these lectures

The goal of these lectures is to describe progress towards a nonperturbative background-independent formulation of string theory. Such a formulation is needed to address fundamental questions such as: What is string theory/M-theory? How is the vacuum of string theory selected? (*i.e.*, Why can the observable low-energy universe be accurately described by the standard model of particle physics in four space-time dimensions with an apparently small but nonzero positive cosmological constant?), and other questions of a cosmological nature. Obviously, aspiring to address these questions is an ambitious undertaking, but we believe that attaining a better understanding of string field theory is a useful step in this direction.

More concretely, in these lectures we will describe recent progress on open string field theory. It may be useful here to recall some basic aspects of open and closed strings and the relationship between them.

Closed strings, which are topologically equivalent to a circle S^1, give rise upon quantization to a massless set of spacetime fields associated with the graviton $g_{\mu\nu}$, the dilaton φ, and the antisymmetric two-form $B_{\mu\nu}$, as well as an infinite family of massive fields. For the supersymmetric closed string, further massless fields associated with the graviton supermultiplet appear—these are the Ramond-Ramond p-form fields $A^{(p)}_{\mu_1\cdots\mu_p}$ and the gravitini $\psi_{\mu\alpha}$. Thus, the quantum theory of closed strings is naturally associated with a theory of gravity in space-time. On the other hand, open strings, which are topologically equivalent to an interval $[0, \pi]$, give rise under quantization to a massless gauge field A_μ in space-time. The supersymmetric open string also has a massless gaugino field ψ_α. It is now understood that generally open strings should be thought of as ending on a Dirichlet p-brane (Dp-brane), and that the massless open string fields describe the fluctuations of the D-brane and the gauge field living on the world-volume of the D-brane.

It may seem, therefore, that open and closed strings are quite distinct, and describe disjoint aspects of the physics in a fixed background space-time containing some family of D-branes. At tree level, the closed strings indeed describe gravitational physics in the bulk space-time, while the open strings describe the D-brane dynamics. At the quantum level, however, the physics of open and closed strings are deeply connected. Indeed, historically open strings were discovered first through the form of their scattering amplitudes [11]. Looking at one-loop processes for open strings led to the first discovery of closed strings, which appeared as *poles* in nonplanar one-loop open string diagrams [12, 13]. The fact that open string diagrams naturally contain closed string intermediate states indicates that in some sense all closed string interactions are implicitly defined through the complete set of open string diagrams. This connection underlies many of the important recent developments in string theory. In particular, the M(atrix) theory and AdS/CFT correspondences between gauge theories and quantum gravity are essentially limits in which closed string physics in a fixed space-time background is captured by a simple limiting Yang-Mills description of an open string theory on a family of branes (D0-branes for M(atrix) theory, D3-branes for the CFT describing AdS$_5 \times S^5$, etc.)

The fact that, in certain fixed space-time backgrounds, quantum gravity theories can be encoded in terms of open string degrees of freedom through the M(atrix) and AdS/CFT correspondences leads to the question of how a global change of the space-time background would appear in the quantum field theory describing the appropriate limit of the open string model in question. If such a change of background could be described in the context of M(atrix) theory or AdS/CFT, it would indicate that these models could be generalized to a background-independent framework. Unfortunately, however, such a change in the background involves adding nonrenormalizable interactions to the field theories in question. At this point in time we do not have the technology to understand generically how a sensible quantum field theory can be described when an infinite number of nonrenormalizable interaction terms are added to the Lagrangian. One example of a special case where this can be done is the addition of a constant background B field in space-time. In the associated Yang-Mills theory, such as that on a system of N D3-branes in the case of the simplest AdS/CFT correspondence, this change in the background field corresponds to replacing products of open string fields with a noncommutative star-product.

The resulting theory is a noncommutative Yang-Mills theory. Such noncommutative theories are the only well-understood example of a situation where adding an infinite number of apparently nonrenormalizable terms to a field theory action leads to a sensible modification of quantum field theory (for a review of noncommutative field theory and its connection to string theory, see [14]).

String field theory is a nonperturbative formulation in target space of an interacting string theory, in which the infinite family of fields associated with string excitations are described by a space-time field theory action. For open strings, this field theory is a natural extension of the low-energy Yang-Mills action describing a system of D-branes, where the entire hierarchy of massive string fields is included in addition to the massless gauge field on the D-brane. Integrating out all the massive fields from the string field theory action gives rise to a nonabelian Born-Infeld action for the D-branes, including an infinite set of higher-order terms arising from string theory corrections to the simple Yang-Mills action. Like the case of noncommutative field theory discussed above, the new terms appearing in this action are apparently nonrenormalizable, but the combination of terms must work together to form a sensible theory.

In the 1980's, a great deal of work was done on formulating string field theory for open and closed, bosonic and supersymmetric string theories. Most of these string field theories are quite complicated. For the open bosonic string, however, Witten [18] constructed an extremely elegant string field theory based on the Chern-Simons action. This cubic bosonic open string field theory (OSFT) is the primary focus of the work described in these lectures. Although Witten's OSFT can be described in a simple abstract language, practical computations with this theory rapidly become extremely complicated. Despite a substantial amount of work on this theory, little insight was gained in the 1980's regarding how this theory could be used to go beyond standard perturbative string methods. Work on this subject stalled out in the late 80's, and little further attention was paid to OSFT until several years ago.

One simple feature of the 26-dimensional bosonic string has been problematic since the early days of string theory: both the open and closed bosonic strings have tachyons in their spectra, indicating that the usual perturbative vacua used for these theories are unstable. In 1999, Ashoke Sen had a remarkable insight into the nature of the open bosonic string tachyon [19]. He observed that the open bosonic string should be thought of as ending on a space-filling D25-brane. He pointed out that this D-brane is unstable in the bosonic theory, as it does not carry any conserved charge, and he suggested that the open bosonic string tachyon should be interpreted as the instability mode of the D25-brane. This led him to conjecture that Witten's open string field theory could be used to precisely determine a new vacuum for the open string, namely one in which the D25-brane is annihilated through condensation of the tachyonic unstable mode. Sen made several precise conjectures regarding the details of the string field theory description of this new open string vacuum. As we describe in these lectures, there is now overwhelming evidence that Sen's picture is correct, demonstrating that string field theory accurately describes the nonperturbative physics of D-branes. This new nonperturbative application of string field theory has sparked a new wave of work on Witten's cubic open string field theory, revealing many remarkable new structures. In particular, string field theory now provides a concrete framework in which disconnected string backgrounds can emerge from the equations of motion of a single underlying theory. Although so far this can only be shown explicitly in the open string context, this·work paves the way for a deeper understanding of background-independence in quantum theories of gravity.

2.   D-branes

In this section we briefly review some basic features of D-branes. The concepts developed here will be useful in describing tachyonic D-brane configurations in the following section. For more detailed reviews of D-branes, see [15, 16].

2.1   D-branes and Ramond-Ramond charges

D-branes can be understood in two ways: *a*) as extended extremal black brane solutions of supergravity carrying conserved charges, and *b*) as hypersurfaces on which strings have Dirichlet boundary conditions.

a) The ten-dimensional type IIA and IIB supergravity theories each have a set of $(p+1)$-form fields $A^{(p+1)}_{\mu_1\cdots\mu_{(p+1)}}$ in the supergraviton multiplet, with p even/odd for type IIA/IIB supergravity. These are the Ramond-Ramond fields in the massless superstring spectrum. For each of these $(p+1)$-form fields, there is a solution of the supergravity field equations which has $(p+1)$-dimensional Lorentz invariance, and which has the form of an extremal black hole solution in the orthogonal $9-p$ space directions plus time (for a review see [17]). These "black p-brane" solutions carry charge under the R-R fields $A^{(p+1)}$, and are BPS states in the supergravity theory, preserving half the supersymmetries of the theory.

b) In type IIA and IIB string theory, it is possible to consider open strings with Dirichlet boundary conditions on some number $9-p$ of the spatial coordinates $x^\mu(\sigma)$. The locus of points defined by such Dirichlet boundary conditions defines a $(p+1)$-dimensional hypersurface Σ_{p+1} in the ten-dimensional spacetime. When p is even/odd in type IIA/IIB string theory, the spectrum of the resulting quantum open string theory contains a massless set of fields $A_\alpha, \alpha = 0, 1, \ldots, p$ and $X^a, a = p+1, \ldots, 9$. These fields can be associated with a gauge field living on the hypersurface Σ_{p+1}, and a set of degrees of freedom describing the transverse fluctuations of this hypersurface in spacetime. Thus, the quantum fluctuations of the open string describe a fluctuating $(p + 1)$-dimensional hypersurface in spacetime — a Dirichlet-brane, or "D-brane".

The remarkable insight of Polchinski in 1995 [20] was the observation that Dirichlet-branes carry Ramond-Ramond charges, and therefore should be described in the low-energy supergravity limit of string theory by precisely the black p-branes discussed in a). This connection between the string and supergravity descriptions of these nonperturbative objects paved the way to a dramatic series of new developments in string theory, including connections between string theory and supersymmetric gauge theories, string constructions of black holes, and new approaches to string phenomenology.

2.2   Born-Infeld and super Yang-Mills D-brane actions

In this subsection we briefly review the low-energy super Yang-Mills description of the dynamics of one or more D-branes. As discussed in the previous subsection, the massless open string modes on a Dp-brane in type IIA or IIB superstring theory describe a $(p+1)$-component gauge field A_α, $9 - p$ transverse scalar fields X^a, and a set of massless fermionic gaugino fields. The scalar fields X^a describe small fluctuations of the D-brane around a flat hypersurface. If the D-brane geometry is sufficiently far from flat, it is useful to describe the D-brane configuration by a general embedding $X^\mu(\xi)$, where ξ^α are $p+1$ coordinates on the Dp-brane world-volume $\Sigma_{(p+1)}$, and X^μ are ten functions giving a map from $\Sigma_{(p+1)}$ into the space-time manifold $\mathbf{R}^{9,1}$. Just as the Einstein equations governing the geometry of spacetime arise from the condition that the one-loop contribution to the closed string beta function vanish, a set of equations of motion for a general Dp-brane geometry and associated world-volume gauge field can be derived from a calculation of the one-loop open string beta function [21]. These equations of motion arise from the classical Born-Infeld action

$$S = -T_p \int d^{p+1}\xi \, e^{-\varphi} \sqrt{-\det(G_{\alpha\beta} + B_{\alpha\beta} + 2\pi\alpha' F_{\alpha\beta})} + S_{\text{CS}} + \text{fermions}$$

$$\text{(1)}$$

where G, B and φ are the pullbacks of the 10D metric, antisymmetric tensor and dilaton to the D-brane world-volume, while F is the field strength of the world-volume $U(1)$ gauge field A_α. S_{CS} represents a set of Chern-Simons terms which will be discussed in the following subsection. This action can be verified by a perturbative string calculation [15], which also gives a precise expression for the brane tension

$$\tau_p = \frac{T_p}{g} = \frac{1}{g\sqrt{\alpha'}} \frac{1}{(2\pi\sqrt{\alpha'})^p} \tag{2}$$

where $g = e^{\langle \varphi \rangle}$ is the string coupling, equal to the exponential of the dilaton expectation value.

A particular limit of the Born-Infeld action (1) is useful for describing many low-energy aspects of D-brane dynamics. Take the background space-time $G_{\mu\nu} = \eta_{\mu\nu}$ to be flat, and all other supergravity fields $(B_{\mu\nu}, A^{(p+1)}_{\mu_1 \cdots \mu_{p+1}})$ to vanish. We then assume that the D-brane is approximately flat, and is close to the hypersurface $X^a = 0, a > p$, so that we may make the static gauge choice $X^\alpha = \xi^\alpha$. We furthermore assume that $\partial_\alpha X^a$ and $2\pi\alpha' F_{\alpha\beta}$ are small and of the same order. In this limit, the action (1) can be expanded as

$$S = -T_p V_p - \frac{1}{4g^2_{YM}} \int d^{p+1}\xi \left(F_{\alpha\beta} F^{\alpha\beta} + \frac{2}{(2\pi\alpha')^2} \partial_\alpha X^a \partial^\alpha X^a \right) + \cdots \quad (3)$$

where V_p is the p-brane world-volume and the coupling g_{YM} is given by

$$g^2_{YM} = \frac{1}{4\pi^2\alpha'^2 T_p} = \frac{g}{\sqrt{\alpha'}}(2\pi\sqrt{\alpha'})^{p-2} . \quad (4)$$

Including fermionic terms, the second term in (3) is simply the dimensional reduction to $(p+1)$ dimensions of the 10D $\mathcal{N} = 1$ super Yang-Mills action

$$S = \frac{1}{g^2_{YM}} \int d^{10}\xi \left(-\frac{1}{4} F_{\mu\nu} F^{\mu\nu} + \frac{i}{2} \bar{\psi} \Gamma^\mu \partial_\mu \psi \right) \quad (5)$$

where for $\alpha, \beta \leq p$, $F_{\alpha\beta}$ is the world-volume U(1) field strength, and for $a > p, \alpha \leq p, F_{\alpha a} \to \partial_\alpha X^a$ (setting $2\pi\alpha' = 1$).

When multiple Dp-branes are present, the D-brane action is modified in a fairly simple fashion [22]. Consider a system of N coincident D-branes. For every pair of branes $\{i, j\}$ there is a set of massless fields

$$(A_\alpha)_i^j, \quad (X^a)_i^j \quad (6)$$

associated with strings stretching from the ith brane to the jth brane; the indices i, j are known as Chan-Paton indices. Treating the fields (6) as matrices, the analogue for multiple branes of the Born-Infeld action (1) takes the form

$$S \sim \int \mathrm{Tr} \sqrt{-\det(G + B + F)} . \quad (7)$$

This action is known as the nonabelian Born-Infeld action (NBI). In order to give a rigorous definition to the nonabelian Born-Infeld action, it is necessary to resolve ordering ambiguities in the expression (7). Since the spacetime coordinates X^a associated with the D-brane positions in space-time become themselves matrix-valued, even evaluating the pullbacks $G_{\alpha\beta}, B_{\alpha\beta}$ involves resolving ordering issues. Much work has been done recently to resolve these ordering ambiguities (see [23] for some recent papers in this direction which contain further references to the literature), but there is still no consistent definition of the nonabelian Born-Infeld theory (7) which is valid to all orders.

The nonabelian Born-Infeld action (7) becomes much simpler in the low-energy limit when the background space-time is flat. In the same limit discussed above for the single D-brane, where we find a low-energy limit giving the U(1) super Yang-Mills theory in $p+1$ dimensions, the inclusion of multiple D-branes simply leads in the low-energy limit to the nonabelian U(N) super Yang-Mills action in $p + 1$ dimensions. This action is the dimensional reduction of the 10D U(N) super Yang-Mills action (analogous to (5), but with an overall trace) to $p + 1$ dimensions. In this reduction, as before, for $\alpha, \beta \leq p$, $F_{\alpha\beta}$ is the world-volume U(1) field strength, and for $a > p, \alpha \leq p$, $F_{\alpha a} \to \partial_\alpha X^a$, where now A_α, X^a, and $F_{\alpha\beta}$ are $N \times N$ matrices. We furthermore have, for $a, b > p$, $F_{ab} \to -i[X^a, X^b]$ in the dimensional reduction.

The low-energy description of a system of N coincident flat D-branes is thus given by $U(N)$ super Yang-Mills theory in the appropriate dimension. This connection between D-brane actions in string theory and super Yang-Mills theory has led to many new developments, including new insights into supersymmetric field theories, the M(atrix) theory and AdS/CFT correspondences, and brane world scenarios.

2.3   Branes from branes

In this subsection we describe a remarkable feature of D-brane systems, namely a mechanism by which one or more D-branes of a fixed dimension can be used to construct additional D-branes of higher or lower dimension.

In our discussion of the D-brane action (1) above, we mentioned a group of terms S_{CS} which we did not describe explicitly. For a single Dp-brane, these Chern-Simons terms can be combined into a single expression of the form

$$S_{\text{CS}} \sim \int_{\Sigma_{p+1}} \mathcal{A} \, e^{F+B} \tag{8}$$

where $\mathcal{A} = \sum_k A^{(k)}$ represents a formal sum over all the Ramond-Ramond fields $A^{(k)}$ of various dimensions. In this integral, for each term $A^{(k)}$, the nonvanishing contribution to (8) is given by expanding the exponential of $F+B$ to order $(p+1-k)/2$, where the dimension of the resulting form saturates the dimension of the brane. For example, on a Dp-brane, there is a coupling of the form

$$\int_{\Sigma_{(p+1)}} A^{(p-1)} \wedge F \,. \tag{9}$$

This coupling implies that the U(1) field strength on the Dp-brane couples to the R-R field associated with $(p-2)$-branes. Thus, we can associate magnetic fields on a Dp-brane with dissolved $(p-2)$-branes living on the Dp-brane. This result generalizes to a system of multiple Dp-branes by simply performing a trace on the RHS of (8) For example, on N compact Dp-branes, the charge

$$\frac{1}{2\pi} \int \mathrm{Tr}\, F_{\alpha\beta}, \tag{10}$$

which is the first Chern class of the U(N) bundle described by the gauge field on the N branes, is quantized and measures the number of units of D$(p-2)$-brane charge living on the Dp-branes, which are encoded in the field strength $F_{\alpha\beta}$. Similarly,

$$\frac{1}{8\pi^2} \int \mathrm{Tr}\, F \wedge F \tag{11}$$

encodes D$(p-4)$-brane charge on the Dp-branes.

Just as lower-dimensional branes can be described in terms of the degrees of freedom associated with a system of N Dp-branes through the field strength $F_{\alpha\beta}$, higher-dimensional branes can be described by a system of N Dp-branes in terms of the commutators of the matrix-valued scalar fields X^a. Just as $\frac{1}{2\pi} F$ measures $(p-2)$-brane charge, the matrix

$$2\pi i [X^a, X^b] \tag{12}$$

measures $(p+2)$-brane charge [16, 24, 25]. The charge (12) should be interpreted as a form of local charge density. The fact that the trace of (12) vanishes for finite sized matrices corresponds to the fact that the net Dp-brane charge of a finite-size brane configuration in flat spacetime vanishes.

A simple example of the mechanism by which a system of multiple Dp-branes form a higher-dimensional brane is given by the matrix sphere. If we take a system of D0-branes with scalar matrices X^a given by

$$X^a = \frac{2r}{N} J^a, \qquad a = 1, 2, 3 \tag{13}$$

where J^a are the generators of SU(2) in the N-dimensional representation, then we have a configuration corresponding to the "matrix sphere". This is a D2-brane of spherical geometry living on the locus of points satisfying $x^2 + y^2 + z^2 = r^2$. The "local" D2-brane charge of this brane is given by (12). The D2-brane configuration given by (13) is rotationally invariant (up to a gauge transformation). The restriction of the brane to the desired locus of points can be seen from the relation $(X^1)^2 + (X^2)^2 + (X^3)^2 = r^2 \mathbb{1} + \mathcal{O}(N^{-2})$.

2.4   T-duality

We conclude our discussion of D-branes with a brief description of T-duality. T-duality is a perturbative symmetry which relates the type IIA and type IIB string theories. This duality symmetry was in fact crucial in the original discovery of D-branes [20]. A more detailed discussion of T-duality can be found in the textbook by Polchinski [26]. Using T-duality, we construct an explicit example of a brane within a brane encoded in super Yang-Mills theory, illustrating the ideas of the previous subsection. This example will be used in the following section to construct an analogous configuration with a tachyon.

Consider type IIA string theory on a spacetime of the form $M^9 \times S^1$ where M^9 is a generic 9-manifold of Lorentz signature, and S^1 is a circle of radius R. T-duality is the statement that this theory is precisely equivalent, at the perturbative level, to type IIB string theory on the spacetime $M^9 \times (S^1)'$, where $(S^1)'$ is a circle of radius $R' = \alpha'/R$.

T-duality is most easily understood in terms of closed strings, where it amounts to an exchange of winding and momentum modes of the string. The string winding modes on S^1 have energy $m = Rw/\alpha'$, where w is the winding number. the T-dual momentum modes on $(S^1)'$ have $m = n/R'$; it is straightforward to check that the spectrum of closed string states is unchanged under T-duality. T-duality can also be understood in terms of open strings. Under T-duality, an open string with Neumann boundary conditions on S^1 is mapped to an open string with Dirichlet boundary conditions on $(S^1)'$, and vice versa. Thus, a Dirichlet p-brane which is wrapped around the circle S^1 is mapped under T-duality to a Dirichlet $(p-1)$-brane of one lower dimension which is localized to a point on the circle $(S^1)'$. At the level of the low-energy theory on the D-brane, the $(p+1)$-dimensional Yang-Mills theory on the p-brane is replaced under T-duality with the p-dimensional Yang-Mills theory on the dual $(p-1)$-brane. Mathematically, the covariant derivative operator in the direction S^1 is replaced under T-duality with an adjoint scalar field X^a. Formally, this adjoint scalar field is an infinite size matrix, containing information about the open strings wrapped an arbitrary number of times around the compact direction $(S^1)'$.

We can summarize the relevant mappings under T-duality in the following table

IIA/S^1	$\leftrightarrow$	IIB/$(S^1)'$
R	$\leftrightarrow$	$R' = \alpha'/R$
Neumann/Dirichlet b.c.'s	$\leftrightarrow$	Dirichlet/Neumann b.c.'s
p-brane	$\leftrightarrow$	$(p \pm 1)$-brane
$2\pi\alpha'(i\partial_a + A_a)$	$\leftrightarrow$	X^a

The phenomena by which field strengths in one brane describe lower- or higher-dimensional branes can be easily understood using T-duality. The following simple example may help to clarify this connection. (For a more detailed discussion from this point of view see [16].)

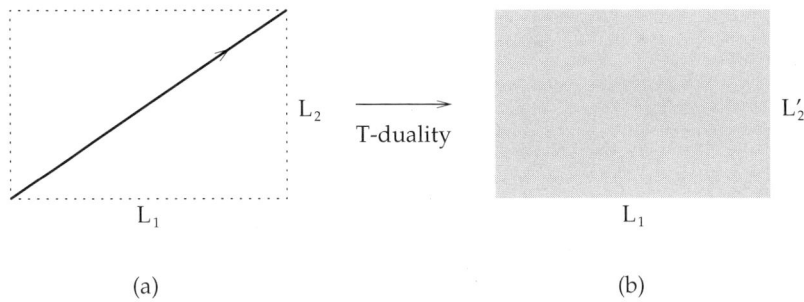

Figure 1.   T-duality takes a diagonal D1-brane on a two-torus (a) to a D2-brane on the dual torus with constant magnetic flux encoding an embedded D0-brane (b).

Consider a D1-brane wrapped diagonally on a two-torus T^2 with sides of length $L_1 = L$ and $L_2 = 2\pi R$. (Figure 1(a)). This configuration is described in terms of the world-volume Yang-Mills theory on a D1-brane stretched in the L_1 direction through a transverse scalar field

$$X^2 = 2\pi R \xi_1 / L . \tag{14}$$

To be technically precise, this scalar field should be treated as an $\infty \times \infty$ matrix [27] whose (n, m) entry is associated with strings connecting the nth and mth images of the D1-brane on the covering space of S^1. The diagonal elements $X^2_{n,n}$ of this infinite matrix are given by $2\pi R(\xi_1 + nL)/L$, while all off-diagonal elements vanish. While the resulting matrix-valued function of ξ_1 is not periodic, it is periodic up to a gauge transformation

$$X^2(L) = V X^2(0) V^{-1} \tag{15}$$

where V is the shift matrix with nonzero elements $V_{n,n+1} = 1$.

Under T-duality in the x^2 direction the infinite matrix X^2_{nm} becomes the Fourier mode representation of a gauge field on a dual D2-brane

$$A_2 = \frac{1}{R'L} \xi_1 . \tag{16}$$

The magnetic flux associated with this gauge field is

$$F_{12} = \frac{1}{R'L} \tag{17}$$

so that

$$\frac{1}{2\pi} \int F_{12} \, d\xi^1 \, d\xi^2 = 1 . \tag{18}$$

Note that the boundary condition (15) on the infinite matrix X^2 transforms under T-duality to the boundary condition on the gauge field

$$
\begin{aligned}
A_2(L, x_2) &= e^{2\pi i \xi_2/L_2'} \left(A_2(0, x_2) + i\partial_2 \right) e^{-2\pi i \xi_2/L_2'} \qquad (19)\\
&= e^{2\pi i \xi_2/L_2'} A_2(0, x_2) e^{-2\pi i \xi_2/L_2'} + \frac{2\pi}{L_2'},
\end{aligned}
$$

where the off-diagonal elements of the shift matrix V in (15) describe winding modes which correspond after T-duality to the first Fourier mode $e^{2\pi i \xi_2/L_2'}$. The boundary condition on the gauge fields in the ξ_2 direction is trivial, which simplifies the T-duality map; a similar construction can be done with a nontrivial boundary condition in both directions, although the configuration looks more complicated in the D1-brane picture.

This construction gives a simple Yang-Mills description of the mapping of D-brane charges under T-duality: the initial configuration described above has charges associated with a single D1-brane wrapped around each of the directions of the 2-torus: $D1_1 + D1_2$. Under T-duality, these D1-branes are mapped to a D2-brane and a D0-brane respectively: $D2_{12} + D0$. The flux integral (18) is the representation in the D2-brane world-volume Yang-Mills theory of the charge associated with a D0-brane which has been uniformly distributed over the surface of the D2-brane, just as in (10).

3.   Tachyons and D-branes

We now turn to the subject of tachyons. Certain D-brane configurations are unstable, both in supersymmetric and nonsupersymmetric string theories. This instability is manifested as a tachyon with $M^2 < 0$ in the spectrum of open strings ending on the D-brane. We will explicitly describe the tachyonic mode in the case of the open bosonic string in Section 4.1; this open bosonic string tachyon will be the focal point of most of the developments described in these notes. In this section we list some elementary D-brane configurations where tachyons arise, and we describe a particular situation in which the tachyon can be seen in the low-energy Yang-Mills description of the D-branes. This Yang-Mills background with a tachyon provides a simple field-theory model of a system analogous to the more complicated string field theory tachyon we describe in the later part of these notes. This simpler model may be useful to keep in mind in the later analysis.

3.1   D-brane configurations with tachyonic instabilities

Some simple examples of unstable D-brane configurations where the open string contains a tachyon include the following:

Brane-antibrane: A pair of parallel Dp-branes with opposite orientation in type IIA or IIB string theory which are separated by a distance $d < l_s$ give rise to a tachyon in the spectrum of open strings stretched between the branes [28]. The difference in orientation of the branes means that the two branes are really a brane and antibrane, carrying equal but opposite R-R charges. Since the net R-R charge is 0, the brane and antibrane can annihilate, leaving an uncharged vacuum configuration.

Wrong-dimension branes: In type IIA/IIB string theory, a Dp-brane of even/odd spatial dimension p is a stable BPS state carrying a nonzero R-R charge. On the other hand, a Dp-brane of the *wrong* dimension (*i.e.,* odd/even for IIA/IIB) carries no charges under the classical IIA/IIB supergravity fields, and has a tachyon in the open string spectrum. Such a brane can annihilate to the vacuum without violating charge conservation.

Bosonic D-branes: Like the wrong-dimension branes of IIA/IIB string theory, a Dp-brane of any dimension in the bosonic string theory carries no conserved charge and has a tachyon in the open string spectrum. Again, such a brane can annihilate to the vacuum without violating charge conservation.

3.2   Example: tachyon in low-energy field theory of two D-branes

As an example of how tachyonic configurations behave physically, we consider in this subsection a simple example where a brane-antibrane tachyon can be seen in the context of the low-energy Yang-Mills theory. This system was originally considered in [29, 30].

The system we want to consider is a simple generalization of the (D2 + D0)-brane configuration we described using Yang-Mills theory in Section 2.4. Consider a pair of D2-branes wrapped on a two-torus, one of which has a D0-brane embedded in it as a constant positive magnetic flux, and the other of which has an anti-D0-brane within it described by a constant negative magnetic flux. We take the two dimensions of the torus to be L_1, L_2. Following the discussion of Section 2.4, this configuration is equivalent under T-duality in the L_2 direction to a pair of crossed D1-branes (see Figure 2). The Born-Infeld energy of this configuration is

$$E_{\mathrm{BI}} \;=\; 2\sqrt{(\tau_2 L_1 L_2)^2 + \tau_0^2}$$

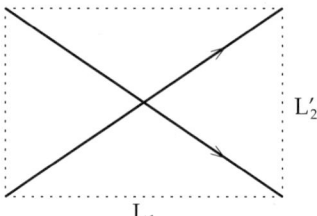

Figure 2.   A pair of crossed D1-branes, T-dual to a pair of D2-branes with uniformly embedded D0- and anti-D0-branes.

$$= \frac{1}{g}\left[\frac{2L_1L_2}{\sqrt{2\pi}} + \frac{(2\pi)^{3/2}}{L_1L_2} + \cdots\right] \tag{20}$$

in units where $2\pi\alpha' = 1$. The second term in the last line corresponds to the Yang-Mills approximation. In this approximation (dropping the D2-brane energy) the energy is

$$E_{\mathrm{YM}} = \frac{\tau_2}{4}\int \mathrm{Tr}\, F_{\alpha\beta}F^{\alpha\beta} = \frac{1}{4\sqrt{2\pi}g}\int \mathrm{Tr}\, F_{\alpha\beta}F^{\alpha\beta}. \tag{21}$$

We are interested in studying this configuration in the Yang-Mills approximation, in which we have a $U(2)$ theory on T^2 with field strength

$$F_{12} = \begin{pmatrix} \frac{2\pi}{L_1L_2} & 0 \\ 0 & -\frac{2\pi}{L_1L_2} \end{pmatrix} = \frac{2\pi}{L_1L_2}\tau_3. \tag{22}$$

This field strength can be realized as the curvature of a linear gauge field

$$A_1 = 0, \qquad A_2 = \frac{2\pi}{L_1L_2}\xi\tau_3 \tag{23}$$

which satisfies the boundary conditions

$$A_j(L,\xi_2) = \Omega(i\partial_j + A_j(0,\xi_2))\Omega^{-1} \tag{24}$$

where

$$\Omega = e^{2\pi i(\xi_1/L_2)\tau_3}. \tag{25}$$

It is easy to check that this configuration indeed satisfies

$$E_{\mathrm{YM}} = \frac{1}{2g}\frac{(2\pi)^{3/2}}{L_1L_2}\mathrm{Tr}\,\tau_3^2 = \frac{1}{g}\frac{(2\pi)^{3/2}}{L_1L_2} \tag{26}$$

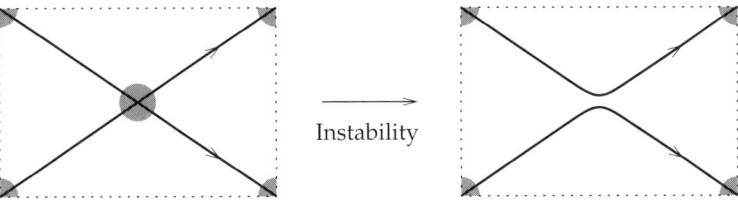

Figure 3. The brane-antibrane instability of a D0-D$\bar{0}$ system embedded in two D2-branes, as seen in the T-dual D1-brane picture.

as desired from (20). Since, however,

$$\text{Tr } F_{\alpha\beta} = 0, \tag{27}$$

the gauge field we are considering is in the same topological equivalence class as $F = 0$. This corresponds to the fact that the D0-brane and anti-D0-brane can annihilate. To understand the appearance of the tachyon, we can consider the spectrum of excitations δA_α around the background (23) [29]. The eigenvectors of the quadratic mass terms in this background are described by theta functions on the torus satisfying boundary conditions related to (24). There are precisely two elements in the spectrum with the negative eigenvalue $-4\pi/L_1 L_2$. These theta functions, given explicitly in [29], are tachyonic modes of the theory which are associated with the annihilation of the positive and negative fluxes encoding the D0- and anti-D0-brane. These tachyonic modes are perhaps easiest to understand in the dual configuration, where they provide a direction of instability in which the two crossed D1-branes reconnect as in Figure 3. In the T-dual picture it is also interesting to note that the two tachyonic modes of the gauge field have support which is localized near the two brane intersection points. These modes have off-diagonal form

$$\delta A_t \sim \begin{pmatrix} 0 & \star \\ \star & 0 \end{pmatrix}. \tag{28}$$

This form of the tachyonic modes naturally encodes our geometric understanding of these modes as reconnecting the two D1-branes near the intersection point.

The full Yang-Mills action around the background (23) can be written as a quartic function of the mass eigenstates around this background. Written in terms of these modes, there are nontrivial cubic and quartic terms which couple the tachyonic modes to all the massive modes in the system. If we integrate out the massive modes, we know from the topological reasoning above that an effective potential arises for the tachyonic mode A_t, with a maximum value of (26) and a minimum value of 0. This system is highly analogous to the bosonic open string tachyon we will discuss in the remainder of these lectures. Our current understanding of the bosonic string through bosonic string field theory is analogous to that of someone who only knows the Yang-Mills theory around the background (23) in terms of a complicated quartic action for an infinite family of modes. Without knowledge of the topological structure of the theory, and given only a list of the coefficients in the quartic action, such an individual would have to systematically calculate the tachyon effective potential by explicitly integrating out all the massive modes one by one. This would give a numerical approximation to the minimum of the effective potential, which could be made arbitrarily good by raising the mass of the cutoff at which the effective action is computed. It may be helpful to keep this example system in mind in the following sections, where an analogous tachyonic system is considered in string field theory. For further discussion of this unstable configuration in Yang-Mills theory, see [29, 30].

4.   Open string field theory and the Sen conjectures

The discussion of the previous sections gives us an overview of string theory, and an example of how tachyons appear in a simple gauge theory context, when an unstable brane-antibrane configuration is embedded in a higher-dimensional brane. We now turn our attention back to string theory, where the appearance of a tachyon necessitates a nonperturbative approach to the theory. In subsection 4.1, we review the BRST quantization approach to the bosonic open string. Subsection 4.2 describes Witten's cubic open string field theory, which gives a nonperturbative off-shell definition to the open bosonic string. In subsection 4.3 we describe Sen's conjectures on tachyon condensation in the open bosonic string.

4.1   The bosonic open string

In this subsection we review the quantization of the open bosonic string. For further details see the textbooks by Green, Schwarz, and Witten [31] and by Polchinski [26]. The bosonic open string can be quantized using the BRST quantization approach starting from the action

$$S = -\frac{1}{4\pi\alpha'} \int \sqrt{-\gamma}\gamma^{ab}\partial_a X^\mu \partial_b X_\mu, \tag{29}$$

where γ is an auxiliary dynamical metric on the world-sheet. This action can be gauge-fixed to conformal gauge $\gamma_{ab} \sim \delta_{ab}$. Using the BRST approach to gauge fixing introduces ghost and antighost fields $c^{\pm}(\sigma), b_{\pm\pm}(\sigma)$. The gauge-fixed action, including ghosts, then becomes

$$S = -\frac{1}{4\pi\alpha'} \int \partial_a X^\mu \partial^a X_\mu + \frac{1}{\pi} \int \left(b_{++}\partial_- c^+ + b_{--}\partial_+ c^- \right) . \tag{30}$$

The matter fields X^μ can be expanded in modes using

$$X^\mu(\sigma, \tau) = x_0^\mu + l_s^2 p^\mu \tau + \sum_{n \neq 0} \frac{il_s}{n} \alpha_n^\mu \cos(n\sigma) e^{-in\tau} . \tag{31}$$

Throughout the remainder of these notes we will use the convention

$$\alpha' = \frac{l_s^2}{2} = 1 , \tag{32}$$

so that $l_s = \sqrt{2}$. In the quantum theory, x_0^μ and p^μ obey the canonical commutation relations

$$[x_0^\mu, p^\nu] = i\eta^{\mu\nu} . \tag{33}$$

The α_n^μ's with negative/positive values of n become raising/lowering operators for the oscillator modes on the string, and satisfy the commutation relations

$$[\alpha_m^\mu, \alpha_n^\nu] = m\eta^{\mu\nu} \delta_{m+n,0} . \tag{34}$$

We will often use the canonically normalized raising and lowering operators

$$a_n^\mu = \frac{1}{\sqrt{|n|}} \alpha_n^\mu \tag{35}$$

which obey the commutation relations

$$[a_m^\mu, a_n^\nu] = \eta^{\mu\nu} \delta_{m+n,0} . \tag{36}$$

The raising and lowering operators satisfy $(\alpha_n^\mu)^\dagger = \alpha_{-n}^\mu, (a_n^\mu)^\dagger = a_{-n}^\mu$. We will also frequently use position modes x_n for $n \neq 0$ and raising and lowering operators $a_0, a_0^\dagger$ for the zero modes. These are related to the modes in (31) through (dropping space-time indices)

$$x_n = \frac{i}{\sqrt{n}}(a_n - a_n^\dagger) \tag{37}$$

$$x_0 = \frac{i}{\sqrt{2}}(a_0 - a_0^\dagger)$$

The ghost and antighost fields can be decomposed into modes through

$$c^{\pm}(\sigma, \tau) \;=\; \sum_n c_n e^{\mp in(\sigma \pm \tau)} \tag{38}$$

$$b_{\pm\pm}(\sigma, \tau) \;=\; \sum_n b_n e^{\mp in(\sigma \pm \tau)} \,.$$

The ghost and antighost modes satisfy the anticommutation relations

$$\{c_n, b_m\} \;=\; \delta_{n+m,0} \tag{39}$$

$$\{c_n, c_m\} = \{b_n, b_m\} \;=\; 0 \,.$$

A general state in the open string Fock space can be written in the form

$$\alpha^{\mu_1}_{-n_1} \cdots \alpha^{\mu_i}_{-n_i} \, c_{-m_1} \cdots c_{-m_j} \, b_{-p_1} \cdots b_{-p_l} \, |0; k\rangle \tag{40}$$

where $|0; k\rangle$ is the SL(2,R) invariant vacuum annihilated by

$$b_n|0; k\rangle \;=\; 0, \quad n \geq -1 \tag{41}$$

$$c_n|0; k\rangle \;=\; 0, \quad n \geq 2 \tag{42}$$

$$\alpha^{\mu}_{-n}|0; k\rangle \;=\; 0, \quad n \geq 1 \tag{43}$$

with momentum

$$p^{\mu}|0; k\rangle = k^{\mu}|0; k\rangle \,. \tag{44}$$

We will often write the zero momentum vacuum $|0; k = 0\rangle$ simply as $|0\rangle$. This vacuum is taken by convention to have ghost number 0, and satisfies

$$\langle 0; k|c_{-1}c_0 c_1|0\rangle = \delta(k) \tag{45}$$

For string field theory we will also find it convenient to work with the vacua of ghost number 1 and 2

$$G = 1: \qquad |0_1\rangle = c_1|0\rangle \tag{46}$$

$$G = 2: \qquad |0_2\rangle = c_0 c_1|0\rangle \,. \tag{47}$$

In the notation of Polchinski [26], these two vacua are written as

$$|0_1\rangle \;=\; |0\rangle_m \otimes |\downarrow\rangle \tag{48}$$

$$|0_2\rangle \;=\; |0\rangle_m \otimes |\uparrow\rangle$$

where $|0\rangle_m$ is the matter vacuum and $|\downarrow\rangle, |\uparrow\rangle$ are the ghost vacua annihilated by b_0, c_0.

The BRST operator of this theory is given by

$$Q_B = \sum_{n=-\infty}^{\infty} c_n L^{(\mathrm{m})}_{-n} + \sum_{n,m=-\infty}^{\infty} \frac{(m-n)}{2} : c_m c_n b_{-m-n} : -c_0 \tag{49}$$

where the matter Virasoro operators are given by

$$
L_q^{(m)} = \begin{cases} \frac{1}{2} \sum_n \alpha_{q-n}^{\mu} \alpha_{\mu\,n}, & q \neq 0 \\ p^2 + \sum_{n=1}^{\infty} \alpha_{-n}^{\mu} \alpha_{\mu\,n}, & q = 0 \end{cases}
\tag{50}
$$

Some useful features of the BRST operator $Q = Q_B$ include:

- $Q^2 = 0$; *i.e.*, the BRST operator is nilpotent. This identity relies on a cancellation between matter and ghost terms which only works in dimension $D = 26$ for the bosonic theory.

- $\{Q, b_0\} = L_0^{(m)} + L_0^{(g)} - 1$.

- Q has ghost number 1, so acting on a state $|s\rangle$ of ghost number G gives a state $Q|s\rangle$ of ghost number $G + 1$.

- The physical states of the theory are given by the cohomology of Q at ghost number 1

$$
\begin{aligned}
\mathcal{H}_{\text{phys}} &= \mathcal{H}_{\text{closed}} / \mathcal{H}_{\text{exact}} \\
&= \{|\psi\rangle : Q|\psi\rangle = 0\} / (|\psi\rangle \sim |\psi\rangle + Q|\chi\rangle)
\end{aligned}
\tag{51}
$$

- Physical states can be chosen as representatives of each cohomology class so that they are all annihilated by b_0.

It is often convenient to separate out the ghost zero-modes, writing $Q = c_0 L_0 + b_0 M + \tilde{Q}$, where (momentarily reinstating α')

$$
L_0 = \sum_{n=1}^{\infty} (\alpha_{-n}\alpha_n + nc_{-n}b_n + nb_{-n}c_n) + \alpha' p^2 - 1
\tag{52}
$$

In this expression the term in parentheses is simply the oscillator number operator, measuring the level of a given state.

Some simple examples of physical states include the tachyon state

$$
|0_1; p\rangle
\tag{53}
$$

which is physical when $p^2 = 1/\alpha' = -M^2$, and the massless gauge boson

$$
\epsilon_\mu \alpha_{-1}^\mu |0_1; p\rangle
\tag{54}
$$

which is physical when $p^2 = M^2 = 0$, for transverse polarizations $p \cdot \epsilon = 0$. Note that the transverse polarization condition follows from the appearance of a term proportional to $c_{-1} p \cdot \alpha_1$ in $\tilde{Q}$, which must annihilate the state (54)

4.2   Witten's cubic bosonic SFT

The discussion of the previous subsection leads to a systematic quantization of the open bosonic string in the conformal field theory framework. Using this approach it is possible, in principle, to calculate an arbitrary perturbative on-shell scattering amplitude for physical string states. To study tachyon condensation in string theory, however, we require a nonperturbative, off-shell formalism for the theory— a string field theory.

A very simple form for the off-shell open bosonic string field theory action was proposed by Witten in 1986 [18]

$$S = -\frac{1}{2} \int \Psi \star Q\Psi - \frac{g}{3} \int \Psi \star \Psi \star \Psi . \qquad (55)$$

This action has the general form of a Chern-Simons theory on a 3-manifold, although for string field theory there is no explicit interpretation of the integration in terms of a concrete 3-manifold. In Eq. (55), g is interpreted as the string coupling constant. The field Ψ is a string field, which takes values in a graded algebra $\mathcal{A}$. Associated with the algebra $\mathcal{A}$ there is a star product

$$\star : \mathcal{A} \otimes \mathcal{A} \to \mathcal{A}, \qquad (56)$$

under which the degree G is additive ($G_{\Psi\star\Phi} = G_\Psi + G_\Phi$). There is also a BRST operator

$$Q : \mathcal{A} \to \mathcal{A}, \qquad (57)$$

of degree one ($G_{Q\Psi} = 1 + G_\Psi$). String fields can be integrated using

$$\int : \mathcal{A} \to \mathbf{C} . \qquad (58)$$

This integral vanishes for all Ψ with degree $G_\Psi \neq 3$.

The elements $Q, \star, \int$ defining the string field theory are assumed to satisfy the following axioms:

(a) Nilpotency of Q: $Q^2\Psi = 0, \quad \forall\Psi \in \mathcal{A}.$

(b) $\int Q\Psi = 0, \quad \forall\Psi \in \mathcal{A}.$

(c) Derivation property of Q:
$$Q(\Psi \star \Phi) = (Q\Psi) \star \Phi + (-1)^{G_\Psi}\Psi \star (Q\Phi), \quad \forall\Psi, \Phi \in \mathcal{A}.$$

(d) Cyclicity: $\int \Psi \star \Phi = (-1)^{G_\Psi G_\Phi} \int \Phi \star \Psi, \quad \forall\Psi, \Phi \in \mathcal{A}.$

(e) Associativity: $(\Phi \star \Psi) \star \Xi = \Phi \star (\Psi \star \Xi), \quad \forall\Phi, \Psi, \Xi \in \mathcal{A}.$

When these axioms are satisfied, the action (55) is invariant under the gauge transformations

$$\delta\Psi = Q\Lambda + \Psi \star \Lambda - \Lambda \star \Psi \qquad (59)$$

for any gauge parameter $\Lambda \in \mathcal{A}$ with ghost number 0.

When the string coupling g is taken to vanish, the equation of motion for the theory defined by (55) simply becomes $Q\Psi = 0$, and the gauge transformations (59) simply become

$$\delta\Psi = Q\Lambda. \tag{60}$$

Thus, when $g = 0$ this string field theory gives precisely the structure needed to describe the free bosonic string. The motivation for introducing the extra structure in (55) was to find a simple interacting extension of the free theory, consistent with the perturbative expansion of open bosonic string theory.

Witten presented this formal structure and argued that all the needed axioms are satisfied when $\mathcal{A}$ is taken to be the space of string fields

$$\mathcal{A} = \{\Psi[x(\sigma); c(\sigma), b(\sigma)]\} \tag{61}$$

which can be described as functionals of the matter, ghost and antighost fields describing an open string in 26 dimensions with $0 \leq \sigma \leq \pi$. Such a string field can be written as a formal sum over open string Fock space states with coefficients given by an infinite family of space-time fields

$$\Psi = \int d^{26}p \left[\phi(p) |0_1; p\rangle + A_\mu(p) \alpha^\mu_{-1}|0_1; p\rangle + \cdots\right] \tag{62}$$

Each Fock space state is associated with a given string functional, just as the states of a harmonic oscillator are associated with wavefunctions of a particle in one dimension. For example, the matter ground state $|0\rangle_m$ annihilated by a_n for all $n \geq 1$ is associated (up to a constant C) with the functional of matter modes

$$|0\rangle_m \to C \exp\left(-\frac{1}{4}\sum_{n>0}^{\infty} nx_n^2\right). \tag{63}$$

For Witten's cubic string field theory, the BRST operator Q in (55) is the usual open string BRST operator Q_B, given in (49). The star product $\star$ acts on a pair of functionals Ψ, Φ by gluing the right half of one string to the left half of the other using a delta function interaction

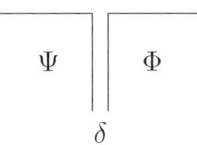

This star product factorizes into separate matter and ghost parts. In the matter sector, the star product is given by the formal functional integral

$$(\Psi \star \Phi)[z(\sigma)] \tag{64}$$

$$\equiv \int \prod_{0\leq\tilde\tau\leq\frac{\pi}{2}} dy(\tilde\tau)\, dx(\pi-\tilde\tau) \prod_{\frac{\pi}{2}\leq\tau\leq\pi} \delta[x(\tau)-y(\pi-\tau)]\, \Psi[x(\tau)]\Phi[y(\tau)]\,,$$

$$x(\tau)=z(\tau) \quad \text{for} \quad 0\leq\tau\leq\frac{\pi}{2}\,,$$

$$y(\tau)=z(\tau) \quad \text{for} \quad \frac{\pi}{2}\leq\tau\leq\pi\,.$$

Similarly, the integral over a string field factorizes into matter and ghost parts, and in the matter sector is given by

$$\int\Psi=\int \prod_{0\leq\sigma\leq\pi} dx(\sigma) \prod_{0\leq\tau\leq\frac{\pi}{2}} \delta[x(\tau)-x(\pi-\tau)]\,\Psi[x(\tau)]\,. \tag{65}$$

This corresponds to gluing the left and right halves of the string together with a delta function interaction

The ghost sector of the theory is defined in a similar fashion, but has an anomaly due to the curvature of the Riemann surface describing the three-string vertex. The ghost sector can be described either in terms of fermionic ghost fields $c(\sigma)$, $b(\sigma)$ or through bosonization in terms of a single bosonic scalar field $\phi_g(\sigma)$. From the functional point of view of Eqs. (64, 65), it is easiest to describe the ghost sector in the bosonized language. In this language, the ghost fields $b(\sigma)$ and $c(\sigma)$ are replaced by the scalar field $\phi_g(\sigma)$, and the star product in the ghost sector is given by (64) with an extra insertion of $\exp(3i\phi_g(\pi/2)/2)$ inside the integral. Similarly, the integration of a string field in the ghost sector is given by (65) with an insertion of $\exp(-3i\phi_g(\pi/2)/2)$ inside the integral. Witten first described the cubic string field theory using bosonized ghosts. While this approach is useful for some purposes, we will use fermionic ghost fields in the remainder of these lecture notes.

The expressions (64, 65) may seem rather formal, as they are written in terms of functional integrals. These expressions, however, can be given precise meaning when described in terms of creation and annihilation operators acting on the string Fock space. In the Fock space language, the integral of a star product of two or three fields is described in terms of two- and three-string vertices

$$\langle V_2|\in \mathcal{H}^*\otimes\mathcal{H}^*,\quad \langle V_3|\in(\mathcal{H}^*)^3 \tag{66}$$

so that

$$\int \Phi \star \Psi \quad \rightarrow \quad \langle V_2 | \left(|\Phi\rangle \otimes |\Psi\rangle \right) \tag{67}$$

$$\int \Psi_1 \star \Psi_2 \star \Psi_3 \quad \rightarrow \quad \langle V_3 | \left(|\Psi_1\rangle \otimes |\Psi_2\rangle \otimes |\Psi_3\rangle \right)$$

In the next section we will give explicit forms for the two- and three-string vertices (66). In terms of these vertices, the string field theory action becomes

$$S = -\frac{1}{2} \langle V_2 | \Psi, Q\Psi \rangle - \frac{g}{3} \langle V_3 | \Psi, \Psi, \Psi \rangle \,. \tag{68}$$

This action is often written using the BPZ dual $\langle \Psi |$ of the string field $|\Psi\rangle$, defined by the conformal map $z \to -1/z$, as

$$S = -\frac{1}{2} \langle \Psi | Q\Psi \rangle - \frac{g}{3} \langle \Psi | \Psi \star \Psi \rangle \,. \tag{69}$$

In the remainder of these lectures, however, we will use the form (68). Using explicit formulae for the vertices (66) and the string field expansion (62) leads to the full string field theory action, given by an off-shell action in the target space-time for an infinite family of fields $\phi(p), A_\mu(p), \ldots$ We discuss this action in more detail in Section 5.

4.3   The Sen conjectures

The existence of the tachyonic mode in the open bosonic string indicates that the standard choice of perturbative vacuum for this theory is unstable. In the early days of the subject, there was some suggestion that this tachyon could condense, leading to a more stable vacuum (see for example [32]). Kostelecky and Samuel argued early on that the stable vacuum could be identified in string field theory in a systematic way [33], however there was no clear physical picture for the significance of this stable vacuum. In 1999, Ashoke Sen reconsidered the problem of tachyons in string field theory. Sen suggested that the open bosonic string should really be thought of as living on a D25-brane, and hence that the perturbative vacuum for this string theory should have a nonzero vacuum energy associated with the tension of this D25-brane. He suggested that the tachyon is simply the instability mode of the D25-brane, which carries no conserved charge and hence is not expected to be stable, as discussed in section 3. Sen furthermore suggested that Witten's cubic open string field theory is a natural framework to use to study this tachyon, and that this string field theory should give an analytic description of the true vacuum. More precisely, Sen made the following 3 conjectures [19]:

   i Witten's classical open string field theory should have a locally stable
     nontrivial vacuum solution. The energy density of this vacuum should

be given by the D25-brane tension

$$\frac{\Delta E}{V} = T_{25} = -\frac{1}{2\pi^2 g^2} . \tag{70}$$

ii Lower-dimensional D-branes should exist as solitonic solutions of SFT which break part of the Lorentz symmetry of the perturbative vacuum.

iii Open strings should decouple from the theory in the nontrivial vacuum, since the D25-brane is absent in this vacuum.

In Section 6 of these lectures we discuss the evidence for these conjectures, focusing particularly on the first and third conjectures. First, however, we need to develop the technical tools to do specific calculations in string field theory.

5.   Basics of SFT

In this section, we give a more detailed discussion of Witten's open bosonic string field theory. Subsection 5.1 is a warmup, in which we review some basic features of the simple harmonic oscillator and discuss squeezed states. In Subsection 5.2 we derive the two-string vertex, and in subsection 5.3 we give an explicit formula for the three-string vertex. In subsection 5.4 we put these pieces together and discuss the calculation of the full SFT action. 5.5 contains a brief description of some more general features of Witten's open bosonic string field theory. For more details about this string field theory, the reader is referred to the reviews [34, 35, 36].

5.1   Squeezed states and the simple harmonic oscillator

Let us consider a simple harmonic oscillator with annihilation operator

$$a = -i \left(\sqrt{\frac{\alpha}{2}} x + \frac{1}{\sqrt{2\alpha}} \partial_x \right) \tag{71}$$

and ground state

$$|0\rangle = \left(\frac{\alpha}{\pi} \right)^{1/4} e^{-\alpha x^2/2} . \tag{72}$$

In the harmonic oscillator basis $|n\rangle$, the Dirac position basis states $|x\rangle$ have a squeezed state form

$$|x\rangle = \left(\frac{\alpha}{\pi} \right)^{1/4} \exp \left(-\frac{\alpha}{2} x^2 - i\sqrt{2\alpha} a^\dagger x + \frac{1}{2} (a^\dagger)^2 \right) |0\rangle . \tag{73}$$

A general wavefunction is associated with a state through the correspondence

$$f(x) \rightarrow \int_{-\infty}^{\infty} dx \, f(x)|x\rangle . \tag{74}$$

In particular, we have

$$\delta(x) \;\to\; \left(\frac{\alpha}{\pi}\right)^{1/4} \exp\left(\frac{1}{2}(a^\dagger)^2\right)|0\rangle \tag{75}$$

$$1 \;\to\; \int dx\,|x\rangle = \left(\frac{4\pi}{\alpha}\right)^{1/4} \exp\left(-\frac{1}{2}(a^\dagger)^2\right)|0\rangle$$

This shows that the delta and constant functions both have squeezed state representations in terms of the harmonic oscillator basis. The norm of a squeezed state

$$|s\rangle = \exp\left(\frac{1}{2}s(a^\dagger)^2\right)|0\rangle \tag{76}$$

is given by

$$\langle s|s\rangle = \frac{1}{\sqrt{1-s^2}} \tag{77}$$

Thus, the states (75) are non-normalizable (as we would expect), however they are right on the border of normalizability. As for the Dirac basis states $|x\rangle$, which are computationally useful although technically not well-defined states in the single-particle Hilbert space, we expect that many calculations using the states (75) will give sensible physical answers.

It will be useful for us to generalize the foregoing considerations in several ways. A particularly simple generalization arises when we consider a pair of degrees of freedom x, y described by a two-harmonic oscillator Fock space basis. In such a basis, repeating the preceding analysis leads us to a function-state correspondence for the delta functions relating x, y of the form

$$\delta(x \pm y) \to \exp\left(\pm\frac{1}{2}a^\dagger_{(x)}a^\dagger_{(y)}\right)\left(|0\rangle_x \otimes |0\rangle_y\right). \tag{78}$$

we will find these squeezed state expressions very useful in describing the two- and three-string vertices of Witten's open string field theory.

5.2   The two-string vertex $|V_2\rangle$

We can immediately apply the oscillator formulae from the preceding section to calculate the two-string vertex. Recall that the matter fields are expanded in modes through

$$x(\sigma) = x_0 + \sqrt{2}\sum_{n=1}^{\infty} x_n \cos n\sigma. \tag{79}$$

(We suppress Lorentz indices in most of this section for clarity.) Using this mode decomposition, we associate the string field functional $\Psi[x(\sigma)]$ with a function $\Psi(\{x_n\})$ of the infinite family of string oscillator mode amplitudes.

The overlap integral combining (65) and (64) can then be expressed in modes as

$$\int \Psi \star \Phi = \int \prod_{n=0}^{\infty} dx_n dy_n \; \delta(x_n - (-1)^n y_n) \Psi(\{x_n\}) \Phi(\{y_n\}) \,. \qquad (80)$$

Geometrically this just encodes the overlap condition $x(\sigma) = y(\pi - \sigma)$ described through

$$\xrightarrow[\Psi]{\Phi}$$

From (78), it follows that we can write the two-string vertex as a squeezed state

$$\langle V_2 |_{\text{matter}} = (\langle 0 | \otimes \langle 0 |) \exp \left(\sum_{n,m=0}^{\infty} -a_n^{(1)} C_{nm} a_m^{(2)} \right) \qquad (81)$$

where $C_{nm} = \delta_{nm}(-1)^n$ is an infinite-size matrix connecting the oscillator modes of the two single-string Fock spaces, and the sum is taken over all oscillator modes including zero. In the expression (81), we have used the formalism in which $|0\rangle$ is the vacuum annihilated by a_0. To translate this expression into a momentum basis, we use only $n, m > 0$, and replace

$$(\langle 0 | \otimes \langle 0 |) \exp \left(-a_0^{(1)} a_0^{(2)} \right) \rightarrow \int d^{26} p \, (\langle 0; p | \otimes \langle 0; -p |) \,. \qquad (82)$$

The extension of this analysis to ghosts is straightforward. For the ghost and antighost respectively, the overlap conditions corresponding with $x_1(\sigma) = x_2(\pi - \sigma)$ are [37] $c_1(\sigma) = -c_2(\pi - \sigma)$ and $b_1(\sigma) = b_2(\pi - \sigma)$. This leads to the overall formula for the two-string vertex

$$\langle V_2 | = \int d^{26} p \, (\langle 0; p | \otimes \langle 0; -p |) \, (c_0^{(1)} + c_0^{(2)}) \exp \left(-\sum_{n=1}^{\infty} (-1)^n [a_n^{(1)} a_n^{(2)} + c_n^{(1)} b_n^{(2)} - \right.$$

$$\qquad (83)$$

This expression for the two-string vertex can also be derived directly from the conformal field theory approach, computing the two-point function of an arbitrary pair of states on the disk.

5.3   The three-string vertex $|V_3\rangle$

The three-string vertex, which is associated with the three-string overlap diagram

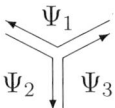

can be computed in a very similar fashion to the two-string vertex above. The details of the calculation, however, are significantly more complicated. There are several different ways to carry out the calculation. One approach is to first rewrite the modes $\cos n\sigma$ on the full string in terms of modes l, r on the two halves of the string with $\sigma < \pi/2, \sigma > \pi/2$. This rewriting can be accomplished using an infinite orthogonal transformation matrix X. The delta function overlap condition can then be applied to the half-string modes as above, giving a squeezed state expression for $|V_3\rangle$ with a squeezing matrix which can be expressed in terms of X. The three-string vertex can also be computed using the conformal field theory approach. The three-string vertex was computed using various versions of these approaches in [37, 38, 39, 40, 41] [1].

In these lectures we will not have time to go through a detailed derivation of the three-string vertex using any of these methods[2]. We simply quote the final result from [37, 66]. Like the two-string vertex, the three-string vertex takes the form of a squeezed state

$$
\langle V_3| = \int d^{26}p^{(1)} d^{26}p^{(2)} d^{26}p^{(3)} \left(\langle 0; p^{(1)}| \otimes \langle 0; p^{(2)}| \otimes \langle 0; p^{(3)}| \right) \delta(p^{(1)} + p^{(2)} + p^{(3)}) c_0^{(1)} c_0^{(2)} c_0^{(3)}
$$
$$
\kappa \exp \left(-\frac{1}{2} \sum_{r,s=1}^{3} [a_m^{(r)} V_{mn}^{rs} a_n^{(s)} + 2 a_m^{(r)} V_{m0}^{rs} p^{(s)} + p^{(r)} V_{00}^{rs} p^{(s)} + c_m^{(r)} X_{mn}^{rs} b_n^{(s)}] \right), \qquad (84)
$$

---

[1] Another interesting approach to understanding the cubic vertex has been explored extensively since these lectures were given. By diagonalizing the Neumann matrices, the star product encoded in the 3-string vertex takes the form of a continuous Moyal product. This simplifies the complexity of the cubic vertex, but at the cost of complicating the propagator. For a recent review of this work and further references, see [42]

[2] A more detailed discussion of the derivation of the Neumann coefficients using CFT and oscillator methods will appear in [4]

where $\kappa = 3^{9/2}/2^6$, and where the Neumann coefficients V_{mn}^{rs}, X_{mn}^{rs} are calculable constants given as follows[3]. Define A_n, B_n for $n \geq 0$ through

$$\left(\frac{1+ix}{1-ix}\right)^{1/3} = \sum_{n\,\text{even}} A_n x^n + i \sum_{m\,\text{odd}} A_m x^m \qquad (85)$$

$$\left(\frac{1+ix}{1-ix}\right)^{2/3} = \sum_{n\,\text{even}} B_n x^n + i \sum_{m\,\text{odd}} B_m x^m$$

These coefficients can be used to define 6-string Neumann coefficients $N_{nm}^{r,\pm s}$ through

$$N_{nm}^{r,\pm r} = \begin{cases} \frac{1}{3(n\pm m)}(-1)^n(A_n B_m \pm B_n A_m), & m+n\,\text{even},\ m \neq n \\ 0, & m+n\,\text{odd} \end{cases} \qquad (8$$

$$N_{nm}^{r,\pm(r+\sigma)} = \begin{cases} \frac{1}{6(n\pm\sigma m)}(-1)^{n+1}(A_n B_m \pm \sigma B_n A_m), & m+n\,\text{even},\ m \neq n \\ \frac{\sqrt{3}}{6(n\pm\sigma m)}(A_n B_m \mp \sigma B_n A_m), & m+n\,\text{odd} \end{cases}$$

where in $N^{r,\pm(r+\sigma)}$, $\sigma = \pm 1$, and $r + \sigma$ is taken modulo 3 to be between 1 and 3. The 3-string matter Neumann coefficients V_{nm}^{rs} are then given by

$$V_{nm}^{rs} = -\sqrt{mn}(N_{nm}^{r,s} + N_{nm}^{r,-s}), \qquad m \neq n,\ \text{and}\ m,n \neq 0$$

$$V_{nn}^{rr} = -\frac{1}{3}\left[2\sum_{k=0}^{n}(-1)^{n-k}A_k^2 - (-1)^n - A_n^2\right], \qquad n \neq 0$$

$$V_{nn}^{r,r+\sigma} = \frac{1}{2}\left[(-1)^n - V_{nn}^{rr}\right], \qquad n \neq 0 \qquad (87)$$

$$V_{0n}^{rs} = -\sqrt{2n}\left(N_{0n}^{r,s} + N_{0n}^{r,-s}\right), \qquad n \neq 0$$

$$V_{00}^{rr} = \ln(27/16)$$

The ghost Neumann coefficients X_{mn}^{rs}, $m \geq 0, n > 0$ are given by

$$X_{mn}^{rr} = \left(-N_{nm}^{r,r} + N_{nm}^{r,-r}\right), \qquad n \neq m$$

$$X_{mn}^{r(r\pm1)} = m\left(\pm N_{nm}^{r,r\mp1} \mp N_{nm}^{r,-(r\mp1)}\right), \qquad n \neq m \qquad (88)$$

---

[3]Note that in some references, signs and various factors in κ and the Neumann coefficients may be slightly different. In some papers, the cubic term in the action is taken to have an overall factor of $g/6$ instead of $g/3$; this choice of normalization gives a 3-tachyon amplitude of g instead of $2g$, and gives a different value for κ. Often, the sign in the exponential of (84) is taken to be positive, which changes the signs of the coefficients V_{nm}^{rs}, X_{nm}^{rs}. When the matter Neumann coefficients are defined with respect to the oscillator modes α_n rather than a_n, the matter Neumann coefficients V_{nm}^{rs}, V_{n0}^{rs} must be divided by $\sqrt{nm}$ and $\sqrt{n}$. Finally, when α' is taken to be $1/2$, an extra factor of $1/\sqrt{2}$ appears for each 0 subscript in the matter Neumann coefficients.

$$X_{nn}^{rr} = \frac{1}{3}\left[-(-1)^n - A_n^2 + 2\sum_{k=0}^{n}(-1)^{n-k}A_k^2 - 2(-1)^n A_n B_n\right]$$

$$X_{nn}^{r(r\pm1)} = -\frac{1}{2}(-1)^n - \frac{1}{2}X_{nn}^{rr}$$

The Neumann coefficients have a number of simple symmetries. There is a cyclic symmetry under $r \to r+1, s \to s+1$, which corresponds to the obvious geometric symmetry of rotating the vertex. The coefficients are also symmetric under the exchange $r \leftrightarrow s, n \leftrightarrow m$. Finally, there is a "twist" symmetry,

$$\begin{aligned} V_{nm}^{rs} &= (-1)^{n+m}V_{nm}^{sr} \\ X_{nm}^{rs} &= (-1)^{n+m}X_{nm}^{sr} \,. \end{aligned} \tag{89}$$

This symmetry follows from the invariance of the 3-vertex under reflection.

5.4 Calculating the SFT action

Given the action (68) and the explicit formulae (83, 84) for the two- and three-string vertices, we can in principle calculate the string field action term by term for each of the fields in the string field expansion

$$\Psi = \int d^{26}p \left[\phi(p) \, |0_1; p\rangle \quad + \quad A_\mu(p) \, \alpha_{-1}^\mu |0_1; p\rangle + \chi(p) b_{-1} c_0 |0_1; p\rangle \right.$$

$$\left. + \quad B_{\mu\nu}(p)\alpha_{-1}^\mu \alpha_{-1}^\nu |0_1; p\rangle + \cdots \right] . \tag{90}$$

Since the resulting action has an enormous gauge invariance given by (59), it is often helpful to fix the gauge before computing the action. A particularly useful gauge choice is the Feynman-Siegel gauge

$$b_0|\Psi\rangle = 0 \,. \tag{91}$$

This is a good gauge choice locally, fixing the linear gauge transformations $\delta|\Psi\rangle = Q|\Lambda\rangle$. This gauge choice is not, however, globally valid; we will return to this point later. In this gauge, all fields in the string field expansion which are associated with states having an antighost zero-mode c_0 are taken to vanish. For example, the field $\chi(p)$ in (90) vanishes. In Feynman-Siegel gauge, the BRST operator takes the simple form

$$Q = c_0 L_0 = c_0(N + p^2 - 1) \tag{92}$$

where N is the total (matter + ghost) oscillator number.

Using (92), it is straightforward to write the quadratic terms in the string field action. They are

$$\frac{1}{2}\langle V_2|\Psi, Q\Psi\rangle = \int d^{26}p \left\{ \phi(-p) \left[\frac{p^2-1}{2}\right] \phi(p) + A_\mu(-p)\left[\frac{p^2}{2}\right]A^\mu(p) + \cdots \right\} .$$

$$\tag{93}$$

The cubic part of the action can also be computed term by term, although the terms are somewhat more complicated. The leading terms in the cubic action are given by

$$\frac{1}{3}\langle V_3|\Psi, \Psi, \Psi\rangle = \tag{94}$$

$$\int d^{26}p\, d^{26}q\, \frac{\kappa g}{3}\, e^{(\ln 16/27)(p^2+q^2+p\cdot q)} \left\{ \phi(-p)\phi(-q)\phi(p+q) + \frac{16}{9}A^\mu(-p)A_\mu(-q)\phi(p+q) \right.$$

$$\left. -\frac{8}{9}(p^\mu + 2q^\mu)(2p^\nu + q^\nu)A^\mu(-p)A_\nu(-q)\phi(p+q) + \cdots \right.$$

In computing the ϕ^3 term we have used

$$V_{00}^{rs} = \delta^{rs}\ln\left(\frac{27}{16}\right) \tag{95}$$

The $A^2\phi$ term uses

$$V_{11}^{rs} = -\frac{16}{27}, \quad r \neq s, \tag{96}$$

while the $(A \cdot p)^2\phi$ term uses

$$V_{10}^{12} = -V_{10}^{13} = -\frac{2\sqrt{2}}{3\sqrt{3}} \tag{97}$$

The most striking feature of this action is that for a generic set of three fields, there is a *nonlocal* cubic interaction term, containing an exponential of a quadratic form in the momenta. This means that the target space formulation of string theory has a dramatically different character from a standard quantum field theory. From the point of view of quantum field theory, string field theory seems to contain an infinite number of nonrenormalizable interactions. Just like the simpler case of noncommutative field theories, however, the magic of string theory seems to combine this infinite set of interactions into a sensible model. [Note, though, that we are working here with the bosonic theory, which becomes problematic quantum mechanically due to the closed string tachyon; the superstring should be better behaved, although a complete understanding of superstring field theory is still lacking despite recent progress [43, 44]]. For the purposes of the remainder of these lectures, however, it will be sufficient for us to restrict attention to the classical action at zero momentum, where the action is quite well-behaved.

5.5   General features of Witten's open bosonic SFT

There are several important aspects of Witten's open bosonic string field theory which are worth reviewing here, although they will not be central to the remainder of these lectures.

The first important aspect of this string field theory is that the perturbative on-shell amplitudes computed using this SFT are in precise agreement with the results of standard perturbative string theory (CFT). This result was shown by Giddings, Martinec, Witten, and Zwiebach in [45, 46, 47]; the basic idea underlying this result is that in Feynman-Siegel gauge, the Feynman diagrams of SFT precisely cover the appropriate moduli space of open string diagrams of an arbitrary genus Riemann surface with boundaries, with the ghost factors contributing the correct measure. The essential feature of this construction is the replacement of the Feynman-Siegel gauge propagator L_0^{-1} with a Schwinger parameter

$$\frac{1}{L_0} = \int_0^\infty dt \; e^{-tL_0}. \tag{98}$$

The Schwinger parameter t plays the role of a modular parameter measuring the length of the strip, for each propagator. This sews the string field theory diagram together into a Riemann surface for each choice of Schwinger parameters; the result of [45, 46, 47] was to show that this parameterization always precisely covers the moduli space correctly. Thus, we know that to arbitrary orders in the string coupling the SFT perturbative expansion agrees with standard string perturbation theory, although string field theory goes beyond the conformal field theory approach since it is a nonperturbative, off-shell formulation of the theory.

A consequence of the perturbative agreement between SFT and standard perturbative string theory is that loop diagrams in open string field theory must include closed string poles at appropriate values of the external momenta. It is well-known that while closed string theory in a fixed space-time background (without D-branes) can be considered as a complete and self-contained theory without including open strings, the same is not true of open string theory. Open strings can always close up in virtual processes to form intermediate closed string states. The closed string poles were found explicitly in the one-loop 2-point function of open string field theory in [48]. The appearance of these poles raises a very important question for open string field theory, namely: Can closed strings appear as asymptotic states in open string field theory? Indeed, standard arguments of unitarity would seem to imply that open string field theory cannot be consistent at the quantum level unless open strings can scatter into outgoing closed string states. This question becomes particularly significant in the context of Sen's tachyon condensation conjectures, where we expect that all open string degrees of freedom disappear from the theory in the nonperturbative locally stable vacuum. We will discuss this issue further in Section 8.

6.   Evidence for the Sen conjectures

Now that we have a more concrete understanding of how to carry out calculations in open string field theory, we can address the conjectures made by Sen regarding tachyon condensation. In subsection 6.1, we discuss evidence for Sen's first conjecture, which states that there exists a stable vacuum with energy density $-T_{25}$. In Subsection 6.2, we discuss physics in the stable vacuum and Sen's third conjecture, which states that open strings decouple completely from the theory in this vacuum. There is also a large body of evidence by now for Sen's second conjecture (see [49, 50, 51] for some of the early papers in this direction), but due to time and space constraints we will not cover this work here[4].

6.1   Level truncation and the stable vacuum

Sen's first conjecture states that the string field theory action should lead to a nontrivial vacuum solution, with energy density

$$-T_{25} = -\frac{1}{2\pi^2 g^2} \,. \tag{99}$$

In this subsection we discuss evidence for the validity of this conjecture.

The string field theory equation of motion is

$$Q\Psi + g\Psi \star \Psi = 0 \,. \tag{100}$$

Despite much work over the last few years, there is still no analytic solution of this equation of motion[5]. There is, however, a systematic approximation scheme, known as level truncation, which can be used to solve this equation numerically. The level (L, I) truncation of the full string field theory involves dropping all fields at level $N > L$, and disregarding any cubic interaction terms between fields whose total level is greater than I. For example, the simplest truncation of the theory is the level $(0, 0)$ truncation. Including only $p = 0$ components of the tachyon field, with the justification that we are looking for a Lorentz-invariant vacuum, the theory in this truncation is simply described by a potential for the tachyon zero-mode

$$V(\phi) = -\frac{1}{2}\phi^2 + g\bar{\kappa}\phi^3 \,. \tag{101}$$

where $\bar{\kappa} = \kappa/3 = 3^{7/2}/2^6$. This cubic function is graphed in Figure 4. Clearly,

---

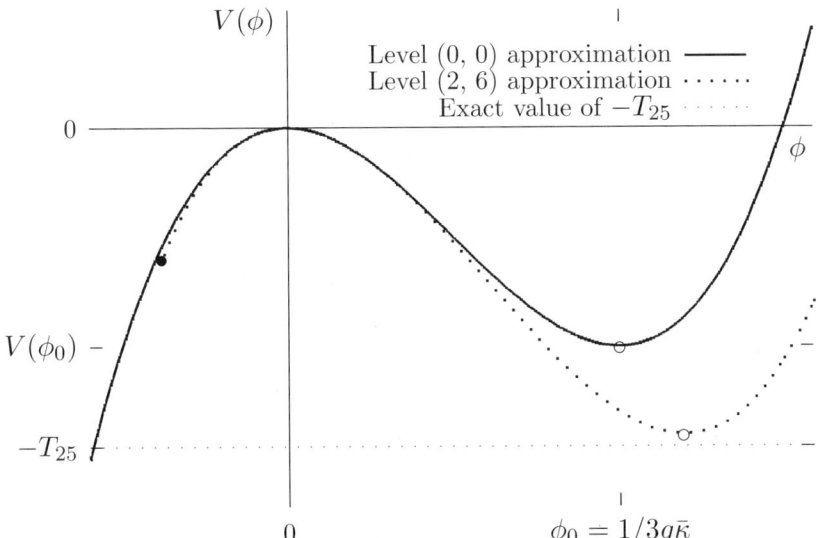

$V(\phi)$

Level $(0, 0)$ approximation ———
Level $(2, 6)$ approximation $\cdots\cdots$
Exact value of $-T_{25}$ $\cdots\cdots$

0

ϕ

$V(\phi_0)$ —

$-T_{25}$

0

$\phi_0 = 1/3g\bar{\kappa}$

Figure 4. The effective tachyon potential in level $(0, 0)$ and $(2, 6)$ truncations. The open circles denote minima in each level truncation. The filled circle denotes a branch point where the level $(2, 6)$ truncation approximation reaches the limit of Feynman-Siegel gauge validity.

this potential has a local minimum at

$$\phi_0 = \frac{1}{3g\bar{\kappa}} . \tag{102}$$

At this point the potential is

$$V(\phi_0) = -\frac{1}{54}\frac{1}{g^2\bar{\kappa}^2} = -\frac{2^{11}}{3^{10}}\frac{1}{g^2} \approx (0.68)\left(-\frac{1}{2\pi^2 g^2}\right) \tag{103}$$

Thus, we see that simply including the tachyon zero-mode gives a nontrivial vacuum with 68% of the vacuum energy density predicted by Sen. This vacuum is denoted by an open circle in Figure 4.

At higher levels of truncation, there are a multitude of fields with various tensor structures. However, again assuming that we are looking for a vacuum which preserves Lorentz symmetry, we can restrict attention to the interactions between scalar fields at $p = 0$. We will work in Feynman-Siegel gauge to simplify calculations. The situation is further simplified by the existence of the "twist" symmetry mentioned in Section 5.3, which guarantees that no cubic vertex between $p = 0$ scalar fields can connect three fields with a total level which is odd. This means that odd fields are not relevant to diagrams with only external tachyons at tree level. Thus, we need only consider even-level scalar fields in looking for Lorentz-preserving solutions to the SFT equations of motion. With these simplifications, in a general level truncation the string field is simply expressed as a sum of a finite number of terms

$$\Psi_s = \sum_i \phi_i |s_i\rangle \tag{104}$$

where ϕ_i are the zero-modes of the scalar fields associated with even-level states $|s_i\rangle$. For example, including fields up to level 2, we have

$$\Psi_s = \phi|0_1\rangle + B\left(\alpha_{-1}\cdot\alpha_{-1}\right)|0_1\rangle + \beta\, b_{-1}c_{-1}|0_1\rangle . \tag{105}$$

The potential for all the scalars appearing in the level-truncated expansion (104) can be simply expressed as a cubic polynomial in the zero-modes of the scalar fields

$$V = \sum_{i,j} d_{ij}\phi_i\phi_j + g\bar{\kappa}\sum_{i,j,k} t_{ijk}\phi_i\phi_j\phi_k . \tag{106}$$

Using the expressions for the Neumann coefficients given in Section 5.3, the potential for all the scalar fields up to level L can be computed in a level (L, I) truncation. For example, the potential in the level $(2, 6)$ truncation is given by

$$V = -\frac{1}{2}\phi^2 + 26B^2 - \frac{1}{2}\beta^2$$

$$+\bar{\kappa}g\left[\phi^3 - \frac{130}{9}\phi^2 B - \frac{11}{9}\phi^2\beta + \frac{30212}{243}\phi B^2 + \frac{2860}{243}\phi B\beta\right.$$

$$\left.+\frac{19}{81}\phi\beta^2 - \frac{2178904}{6561}B^3 - \frac{332332}{6561}B^2\beta - \frac{2470}{2187}B\beta^2 - \frac{1}{81}\beta^3\right]. \tag{107}$$

As an example of how these terms arise, consider the $\phi^2 B$ term. The coefficient in this term is given by

$$g\left\langle V_3 | (|0_1\rangle \otimes |0_1\rangle \otimes \alpha_{-1}\cdot\alpha_{-1}|0_1\rangle)\right. = -g\bar{\kappa}\,(3\cdot 26)\,V_{11}^{11} \tag{108}$$

$$= -g\bar{\kappa}\frac{130}{9}$$

where we have used $V_{11}^{11} = 5/27$.

In the level $(2, 6)$ truncation of the theory, with potential (107), the nontrivial vacuum is found by simultaneously solving the three quadratic equations found by setting to zero the derivatives of (107) with respect to ϕ, B, and β. There are a number of different solutions to these equations, but only one is in the vicinity of $\phi = 1/3g\bar{\kappa}$. The solution of interest is

$$\phi \approx 0.39766\,\frac{1}{g\bar{\kappa}}$$

$$B \approx 0.02045\,\frac{1}{g\bar{\kappa}} \tag{109}$$

$$\beta \approx -0.13897\,\frac{1}{g\bar{\kappa}}$$

Plugging these values into the potential gives

$$E_{(2,6)} = -0.95938\,T_{25}, \tag{110}$$

or 95.9% of the result predicted by Sen. This vacuum is denoted by an open circle in Figure 4.

It is a straightforward, although computationally intensive, project to generalize this calculation to higher levels of truncation. This calculation was carried out to level $(4, 8)$ by Kostelecky and Samuel [33] many years ago. They noted that the vacuum seemed to be converging, but they lacked any physical picture giving meaning to this vacuum. Following Sen's conjectures, the level $(4, 8)$ calculation was done again using somewhat different methods by Sen and Zwiebach [52], who showed that the energy at this level is $-0.986\,T_{25}$. The calculation was automated by Moeller and Taylor [53], who calculated up to level $(10, 20)$, where there are 252 scalar fields. Up to this level, the vacuum energy converges monotonically, as shown in Table 1. These numerical calculations

level	$g\bar{\kappa}\langle\phi\rangle$	V/T_{25}
(0, 0)	0.3333	-0.68462
(2, 4)	0.3957	-0.94855
(2, 6)	0.3977	-0.95938
(4, 8)	0.4005	-0.98640
(4, 12)	0.4007	-0.98782
(6, 12)	0.4004	-0.99514
(6, 18)	0.4004	-0.99518
(8, 16)	0.3999	-0.99777
(8, 20)	0.3997	-0.99793
(10, 20)	0.3992	-0.99912

Table 1.   Tachyon VEV and vacuum energy in stable vacua of level-truncated theory

indicate that level truncation of string field theory leads to a good systematic approximation scheme for computing the nonperturbative tachyon vacuum [6].

It is interesting to consider the tachyon condensation problem from the point of view of the effective tachyon potential. If instead of trying to solve the quadratic equations for all N of the fields appearing in (106), we instead fix the tachyon field ϕ and solve the quadratic equations for the remaining $N - 1$ fields, we can determine a effective potential $V(\phi)$ for the tachyon field. This was done numerically up to level $(10, 20)$ in [53][6]. At each level, the tachyon effective potential smoothly interpolates between the perturbative vacuum and the nonperturbative vacuum near $\phi = 0.4/g\bar{\kappa}$. For example, the tachyon effective potential at level $(2, 6)$ is graphed in Figure 4. In all level truncations other than $(0, 0)$ and $(2, 4)$, the tachyon effective potential has two branch point singularities at which the continuous solution for the other fields breaks down; for the level $(2, 6)$ truncation, these branch points occur at $\phi \approx -0.127/g\bar{\kappa}$ and $\phi \approx 2.293/g\bar{\kappa}$; the lower branch point is denoted by a solid circle in Figure 4. As a result of these branch points, the tachyon effective potential is only valid for a finite range of ϕ, ranging between approximately $-0.1/g\bar{\kappa}$ and $0.6/g\bar{\kappa}$. In [58] it was demonstrated numerically that the branch points in the tachyon effective

---

[6]These were the best values for the vacuum energy and effective potential at the time of the lectures. At strings 2002, Gaiotto and Rastelli reported results up to level $(18, 54)$ [54]. They found the surprising result that while the energy monotonically approaches $-T_{25}$ up to level 12, at level $(14, 42)$ the energy becomes $-1.0002\,T_{25}$, and that the energy continues to decrease, reaching $-1.0005\,T_{25}$ at level $(18, 54)$. In [55], it was shown that this calculation could be theoretically extrapolated to higher levels using the result found in [56] that perturbative amplitudes converge in level truncation with errors described by a power series in $1/L$. This extrapolation suggests that the energy turns around again near $L = 28$, and then increases again, asymptotically approaching $-T_{25}$ as $L \to \infty$. Further analysis supporting this conclusion was given in [57], where the effective tachyon potential was extrapolated to higher order using results calculated up to level 18.

potential arise because the trajectory in field space associated with this potential encounters the boundary of the region of Feynman-Siegel gauge validity. As mentioned earlier, Feynman-Siegel gauge is only valid in a finite-size region around the perturbative vacuum. It seems almost to be a fortunate accident that the nonperturbative vacuum lies within the region of validity of this gauge choice. It is also worth mentioning here that in the "background-independent" formulation of SFT, the tachyon potential can be computed exactly [59]. In this formulation, there is no branch point in the effective potential, which is unbounded below for negative values of the tachyon. On the other hand, the nontrivial vacuum in the background-independent approach arises only as the tachyon field goes to infinity, so it is harder to study the physics of the stable vacuum from this point of view.

Another interesting perspective on the tachyon effective potential is found by performing a perturbative computation of the coefficients in this effective potential in the level-truncated theory. This gives a power series expansion of the effective tachyon potential

$$V(\phi) = \sum_{n=2}^{\infty} c_n (\bar{\kappa} g)^{n-2} \phi^n \tag{111}$$

$$= -\frac{1}{2}\phi^2 + (\bar{\kappa} g)\phi^3 + c_4(\bar{\kappa} g)^2\phi^4 + c_5(\bar{\kappa} g)^3\phi^5 + \cdots$$

In [53], the coefficients up to c_{60} were computed in the level truncations up to $(10, 20)$. Because of the branch point singularity near $\phi = -0.1/g\bar{\kappa}$, this series has a radius of convergence much smaller than the value of ϕ at the nonperturbative vacuum. Thus, the energy at the stable vacuum lies outside the naive range of perturbation theory[7].

6.2   Physics in the stable vacuum

We have seen that numerical results from level-truncated string field theory strongly suggest the existence of a classically stable vacuum solution Ψ_0 to the string field theory equation of motion (100). The state Ψ_0, while still unknown analytically, has been determined numerically to a high degree of precision. This state seems like a very well-behaved string field configuration. While there is no positive-definite inner product on the string field Fock space, the state Ψ_0 certainly has finite norm under the natural inner product $\langle V_2|\Psi_0, c_0 L_0 \Psi_0\rangle$, and is even better behaved under the product $\langle V_2|\Psi_0, c_0 \Psi_0\rangle$. Thus, it is natural to assume that Ψ_0 defines a classically stable vacuum for the theory, around which

---

[7]In [55], however, it was shown that the method of Padé approximants enables us to compute the vacuum energy to excellent precision given a reasonably small number of the coefficients c_n. Thus, the stable vacuum is in some sense accessible from purely perturbative calculations.

we can expand the action to find a new "vacuum string field theory". Expanding

$$\Psi = \Psi_0 + \tilde{\Psi},$$
(112)

we get the action

$$\tilde{S}(\tilde{\Psi}) = S(\Psi_0 + \tilde{\Psi}) = S_0 - \frac{1}{2} \int \tilde{\Psi} \star \tilde{Q}\tilde{\Psi} - \frac{g}{3} \int \tilde{\Psi} \star \tilde{\Psi} \star \tilde{\Psi}.$$
(113)

where

$$\tilde{Q}\Phi = Q\Phi + g(\Psi_0 \star \Phi + \Phi \star \Psi_0).$$
(114)

This string field theory around the stable vacuum has precisely the same form as Witten's original cubic string field theory, only with a different BRST operator $\tilde{Q}$, which so far is only determined numerically. Note that this formulation of vacuum string field theory is distinct from the VSFT model of Rastelli, Sen, and Zwiebach (RSZ) [60]. These authors make an Ansatz that the BRST operator takes a pure ghost form, along the lines of $Q \to c_0$, and they conjecture that the theory with such a BRST operator is equivalent to the VSFT model given by the BRST operator (114). We discuss the RSZ model again briefly in the next section.

Sen's third conjecture states that there should be no open string excitations of the theory around $\Psi = \Psi_0$. This implies that there should be no solutions of the linearized equation $\tilde{Q}\tilde{\Psi}$ in the VSFT (113) other than pure gauge states of the form $\tilde{\Psi} = \tilde{Q}\tilde{\Lambda}$. In this subsection we discuss evidence for this conjecture.

It may seem surprising to imagine that *all* the perturbative open string degrees of freedom will vanish at a particular point in field space, since these are all the degrees of freedom available in the theory. This is not a familiar phenomenon from quantum field theory. To understand how the open strings can decouple, it may be helpful to begin by considering the simple example of the (0, 0) level-truncated theory. In this theory, the quadratic terms in the action become

$$-\int d^{26}p\, \phi(-p) \left[\frac{p^2 - 1}{2} + g\bar{\kappa} \left(\frac{16}{27}\right)^{p^2} \cdot 3\langle\phi\rangle \right] \phi(p).$$
(115)

Taking $\langle\phi\rangle = 1/3\bar{\kappa}g$, we find that the quadratic term is a transcendental expression which does not vanish for any real value of p^2. Thus, this theory has no poles, and the tachyon has decoupled from the theory. Of course, this is not the full story, as there are still finite complex poles. It does, however suggest a mechanism by which the nonlocal parts of the action (encoded in the exponential of p^2) can remove physical poles.

To get the full story, it is necessary to continue the analysis to higher level. At level 2, there are 7 scalar fields, the tachyon and the 6 fields associated with the Fock space states

$$
\begin{array}{ll}
(\alpha_{-1} \cdot \alpha_{-1})|0_1, p\rangle & b_{-1} \cdot c_{-1}|0_1, p\rangle \\
c_0 \cdot b_{-1}|0_1, p\rangle & (p \cdot \alpha_{-2})|0_1, p\rangle \\
(p \cdot \alpha_{-1})^2|0_1, p\rangle & (p \cdot \alpha_{-1})c_0 b_{-1}|0_1, p\rangle
\end{array}
\tag{116}
$$

Note that in this analysis we cannot fix Feynman-Siegel gauge, as we only believe that this gauge is valid for the zero-modes of the scalar fields in the vacuum Ψ_0. An attempt at analyzing the spectrum of the theory in Feynman-Siegel gauge using level truncation was made in [33], with no sensible results. Diagonalizing the quadratic term in the action on the full set of 7 fields of level ≤ 2, we find that poles develop at $M^2 = 0.9$ and $M^2 = 2.0$ (in string units, where the tachyon has $M^2 = -1$) [61]. These poles correspond to states satisfying $\tilde{Q}\tilde{\Psi} = 0$. The question now is, are these states physical? If they are exact states, of the form $\tilde{\Psi} = \tilde{Q}\tilde{\Lambda}$, then they are simply gauge degrees of freedom. If not, however, then they are states in the cohomology of $\tilde{Q}$ and should be associated with physical degrees of freedom. Unfortunately, we cannot precisely determine whether the poles we find in level truncation are due to exact states, as the level-truncation procedure breaks the condition $\tilde{Q}^2 = 0$. Thus, we can only measure *approximately* whether a state is exact. A detailed analysis of this question was carried out in [61]. In this paper, all terms in the SFT action of the form $\phi_i \, \psi_j(p) \, \psi_k(-p)$ were determined, where ϕ_i is a scalar zero-mode, and $\psi_{j,k}$ are nonzero-momentum scalars. In addition, all gauge transformations involving at least one zero-momentum field were computed up to level (6, 12). At each level up to $L = 6$, the ghost number 1 states in the kernel Ker $\tilde{Q}^{(1)}_{(L,2L)}$ were computed. The extent to which each of these states lies in the exact subspace was measured using the formula

$$
\% \text{ exactness} = \sum_i \frac{(s \cdot e_i)^2}{(s \cdot s)}
\tag{117}
$$

where $\{e_i\}$ are an orthonormal basis for Im $\tilde{Q}^{(0)}_{(L,2L)}$, the image of $\tilde{Q}$ acting on the space of ghost number 0 states in the appropriate level truncation. (Note that this measure involves a choice of inner product on the Fock space; several natural inner products were tried, giving roughly equivalent results). The result of this analysis was that up to the mass scale of the level truncation, $M^2 \leq L-1$, all the states in the kernel of $\tilde{Q}^{(1)}$ were $\geq 99.9\%$ within the exact subspace, for $L \geq 4$. This result seems to give very strong evidence for Sen's third conjecture that there are no perturbative open string excitations around the stable classical vacuum Ψ_0. This analysis was only carried out for even level scalar fields; it would be nice to check that a similar result holds for odd-level fields and for tensor fields of arbitrary rank.

Another more abstract argument that there are no open string states in the stable vacuum was given by Ellwood, Feng, He and Moeller [62]. These authors argued that in the stable vacuum, the identity state $|I\rangle$ in the SFT star algebra, which satisfies $I \star A = A$ for a very general class of string fields A, seems to be an exact state,

$$|I\rangle = \tilde{Q}|\Lambda\rangle. \tag{118}$$

If indeed the identity is exact, then it follows immediately that the cohomology of $\tilde{Q}$ is empty, since $\tilde{Q}A = 0$ then implies that

$$\begin{aligned} A &= (\tilde{Q}\Lambda) \star A \\ &= \tilde{Q}(\Lambda \star A) - \Lambda \star \tilde{Q}A \\ &= \tilde{Q}(\Lambda \star A). \end{aligned} \tag{119}$$

So to prove that the cohomology of $\tilde{Q}$ is trivial, it suffices to show that $\tilde{Q}|\Lambda\rangle = |I\rangle$. While there are some subtleties involved with the identity string field, Ellwood *et al.* found a very elegant expression for this field,

$$|I\rangle = \left(\cdots e^{\frac{1}{8}L_{-16}} e^{\frac{1}{4}L_{-8}} e^{\frac{1}{2}L_{-4}} \right) e^{L_{-2}}|0\rangle. \tag{120}$$

(Recall that $|0\rangle = b_{-1}|0_1\rangle$.) They then looked numerically for a state $|\Lambda\rangle$ satisfying (118). For example, truncating at level $L = 3$,

$$\begin{aligned} |I\rangle &= |0\rangle + L_{-2}|0\rangle + \cdots \\ &= |0\rangle - b_{-3}c_1|0\rangle - 2b_{-2}c_0|0\rangle + \frac{1}{2}(\alpha_{-1} \cdot \alpha_{-1})|0\rangle + \cdots \end{aligned} \tag{121}$$

while the only candidate for $|\Lambda\rangle$ is

$$|\Lambda\rangle = \alpha\, b_{-2}|0\rangle, \tag{122}$$

for some constant α. The authors of [62] showed that the state (121) is best approximated as exact when $\alpha \sim 1.12$; for this value, their measure of exactness becomes

$$\frac{\left| \tilde{Q}|\Lambda\rangle - |I\rangle \right|}{|I|} \to 0.17, \tag{123}$$

which the authors interpreted as a 17% deviation from exactness. Generalizing this analysis to higher levels, they found at levels 5, 7, and 9, a deviation from exactness of 11%, 4.5% and 3.5% respectively. At level 9, for example, the identity field has 118 components, and there are only 43 gauge parameters, so this is a highly nontrivial check on the exactness of the identity. Like the results of [61], these results strongly support the conclusion that the cohomology of the theory is trivial in the stable vacuum. In this case, the result applies to fields of all spins and all ghost numbers.

Given that the Witten string field theory seems to have a classical solution with no perturbative open string excitations, in accordance with Sen's conjectures, it is quite interesting to ask what the physics of the vacuum string field theory (113) should describe. One natural assumption might be that this theory should include closed string states in its quantum spectrum. We discuss this question again briefly in the final section.

7.   Further developments

In this section we review briefly some further developments which we do not have time to explore in great detail in these lectures. In Subsection 7.1 we discuss the pure ghost BRST operator Ansatz of RSZ (Rastelli, Sen, and Zwiebach) for vacuum string field theory. In Subsection 7.2 we discuss "sliver" states and related states; these states are projectors in the SFT star algebra, and are closely related to D-branes in the RSZ VSFT model. These topics will be discussed in further detail in [4]

7.1   The vacuum string field theory model of RSZ

In [60], Rastelli, Sen, and Zwiebach made an intriguing proposal regarding the form of Witten's string field theory around the stable tachyon vacuum. Since the exact form of the BRST operator $\tilde{Q}$ given by (114) is not known analytically, and is difficult to work with numerically, these authors suggested that it might be possible to "guess" an appropriate form for this operator (after suitable field redefinition), using the properties expected of the BRST operator in any vacuum. They suggested a simple class of BRST operators $\hat{Q}$ which satisfy the properties (a-c) described in Section 4.2 (actually, they impose the slightly weaker but still sufficient condition $\int (\hat{Q}\Psi \star \Phi + (-1)^{G_\Psi}\Psi \star \hat{Q}\Phi)$ instead of condition (b)). In particular, they propose that after a field redefinition, the BRST operator of the string field theory in the classically stable vacuum should be an operator $\hat{Q}$ expressable purely in terms of ghost operators. For example, the simplest operator in the class they suggest is $\hat{Q} = c_0$, which clearly satisfies $\hat{Q}^2$, and which also satisfies condition (c) and the weaker form of condition (b) mentioned above.

The RSZ model of vacuum string field theory has a number of attractive features.

- This model satisfies all the axioms of string field theory, and has a BRST operator with vanishing cohomology.

- In the RSZ model, the equation of motion factorizes into the usual equation of motion

$$\hat{Q}\Psi_{\text{ghost}} + g\Psi_{\text{ghost}} \star \Psi_{\text{ghost}} = 0 \tag{124}$$

for the ghost part of the field, and a projection equation

$$\Psi_{\text{matter}} = \Psi_{\text{matter}} \star \Psi_{\text{matter}} \tag{125}$$

for the matter part of the field, where the full string field is given by

$$\Psi = \Psi_{\text{ghost}} \otimes \Psi_{\text{matter}}. \tag{126}$$

Thus, finding a solution of the equation of motion reduces to the problem of solving the equation of motion in the ghost sector and identifying projection operators in the string field star algebra. It was also recently shown [63, 64, 65] that by taking the BRST operator $\hat{Q}$ to be given by a ghost insertion localized at the string midpoint, the ghost equation also has essentially the form of the projection equation. Thus, this seems to be a very natural choice for the BRST operator of the RSZ model.

- A number of projection operators have been identified in the string field star algebra. These projection operators have many of the properties desired of D-branes. We will briefly review some aspects of these projection operators in the next subsection.

- Given the projection operators just mentioned, the ratio of tensions between D-branes of different dimensionality can be computed and has the correct value [66][8].

Despite the successes of the RSZ model, there are some difficult technical aspects of this picture. First, it seems very difficult to actually prove that this model is related to the VSFT around the stable vacuum in the Witten model, not least because we lack any analytic control over the Witten theory. Second, the RSZ model seems to have a somewhat singular structure in several respects. Formally, the action on any well-behaved Fock space state satisfying the equation of motion will vanish [69]. Further, the natural solutions of the projection equation corresponding to the matter sector of the equation of motion have rather singular properties [70]. Some of these singular properties are related to the fact that some of the physics in the RSZ model seems to have been "pushed" into the midpoint of the string. In the Witten model, the condition

---

[8]This result was known at the time of the lectures. There was quite a bit of recent work on the problem of computing the exact D-brane tension [67]. A very nice recent paper by Okawa [68] resolved the question and demonstrated that not only the ratio of tensions, but also the tension of an individual brane, is correctly reproduced in the RSZ VSFT theory when singularities are correctly controlled.

that, for example, $Q^2 = 0$ involves a fairly subtle anomaly cancellation between the matter and ghost sectors at the midpoint. In the RSZ model, the matter and ghost sectors are essentially decoupled, so that the theory seems to have separate singularities in each sector, which cancel when the sectors are combined. These are all indications of a theory with problematic singularities. While the Witten theory seems to be free of singularities of this type, it remains to be seen whether resolving the singularities of the RSZ model or finding an analytic approach to the Witten theory will be a more difficult problem to solve.

7.2 Projection operators in SFT

From the point of view of the RSZ model of VSFT just discussed, projection operators in the matter sector of the star algebra play a crucial role in constructing solutions of the equations of motion. Such projection operators may also be useful in understanding solutions in the original Witten theory. Quite a bit of work has been done on constructing and analyzing projectors in the star algebra since the RSZ model was originally proposed. Without going into the technical details, we now briefly review some of the important features of matter projectors.

The first matter projector which was explicitly constructed is the "sliver" state. This state was identified in conformal field theory in [71], and then constructed explicitly using matter oscillators in [72]. The sliver state takes the form of a squeezed state

$$\exp \left[\frac{1}{2} a^\dagger \cdot S \cdot a^\dagger \right] |0\rangle . \tag{127}$$

By requiring that such a state satisfy the projection equation $\Psi \star \Psi = \Psi$, and by making some further assumptions about the nature of the state, an explicit formula for the matrix S was found in [72].

Projectors like the sliver have many properties which are reminiscent of D-branes. This relationship between projection operators and D-branes is familiar from noncommutative field theory, where projectors also play the role of D-brane solitons [73] (for a review of noncommutative field theory, see [14]). In the RSZ model, by tensoring an arbitrary matter projector with a fixed ghost state satisfying the ghost equation of motion (124), states corresponding to an arbitrary configuration of D-branes can be constructed. Particular projectors like the sliver can be constructed which are localized in any number of space-time dimensions, corresponding to the codimension of a D-brane. Under gauge transformations, a rank 1 projector can be rotated into an orthogonal rank 1 projector, so that configurations containing multiple branes can be constructed as higher rank projectors formed from the sum of orthogonal rank one projectors [75, 76]. This gives a very suggestive picture of how arbitrary D-brane configurations can be constructed in string field theory. While this picture is quite compelling, however, there are a number of technical obstacles which make this still a somewhat incomplete story. As mentioned above, in the RSZ model, many singularities appear due to the separation of the matter and ghost sectors. In the context of the matter projectors, these singularities manifest as singular properties of the projectors. For example, the sliver state described above has a matrix S which has eigenvalues of ± 1 for any dimension of D-brane [70, 77]. Such eigenvalues cause the state to be nonnormalizable elements of the matter Fock space. In the Dirichlet directions, this lack of normalizability occurs because the state is essentially localized to a point and is analogous to a delta function. In the Neumann directions, the singularity manifests as a "breaking" of the strings composing the D-brane, so that the functional describing the projector state is a product of a function of the string configurations on the left and right halves of the string, with no connection mediated through the midpoint. These geometric singularities seem to be generic features of matter projectors, not just of the sliver state [78, 77]. These singular geometric features are one of the things which makes direct calculation in the RSZ model somewhat complicated, as all these singularities must be sensibly regulated. These singularities do not seem to appear in the Witten theory, where the BRST operator and numerically calculated solutions seem to behave smoothly at the string midpoint. On the other hand, it may be that further study of the matter projection operators and their cousins in the ghost sector which satisfy (124) will lead to analytic progress on the Witten theory.

8.   Conclusions and open problems

The work described in these lectures has brought our understanding of string field theory to a new level. We now have fairly conclusive evidence that open string field theory can successfully describe distinct vacua with very different geometrical properties, which are not related to one another through a marginal deformation. The resulting picture, in which a complicated set of degrees of freedom defined primarily through an algebraic structure, can produce different geometrical backgrounds at different solutions of the equations of motion, represents an important step beyond perturbative string theory. Such an approach, where different backgrounds with different low-energy degrees of freedom arise from a single underlying formalism, is clearly necessary to discuss questions of a cosmological nature in string theory. It is clearly essential, however, to generalize from the work described here in which the theory describes distinct *open* string backgrounds, to a formalism where different *closed* string backgrounds appear as solutions to an equation of motion for a single set of degrees of freedom.

Clearly, it is an important goal to have a formulation of string/M-theory in which all the currently understood vacua can arise in terms of a single well-defined set of degrees of freedom. It is not yet clear, however, how far it is possible go towards this goal using the current formulations of string field theory. It may be that the correct lesson to take from the work described here is simply that there *are* nonperturbative formulations in which distinct vacua can be brought together as solutions of a single classical theory, and that one should search for some deeper fundamental algebraic formulation where geometry, and even the dimension of space-time emerge from the fundamental degrees of freedom in the same way that D-brane geometry emerges from the degrees of freedom of Witten's open string field theory. A more conservative scenario, however, might be that we could perhaps use the current framework of string field theory, or some limited refinement thereof, to achieve this goal of providing a universal nonperturbative definition of string theory and M-theory. Following this latter scenario, we propose here a series of questions aimed at continuing the recent developments in open string field theory as far as possible towards this ultimate goal. It is not certain that this research program can be carried to its conclusion, but it will be very interesting to see how far open string field theory can go in reproducing important nonperturbative aspects of string theory.

Some open problems:

1) The first important unsolved problem in this area is to find an analytic description of the tachyonic vacuum. Despite several years of work on this problem, great success with numerical approximations, and some insight from the RSZ vacuum string field theory model, we still have no good analytic understanding of the difference between the D-brane vacuum and the empty vacuum in Witten's open cubic string field theory. It seems almost unbelievable that there is not some elegant analytic solution to this problem. Finding such an analytic solution would almost certainly greatly enhance our understanding of this theory, and would probably lead to other significant advances.

2) Another interesting and important unsolved problem is to find, either analytically or numerically, a solution of the Witten theory describing *two* D25-branes. If open string field theory is truly a background-independent theory, at least in the open string sense, it should be just as feasible to go from a vacuum with one D-brane to a vacuum with two D-branes as it is to go from a vacuum with one D-brane to the empty vacuum (or from the vacuum with two D-branes to the vacuum with one D-brane, which is essentially the same problem as going from one to none). Despite some work on this problem [79], there is as yet no evidence that a double D-brane solution exists for the Witten theory on a single D-brane. Several approaches which have been tried (and will be described in more detail in [79]) include: *i*) following a positive mass field upward, looking for a stable point; this method seems to fail because of

gauge-fixing problems—the effective potential often develops a singularity before reaching the energy $+T_{25}$, *ii*) following the intuition of the RSZ model and constructing a gauge transform of the original D-brane solution which is $\star$—orthogonal to the original D-brane vacuum. It can be shown formally that such a state, when added to the original D-brane vacuum gives a new solution with the correct energy for a double D-brane; unfortunately, however, we have been unable to identify such a state numerically in level truncation.

There are several other problems closely related to the double D-brane problem. One related problem is the problem of studying a D0-brane lump solution from the tachyon field on a D1-brane wrapped on a small circle. When the circle is sufficiently small, the mass of the D0-brane is larger than that of the wrapped D1-brane. In this case, it seems much more difficult to construct the D0-brane lump solution than it is when the D0-brane has mass smaller than the D1-brane [80]. Another possibly related problem is the problem of translating a single D-brane of less than maximal dimension in a transverse direction. It was shown by Sen and Zwiebach [81] (in a T-dual picture) that after moving a D-brane a finite distance of order of the string length in a transverse direction, the level-truncated string field theory equations develop a singularity. Thus, in level truncation it does not seem possible to move a D-brane a macroscopic distance in a transverse direction[9]. In this case, a toy model [83] suggests that the problem is that the infinitesimal marginal parameter for the brane translation ceases to parameterize the marginal trajectory in field space after a finite distance, just as the coordinate x ceases to parameterize the circle $x^2 + y^2 = 1$ near $x = 1$. This is similar in spirit to the breakdown of Feynman-Siegel gauge along the tachyon potential discussed in section 6.1.

To show that open string field theory is sufficiently general to address arbitrary questions involving different vacua, it is clearly necessary to show that the formalism is powerful enough to describe multiple brane vacua, the D0-brane lump on an arbitrary radius circle, and translated brane vacua. It is currently unclear whether the obstacles to finding these vacua are conceptual or technical. It may be that the level-truncation approach is not well-suited to finding these vacua. If this is true, however, we may need a clearer mathematical formalism for describing the theory. There is currently some ambiguity in the definition of the theory, in terms of precisely which states are allowed in the string field. Level-truncation

---

[9]although this can be done formally [82], it is unclear how the formal solution relates to an explicit expression in the oscillator language

in some sense gives a regularization of, and a concrete definition to, the theory. Without level truncation, we would need some more definitive mathematical tools for analyzing various features of the theory, such as the other vacua mentioned here.

3) Another open question involves the role that closed strings play in open string field theory. As has been known since the earliest days of the subject, closed strings appear as poles in perturbative open string scattering amplitudes. This was shown explicitly for Witten's SFT in [48], where it was shown that closed string poles arise in the one-loop 2-point function. If Witten's theory is well-defined as a quantum theory, it would follow from unitarity that the closed string states should also arise in some natural sense as asymptotic states of the quantum open string field theory. It is currently rather unclear, however, whether, and if so how, this might be realized. There are subtleties in the quantum formulation of the theory which have never completely been resolved [35]. Both older SFT literature [84, 85] and recent work [86, 70, 64, 87, 88] have suggested ways in which closed strings might be incorporated into the open string field theory formalism, but a definitive resolution of this question is still not available. If it is indeed possible to encode closed string degrees of freedom in some way in the quantum open string field theory, it suggests that one could use the Witten formalism in principle to not only compute general closed string scattering amplitudes, but perhaps even to address questions of closed string vacua. This is clearly an optimistic scenario, but one can imagine that the open string theory might really contain all of closed string physics as well as open string physics. This scenario is perhaps not so farfetched, as it really represents simply a lifting to the level of string field theory of the AdS/CFT story, where the massive as well as the massless modes are included. Furthermore, the fact that, as discussed in Section 5.5, the open string diagrams precisely cover the moduli space of Riemann surfaces with an arbitrary number of handles (and at least one boundary), suggests that by shrinking the boundaries to closed strings, one might neatly describe all perturbative closed string amplitudes in the open string language. On the other hand, it seems quite possible that the closed string sector of the theory is encoded in a singular fashion (like the encoding of the D-brane in the RSZ VSFT model), so that extracting the closed string physics from the open string field theory may involve such complicated manipulations that one is better off directly working with a closed string formalism. It would certainly be nice to have a clearer picture of how far one can go in this direction purely from the open string point of view.

4) Another obvious, but crucial, question is how this whole story can be generalized to superstrings. The naive Witten cubic superstring field theory has technical problems arising from contact terms between picture-changing operators [89, 90]. It has been suggested that these problems can be resolved directly in the cubic theory [43]. Berkovits has also suggested a new non-polynomial string field theory framework which seems to deal successfully with the contact term problem, at least in the NS-NS sector [44]. Some preliminary work indicates that numerical calculations on the tachyon condensation problem for the open superstring can be carried out in the Berkovits model with analogous results to those described here for the bosonic open string, although the results to date for the superstring are much more limited [91]. It would be nice to have a more complete picture for the superstring, and some sense of how issues like the closed string question would work in the supersymmetric framework.

5) Perhaps the most important lesson we have learned from the body of work discussed in these lectures is that open string field theory is a consistent framework in which geometrically distinct open string backgrounds can arise as classical solutions of a single theory. A fundamental outstanding problem in string theory is to find a framework in which different closed string backgrounds arise in a similar fashion from some fixed set of degrees of freedom within a single well-defined theory. In principle, we would hope that all the different closed string backgrounds would arise as solutions of the equations of motion for the fundamental underlying degrees of freedom of string field theory, either by incorporating closed strings into the open string field theory framework as described above, or by working directly in some formulation of closed string field theory. It is quite challenging to imagine a single set of degrees of freedom which would encode, in different phases, all the possible string backgrounds we are familiar with. A particularly pressing case is that of M-theory. In principle, a nonperturbative background-independent formulation of type II string theory should allow one to take the string coupling to infinity in such a way that the fundamental degrees of freedom of the theory are still actually at some finite point in their configuration space in the limit. This would lead to the vacuum associated with M-theory in flat space-time. It would be quite remarkable if this can be achieved in the framework of string field theory. Given the nontrivial relationship between string fields and low-energy effective degrees of freedom, however, such a result cannot be ruled out. If this picture could be successfully implemented, it would give a very satisfying understanding of how the complicated network of dualities of string and M-theory could be represented in terms of a single underlying set of degrees of freedom.

Acknowledgments

I would like to thank CECS and the organizers of the School on Quantum Gravity for their support and hospitality, and for an extremely enjoyable summer school experience. Thanks to Erasmo Coletti, Ian Ellwood, David Gross, Nicolas Moeller, Joe Minahan, Greg Moore, Leonardo Rastelli, Martin Schnabl, Ashoke Sen, Jessie Shelton, Ilya Sigalov, and Barton Zwiebach, for many discussions and collaborations which provided the material for these lectures. Thanks also to TASI '01, where some of this material was presented prior to this School. Particular thanks to Barton Zwiebach for suggestions and contributions to these lecture notes, which have substantial overlap with a more extensive set of lecture notes based on lectures by Zwiebach and myself at TASI '01, which will appear presently [4]. This work was supported by the DOE through contract #DE-FC02-94ER40818.

References

[1] K. Ohmori, "A review on tachyon condensation in open string field theories," hep-th/0102085.

[2] P. J. De Smet, "Tachyon condensation: Calculations in string field theory," hep-th/0109182.

[3] I. Y. Aref'eva, D. M. Belov, A. A. Giryavets, A. S. Koshelev and P. B. Medvedev, "Noncommutative field theories and (super)string field theories," hep-th/0111208.

[4] W. Taylor and B. Zwiebach, "TASI '01 lectures on D-branes, tachyons, and string field theory," to appear.

[5] Hull, C. M., and P. K. Townsend, "Unity of superstring dualities," Nucl. Phys. **B 438**, 109, 1995; hep-th/9410167.

[6] Witten, E. "String Theory Dynamics in Various Dimensions," *Nucl. Phys.* **B 443**; 85 1995; hep-th/9503124.

[7] Townsend, P. K., "The eleven-dimensional supermembrane revisited," Phys. Lett. **B 350**, 184, 1995; hep-th/9501068.

[8] Hořava, P. and E. Witten, "Heterotic and type I string dynamics from eleven dimensions," Nucl. Phys. **B 460**, 506, 1996; hep-th/9510209.

[9] W. Taylor, "M(atrix) theory: matrix quantum mechanics as a fundamental theory," Rev. Mod. Phys. **73**, 419, 2001; hep-th/0101126

[10] Aharony, O., S. S. Gubser, J. Maldacena, H. Ooguri, and Y. Oz, "Large N field theories, string theory and gravity," Phys. Rept. **323**, 183, 2000; hep-th/9905111.

[11] G. Veneziano, "Construction of a crossing-symmetric, Regge-behaved amplitude for linearly rising trajectories," Nuovo Cim. **57A** 190 (1968).

[12] D. J. Gross, A. Neveu, J. Scherk and J. H. Schwarz, "Renormalization and unitarity in the dual resonance model," *Phys. Rev.* **D2** (1970) 697.

[13] C. Lovelace, "Pomeron form factors and dual Regge cuts," *Phys. Lett.* **B34** (1971) 500.

[14] M. R. Douglas and N. A. Nekrasov, "Noncommutative field theory," Rev. Mod. Phys. **73**, 977 (2001), hep-th/0106048.

[15] J. Polchinski, "TASI Lectures on D-branes," hep-th/9611050.

[16] W. Taylor, "Trieste lectures on D-branes, gauge theory and M(atrices)," hep-th/9801182.

[17] M. J. Duff, R. R. Khuri and J. X. Lu, "String solitons," Phys. Rept. **259**, 213 (1995), hep-th/9412184.

[18] E. Witten, "Non-commutative geometry and string field theory," *Nucl. Phys.* **B268** 253, (1986).

[19] A. Sen, "Universality of the tachyon potential," JHEP **9912**, 027 (1999), hep-th/9911116.

[20] J. Polchinski, "Dirichlet-Branes and Ramond-Ramond Charges," *Phys. Rev. Lett.* **75** (1995), 4724, hep-th/9510017.

[21] R. G. Leigh, "Dirac-Born-Infeld action from Dirichlet sigma model," *Mod. Phys. Lett.* **A4** (1989), 2767.

[22] E. Witten, "Bound States of Strings and p-Branes," *Nucl. Phys.* **B460** (1996), 335, hep-th/9510135.

[23] A. A. Tseytlin, "Born-Infeld action, supersymmetry and string theory," hep-th/9908105;

P. Koerber and A. Sevrin, "The non-abelian D-brane effective action through order alpha'**4," JHEP **0210**, 046 (2002), hep-th/0208044;

S. Stieberger and T. R. Taylor, "Non-Abelian Born-Infeld action and type I - heterotic duality. II: Nonrenormalization theorems," Nucl. Phys. B **648**, 3 (2003), hep-th/0209064;

D. T. Grasso, "Higher order contributions to the effective action of N = 4 super Yang-Mills," JHEP **0211**, 012 (2002), hep-th/0210146.

[24] W. Taylor and M. Van Raamsdonk, "Multiple D0-branes in weakly curved backgrounds," *Phys. Lett.* **B558** (1999), 63, hep-th/9904095; "Multiple Dp-branes in weak background fields," *Nucl. Phys.* **B573**, 703 (2000), hep-th/9910052.

[25] R. C. Myers, "Dielectric-branes," *JHEP* **9912** (1999), 022, hep-th/9910053.

[26] Polchinski, J. *String theory* (Cambridge University Press, Cambridge, England, 1998).

[27] W. Taylor, "D-brane Field Theory on Compact Spaces," *Phys. Lett.* **B394** (1997), 283; hep-th/9611042.

[28] T. Banks and L. Susskind, "Brane - Antibrane Forces," hep-th/9511194.

[29] A. Hashimoto and W. Taylor, "Fluctuation Spectra of Tilted and Intersecting D-branes from the Born-Infeld Action," *Nucl. Phys.* **B503** (1997), 193-219; hep-th/9703217.

[30] E. Gava, K. S. Narain and M. H. Sarmadi, "On the bound states of p- and (p+2),-branes," Nucl. Phys. B **504**, 214 (1997), hep-th/9704006.

[31] Green, M. B., J. H. Schwarz, and E. Witten, *Superstring theory* (Cambridge University Press, Cambridge, England, 1987).

[32] K. Bardakci, "Spontaneous symmetry breaking in the standard dual string model," *Nucl. Phys.* **B133** (1978), 297.

[33] V. A. Kostelecky and S. Samuel, "On a nonperturbative vacuum for the open bosonic string," *Nucl. Phys.* **B336** (1990), 263-296.

[34] A. Leclair, M. E. Peskin and C. R. Preitschopf, "String field theory on the conformal plane (I)," *Nucl. Phys.* **B317** (1989), 411-463.

[35] C. Thorn, "String field theory," *Phys. Rep.* **175** (1989), 1.

[36] M. R. Gaberdiel and B. Zwiebach, "Tensor constructions of open string theories 1., 2.," *Nucl. Phys.* **B505** (1997), 569, hep-th/9705038; *Phys. Lett.* **B410** (1997), 151, hep-th/9707051.

[37] D. J. Gross and A. Jevicki, "Operator formulation of interacting string field theory (I), (II)," *Nucl. Phys.* **B283** (1987), 1; *Nucl. Phys.* **B287** (1987), 225.

[38] E. Cremmer, A. Schwimmer and C. Thorn, "The vertex function in Witten's formulation of string field theory" *Phys. Lett.* **B179** 57 (1986).

[39] S. Samuel, "The physical and ghost vertices in Witten's string field theory," *Phys. Lett.* **B181** 255 (1986).

[40] N. Ohta, "Covariant interacting string field theory in the Fock space representation," *Phys. Rev.* **D34** (1986), 3785; *Phys. Rev.* **D35** (1987), 2627 (E).

[41] J. Shelton, to appear.

[42] I. Bars, "MSFT: Moyal star formulation of string field theory," hep-th/0211238.

[43] I. Y. Aref'eva, A. S. Koshelev, D. M. Belov and P. B. Medvedev, "Tachyon condensation in cubic superstring field theory," Nucl. Phys. B **638**, 3 (2002), hep-th/0011117.

G. Bandelloni and S. Lazzarini, "The geometry of W3 algebra: A twofold way for the rebirth of a symmetry," Nucl. Phys. B **594**, 477 (2001), hep-th/0011208.

[44] N. Berkovits, "Super-Poincare invariant superstring field theory," *Nucl. Phys.* **B450** (1995), 90, hep-th/9503099; "Review of open superstring field theory," hep-th/0105230; "The Ramond sector of open superstring field theory," JHEP **0111**, 047 (2001), hep-th/0109100.

[45] S. B. Giddings and E. J. Martinec, "Conformal Geometry and String Field Theory," *Nucl. Phys.* **B278**, 91 (1986).

[46] S. B. Giddings, E. J. Martinec and E. Witten, "Modular Invariance In String Field Theory," Phys. Lett. B **176**, 362 (1986).

[47] B. Zwiebach, "A Proof That Witten's Open String Theory Gives A Single Cover Of Moduli Space," Commun. Math. Phys. **142**, 193 (1991).

[48] D. Z. Freedman, S. B. Giddings, J. A. Shapiro and C. B. Thorn, "The Nonplanar One Loop Amplitude In Witten's String Field Theory," Nucl. Phys. B **298**, 253 (1988).

[49] J. A. Harvey and P. Kraus, "D-branes as unstable lumps in bosonic open string field theory," JHEP **0004**, 012 (2000), hep-th/0002117;

[50] R. de Mello Koch, A. Jevicki, M. Mihailescu and R. Tatar, "Lumps and p-branes in open string field theory," Phys. Lett. B **482**, 249 (2000), hep-th/0003031.

[51] N. Moeller, A. Sen and B. Zwiebach, "D-branes as tachyon lumps in string field theory," JHEP **0008**, 039 (2000), hep-th/0005036.

[52] A. Sen and B. Zwiebach, "Tachyon condensation in string field theory," JHEP **0003**, 002 (2000), hep-th/9912249.

[53] N. Moeller and W. Taylor, "Level truncation and the tachyon in open bosonic string field theory," Nucl. Phys. B **583**, 105 (2000), hep-th/0002237.

[54] D. Gaiotto and L. Rastelli, "Progress in open string field theory," Presentation by L. Rastelli at Strings 2002, Cambridge, England; http://www.damtp.cam.ac.uk/strings02/avt/rastelli/.

[55] W. Taylor, "A Perturbative Analysis of Tachyon Condensation," hep-th/0208149.

[56] W. Taylor, "Perturbative diagrams in string field theory," hep-th/0207132.

[57] D. Gaiotto and L. Rastelli, "Experimental string field theory," hep-th/0211012.

[58] I. Ellwood and W. Taylor, "Gauge invariance and tachyon condensation in open string field theory," hep-th/0105156.

[59] A. A. Gerasimov and S. L. Shatashvili, "On exact tachyon potential in open string field theory," JHEP **0010**, 034 (2000), hep-th/0009103; D. Kutasov, M. Marino and G. Moore, "Some exact results on tachyon condensation in string field theory," JHEP **0010**, 045 (2000), hep-th/0009148; D. Ghoshal and A. Sen, "Normalisation of the background independent open string field JHEP **0011**, 021 (2000), hep-th/0009191; D. Kutasov, M. Marino and G. Moore, D. Kutasov, M. Marino and G. W. Moore, "Remarks on tachyon condensation in superstring field theory,", hep-th/0010108; S. Moriyama and S. Naka-mura, "Descent relation of tachyon condensation from boundary string Phys. Lett. B **506**, 161 (2001), hep-th/0011002; A. A. Gerasimov and S. L. Shatashvili, "Stringy Higgs mechanism and the fate of open strings," JHEP **0101**, 019 (2001), hep-th/0011009; I. Y. Aref'eva, A. S. Koshelev, D. M. Belov and P. B. Medvedev, "Tachyon condensation in cubic superstring field theory," Nucl. Phys. B **638**, 3 (2002), hep-th/0011117; P. Kraus and F. Larsen, "Boundary string field theory of the D D-bar system," Phys. Rev. D **63**, 106004 (2001), hep-th/0012198; M. Alishahiha, H. Ita and Y. Oz, "On superconnec-tions and the tachyon effective action," Phys. Lett. B **503**, 181 (2001), hep-th/0012222; G. Chalmers, "Open string decoupling and tachyon condensation," JHEP **0106**, 012 (2001), hep-th/0103056; M. Marino, "On the BV formulation of boundary superstring field theory," JHEP **0106**, 059 (2001), hep-th/0103089; V. Niarchos and N. Prezas, "Boundary superstring field theory," Nucl. Phys. B **619**, 51 (2001), hep-th/0103102; K. S. Viswanathan and Y. Yang, "Tachyon condensation and background independent su-perstring field theory," Phys. Rev. D **64**, 106007 (2001), hep-th/0104099; M. Alishahiha, "One-loop correction of the tachyon action in boundary superstring field theory," Phys. Lett. B **510**, 285 (2001), hep-th/0104164.

[60] L. Rastelli, A. Sen and B. Zwiebach, "String field theory around the tachyon vacuum," Adv. Theor. Math. Phys. **5**, 353 (2002), hep-th/0012251.

[61] I. Ellwood and W. Taylor, "Open string field theory without open strings," Phys. Lett. B **512**, 181 (2001), hep-th/0103085.

[62] I. Ellwood, B. Feng, Y. H. He and N. Moeller, "The identity string field and the tachyon vacuum," JHEP **0107**, 016 (2001), hep-th/0105024.

[63] H. Hata and T. Kawano, "Open string states around a classical solution in vacuum string field theory," JHEP **0111**, 038 (2001), hep-th/0108150.

[64] D. Gaiotto, L. Rastelli, A. Sen and B. Zwiebach, "Ghost structure and closed strings in vacuum string field theory," hep-th/0111129.

[65] K. Okuyama, "Ghost kinetic operator of vacuum string field theory," JHEP **0201**, 027 (2002), hep-th/0201015.

[66] L. Rastelli, A. Sen and B. Zwiebach, "Classical solutions in string field theory around the tachyon vacuum," Adv. Theor. Math. Phys. **5**, 393 (2002), hep-th/0102112.

[67] H. Hata and S. Moriyama, "Observables as twist anomaly in vacuum string field theory," JHEP **0201**, 042 (2002), hep-th/0111034.

L. Rastelli, A. Sen and B. Zwiebach, "A note on a proposal for the tachyon state in vacuum string field theory," JHEP **0202**, 034 (2002), hep-th/0111153.

H. Hata, S. Moriyama and S. Teraguchi, "Exact results on twist anomaly," JHEP **0202**, 036 (2002), hep-th/0201177.

[68] Y. Okawa, "Open string states and D-brane tension from vacuum string field theory," JHEP **0207**, 003 (2002), hep-th/0204012.

[69] D. J. Gross and W. Taylor, "Split string field theory II," JHEP **0108**, 010 (2001), hep-th/0106036.

[70] G. Moore and W. Taylor, "The singular geometry of the sliver," JHEP **0201**, 004 (2002), hep-th/0111069.

[71] L. Rastelli and B. Zwiebach, "Tachyon potentials, star products and universality," JHEP **0109**, 038 (2001), hep-th/0006240.

[72] V. A. Kostelecky and R. Potting, "Analytical construction of a nonperturbative vacuum for the open bosonic string," Phys. Rev. D **63**, 046007 (2001), hep-th/0008252.

[73] R. Gopakumar, S. Minwalla and A. Strominger, "Noncommutative solitons," JHEP **0005**, 020 (2000), hep-th/0003160.

[74] M. R. Douglas and N. A. Nekrasov, "Noncommutative field theory," Rev. Mod. Phys. **73**, 977 (2001), hep-th/0106048.

[75] L. Rastelli, A. Sen and B. Zwiebach, "Half strings, projectors, and multiple D-branes in vacuum string field theory," JHEP **0111**, 035 (2001), hep-th/0105058.

[76] D. J. Gross and W. Taylor, "Split string field theory I," JHEP **0108**, 009 (2001), hep-th/0105059.

[77] D. Gaiotto, L. Rastelli, A. Sen and B. Zwiebach, "Star algebra projectors," JHEP **0204**, 060 (2002), hep-th/0202151.

[78] M. Schnabl, "Wedge states in string field theory," JHEP **0301**, 004 (2003), hep-th/0201095; "Anomalous reparametrizations and butterfly states in string field theory," Nucl. Phys. B **649**, 101 (2003), ep-th/0202139.

[79] I. Ellwood and W. Taylor, to appear.

[80] N. Moeller, A. Sen and B. Zwiebach, "D-branes as tachyon lumps in string field theory," JHEP **0008**, 039 (2000), ep-th/0005036.

[81] A. Sen and B. Zwiebach, "Large marginal deformations in string field theory," JHEP **0010**, 009 (2000), hep-th/0007153.

[82] J. Kluson, "Marginal deformations in the open bosonic string field theory for N D0-branes," hep-th/0203089; "Exact solutions in open bosonic string field theory and marginal deformation in CFT," hep-th/0209255.

[83] B. Zwiebach, "A solvable toy model for tachyon condensation in string field theory," JHEP **0009**, 028 (2000), hep-th/0008227.

[84] A. Strominger, "Closed strings in open string field theory," *Phys. Rev. Lett.* **58** 629 (1987).

[85] J. A. Shapiro and C. B. Thorn, "BRST invariant transitions between open and closed strings," *Phys. Rev.* **D36** 432 (1987); "closed string-open string transitions in Witten's string field theory," Phys. Lett. B **194**, 43 (1987).

[86] A. A. Gerasimov and S. L. Shatashvili, "Stringy Higgs mechanism and the fate of open strings," JHEP **0101**, 019 (2001), hep-th/0011009.

S. L. Shatashvili, "On field theory of open strings, tachyon condensation and closed strings," hep-th/0105076.

[87] A. Hashimoto and N. Itzhaki, "Observables of string field theory," JHEP **0201**, 028 (2002), hep-th/0111092.

[88] M. Alishahiha and M. R. Garousi, "Gauge invariant operators and closed string scattering in open string field theory," Phys. Lett. B **536**, 129 (2002), hep-th/0201249.

[89] C. Wendt, "Scattering amplitudes and contact interactions in Witten's superstring field theory," *Nucl. Phys.* **B314** (1989) 209.

[90] J. Greensite and F. R. Klinkhamer, "Superstring amplitudes and contact interactions," *Phys. Lett.* **B304** (1988) 108.

[91] N. Berkovits, A. Sen and B. Zwiebach, "Tachyon condensation in superstring field theory," Nucl. Phys. B **587**, 147 (2000), hep-th/0002211;

P. De Smet and J. Raeymaekers, "Level four approximation to the tachyon potential in superstring JHEP **0005**, 051 (2000), hep-th/0003220;

A. Iqbal and A. Naqvi, "Tachyon condensation on a non-BPS D-brane,", hep-th/0004015;

P. De Smet and J. Raeymaekers, "The tachyon potential in Witten's superstring field theory," JHEP **0008**, 020 (2000), hep-th/0004112;

BILLIARD DYNAMICS OF EINSTEIN-MATTER SYSTEMS NEAR A SPACELIKE SINGULARITY

Thibault Damour
Institut des Hautes Etudes Scientifiques, 35, Route de Chartres
F-91440 Bures-sur-Yvette
France

Marc Henneaux
Physique Théorique et Mathématique, Université Libre de Bruxelles
C.P. 231, B-1050, Bruxelles
Belgium
and
Centro de Estudios Científicos, Valdivia,
Chile

Hermann Nicolai
Max-Planck-Institut für Gravitationsphysik, Albert-Einstein-Institut
Am Mühlenberg 1, D-14476 Golm
Germany

Abstract   It is shown in detail that the dynamics in the vicinity of a spacelike singularity of the D-dimentional Einstein-dilaton-p-form system can be described, at each spatial point, as a billiard motion in a region of hyperbolic space. This is done within the Hamiltonian formalism. A key rôle is played in the derivation by the Iwasawa decomposition of the spatial metric. We also comment on the strong coupling/small tension limit of the theory.

1.   Introduction

1.1   BKL analysis in four and higher spacetime dimensions

The non-linearities of the Einstein equations are notably known to prevent the construction of an exact, general solution. Only peculiar solutions, corresponding to idealized situations, have been explicitly derived. From this perspective, the work of Belinskii, Khalatnikov and Lifshitz [BKL70, BKL82] – known as "BKL" –, which describes the asymptotic, general behaviour of the gravitational field in four spacetime dimensions as one approaches a spacelike singularity, appears to be quite remarkable. As argued by these authors, the spatial points decouple in this regime, in the sense that the dynamical evolution of the spatial metric at each spatial point is asymptotically governed by a set of ordinary differential equations with respect to time involving only the metric and its time derivatives at that point. These differential equations are the same as those that arise in some spatially homogeneous cosmological models, which provide therefore invaluable insight into the qualitative features of the general solution.

In the vacuum case, the spatially homogeneous models that capture the behaviour of the general solution are of "Bianchi types" IX or VIII (homogeneity groups equal to $SU(2)$ or $SL(2, R)$). The asymptotic evolution of the metric can then be pictured as an infinite sequence of oscillations of the scale factors along independent spatial directions [BKL70, BKL82]. This regime is called "oscillatory" or "of mixmaster type" [M69a] and exhibits strong chaotic features [LLK71, CB83]. The coupling to matter fields does not change the picture, except if one includes a massless scalar field (equivalent to a perfect fluid with "stiff" equation of state $p = \rho$), in which case the relevant homogeneity group is simply the abelian group of translations in R^3 ("Bianchi I" model). The chaotic evolution is replaced by a monotone expansion or contraction of the scale factors [BK73], mimicking at each spatial point the Kasner solution. This behaviour is called "Kasner-like".

It is natural to investigate whether the BKL analysis remains valid in higher dimensions. The study of this question was started some time ago in the context of pure gravity (with no symmetry assumption) in [DHS85, DHHST86], where it was shown that the general BKL ideas remain valid: spatial points decouple as one approaches a spacelike singularity, i.e., the dynamical evolution at each spatial point of the scale factors is again governed by ordinary differential equations. However, while the general behaviour of solutions of the vacuum Einstein equations remains oscillatory for spacetime dimensions $D \leq 10$, it ceases to be so for spacetime dimensions $D \geq 11$, where it becomes Kasner-like. Furthermore, just as in four spacetime dimensions, the coupling to a massless scalar field suppresses the chaotic behaviour in any number of spacetime dimensions (when it is present), and makes the solution monotone.

The authors of [DHS85] did not consider the inclusion of massless p-forms, which are part of the low-energy bosonic sector of superstring/M-theory models. This task was undertaken in [DH00a, DH00b], with the finding that these p-forms play a crucial role and can reinstate chaos when it is otherwise suppressed. In particular, even though pure gravity is non chaotic in $D = 11$ spacetime dimensions, the 3-form of 11-dimensional supergravity renders the system chaotic. Similarly, the bosonic sectors of all ten-dimensional supergravities related to string models define chaotic dynamical systems, thanks again to the p-forms.

The best way to grasp the asymptotic behaviour of the fields as one approaches a spacelike singularity is based on the qualitative Hamiltonian methods initiated by Misner [M69b]. These naturally lead, in four dimensions, to a billiard description of the asymptotic evolution, in which the spatial scale factors define a geodesic motion in a region of the Lobachevsky plane H_2, interrupted by geometric reflections against the walls bounding this region [C72, M94]. Chaos follows from the fact that the billiard has finite volume[1].

It turns out that this useful billiard description is quite general and can be extended to higher spacetime dimensions, with p-forms and dilaton. If $d = D - 1$ is the number of spatial dimensions, the billiard is a region of hyperbolic space H_d in the presence of the dilaton, or H_{d-1} if no dilaton is present. [If there are k dilatons, the dimension of the relevant hyperbolic space is $d+k-1$.] Besides the dimension of the hyperbolic billiard, the other ingredients that enter its definition are the walls that bound it. These walls can be of different types [DH00a, DH01]: symmetry walls related to the off-diagonal components of the spatial metric, gravitational walls related to the spatial curvature, and p-form walls (electric and magnetic) arising from the p-form energy-density. All these walls are hyperplanes and the billiard is a convex polyhedron with finitely many vertices, some of which are at infinity.

The fact that the asymptotic dynamics admits a billiard description was announced in [DH00a, DH01], where it was derived using the somewhat heuristic arguments standard in BKL-type discussions. The purpose of this paper is to provide a more complete derivation of the billiard picture, in the general (inhomogeneous) D-dimensional situation, with dilaton and p-form gauge fields. For that purpose, we shall rely on the Iwasawa decomposition (see e.g. [H78]) of the spatial metric. This provides an efficient derivation of the symmetry walls in any number of spacetime dimensions, which we do by working out explicitly the Hamiltonian that governs the dynamics in the "BKL" or "small volume" limit.

1.2 Organization of the paper

After fixing the conventions, we discuss in the next two sections the Kasner (= invariant under spatial translations) solution in D spacetime dimensions, with a dilaton. This solution plays a particularly important rôle because it turns out to describe the free motion between collisions. First, we consider diagonal

---

[1]Throughout this paper, the word *billiard* used as a noun in the singular will denote the dynamical system consisting of a ball moving freely on a "table" (region in some Riemannian space), with elastic bounces against the edges. *Billiard* will also sometimes mean the table itself.

metrics and develop the geometry of the scale factors (section 2). Next, we do not assume that the spatial metric is diagonal and show, using the Iwasawa decomposition of the spatial metric, that the off-diagonal components induce "symmetry walls" in the BKL limit (section 3). In fact, many of the billiard ideas can already be introduced in this simple context.

In section 4, we derive the billiard picture in full generality, without imposing any homogeneity conditions on the metric and the matter fields. We show how the gravitational and p-form walls emerge (in addition to the symmetry walls) near a spacelike singularity when one takes the BKL limit. We also prove that Yang-Mills or Chern-Simons couplings do not affect the billiard picture.

In section 5, we relate the BKL limit to other limits that have been considered in the past ("strong coupling limit" [I76], "zero velocity of light limit" [T78]). We also discuss the connection with the so-called "velocity-dominated" behavior [E72]. The last section (section 6) provides a brief conclusion as well as a short survey of some related developments (emergence of Kac-Moody algebras for specified models, lists of chaotic and non-chaotic models).

Four appendices illustrate and discuss further some aspects of the problem. Appendix A analyzes the Iwasawa decomposition and the asymptotics of the off-diagonal Kasner metric in the case of three spatial dimensions. Appendices B and C shed further light on the freezing off of the off-diagonal components and of the electromagnetic variables (in a sense made precise in the text) as one goes to the singularity. Finally, appendix D is devoted to the Hamiltonian reduction of the system in the general case.

We should stress that our analysis is purely classical and is accordingly, as it stands, valid only up to the Planck scale. We also drop the fermionic fields. It is reasonable to expect, however, that some of the ideas discussed here will remain relevant in a more general context, at least qualitatively.

The subject of Hamiltonian cosmology has a long history in the context of four-dimensional, spatially homogeneous spacetimes and provides useful insight on the general discussion presented here. Excellent reviews on this subject, with an extensive bibliography, are [R72b, RS75, J01].

1.3   Models and Conventions

We consider models of the general form

$$S[g_{\mu\nu}, \phi, A^{(p)}] = \int d^D x \sqrt{-{}^{(D)}g} \left[R - \partial_\mu \phi \partial^\mu \phi \right.$$

$$\left. -\frac{1}{2} \sum_p \frac{1}{(p+1)!} e^{\lambda^{(p)}\phi} F^{(p)}_{\mu_1 \cdots \mu_{p+1}} F^{(p)\,\mu_1 \cdots \mu_{p+1}} \right] + \text{"more"} \quad (1)$$

where we have chosen units such that $16\pi G = 1$. The spacetime dimension is left unspecified. We work (as a convenient common formulation) in the Einstein conformal frame, and we normalize the kinetic term of the "dilaton" ϕ with a weight 1 with respect to the Ricci scalar. The Einstein metric $g_{\mu\nu}$ has Lorentz signature $(-, +, \cdots, +)$ and is used to lower or raise the indices. Its determinant is $^{(D)}g$, where the index D is used to avoid any confusion with the determinant of the spatial metric introduced below. The integer $p \geq 0$ labels the various p-forms $A^{(p)}$ present in the theory, with field strengths $F^{(p)}$ equal to $dA^{(p)}$,

$$F^{(p)}_{\mu_1 \cdots \mu_{p+1}} = \partial_{\mu_1} A^{(p)}_{\mu_2 \cdots \mu_{p+1}} \pm p \text{ permutations}. \qquad (2)$$

In fact, the field strength could be modified by additional coupling terms of Yang-Mills or Chapline-Manton type [BRWN82, CM83] (e.g., $F_C = dC^{(2)} - C^{(0)}dB^{(2)}$ for two 2-forms $C^{(2)}$ and $B^{(2)}$ and a 0-form $C^{(0)}$, as it occurs in ten-dimensional type IIB supergravity), but we include these additional contributions to the action in "more". Similarly, "more" might contain Chern-Simons terms, as in the action for eleven-dimensional supergravity [CJS78]. If there are many p-form gauge fields with same form degree p, we use different letters A, B, C, ... to distinguish them, as we just did. The real parameter $\lambda^{(p)}$ measures the strength of the coupling to the dilaton. When $p = 0$, we assume that $\lambda^{(0)} \neq 0$, so that there is only one dilaton. This is done mostly for notational convenience. If there were other dilatons among the 0-forms, these should be separated off from the p-forms because they play a distinct rôle. They would appear as additional scale factors and would increase the dimensions of the relevant hyperbolic billiard (they define additional spacelike directions in the space of the scale factors: see Eq. (13) below, in which $(d\phi)^2$ should be replaced by $\sum_i (d\phi^i)^2$, where i runs over the dilatons).

Our work applies both to past and future singularities. In particular, it applies to Schwarzschild-type of singularities inside black holes. To follow historical usage, we shall assume for definiteness that the spacelike singularity occurs in the past, at $t = 0$ ($t \to 0^+$).

2. Kasner solution – Diagonal case

2.1 Diagonal case

One of the basic ingredients in the asymptotic description of the fields as one goes to the singularity is the Kasner solution (generalized to include the dilaton, in D spacetime dimensions).

The Kasner solution is the general solution of the Einstein-dilaton equations (no p-form) in the spatially homogeneous context, where all the fields depend only on the time coordinate ("Bianchi I" models). Assuming the initial data (spatial metric and extrinsic curvature of initial slice) to be diagonal ("diagonal case"), one easily finds that the metric and the dilaton are given by

$$ds^2 \;=\; -dt^2 + \sum_{i=1}^{d} a_i^2(t)(dx^i)^2, \;\; a_i(t) = t^{p^i} \tag{1}$$

$$\phi \;=\; -p^\phi \ln t + C \tag{2}$$

where p^i, p^ϕ and C are integration constants subject to the conditions

$$\sum_{i=1}^{d} p^i = 1 \tag{3}$$

and

$$\sum_{i=1}^{d} (p^i)^2 - (\sum_{i=1}^{d} p^i)^2 + (p^\phi)^2 = 0. \tag{4}$$

(the minus sign in front of p^ϕ in (2) is included for the sake of uniformity in the formulas below). The second condition is the Hamiltonian constraint $G_{00} - (1/2)T_{00} = 0$. The (p^i, p^ϕ) are known as the Kasner exponents. In the absence of the dilaton, one must simply set p^ϕ (and C) equal to zero in the above expressions. In this latter case, there is at least one Kasner exponent p^i that is negative, so that at least one of the scale factors a_i blows up as $t \to 0$. In contrast, the scale factors associated with positive Kasner exponents monotonously contract to zero. In the presence of the dilaton, however, all the p^i can be positive simultaneously. In all cases there is an overall contraction since the determinant g of the spatial metric tends to zero,

$$g = t^2 \to 0 \;\; \text{for } t \to 0. \tag{5}$$

It is convenient to redefine the time coordinate t as

$$t \to \tau = -\ln t. \tag{6}$$

This puts the singularity $t \to 0^+$ at $\tau \to +\infty$. Similarly, as $t \to +\infty$, we have $\tau \to -\infty$. In terms of τ, the logarithms of the scale factors

$$\beta^i = -\ln a_i \tag{7}$$

and the dilaton behave linearly,

$$\beta^i = p^i \tau, \;\; \phi = p^\phi \tau. \tag{8}$$

The Kasner exponents appear therefore as the τ-time velocities of the β^i, ϕ.

2.2 Geometry of the space of the scale factors

2.2.1 Supermetric and Hamiltonian. The Kasner solution can be viewed as a null geodesic in the "superspace" coordinatized by the metric components g_{ij} and by ϕ. Indeed, the action that governs the time-evolution of spatially-homogeneous models of type I, which is obtained from the action (1) by setting $A^{(p)} = 0$ and assuming the other fields to depend on time only, reads, in terms of an arbitrary reparametrized time $x^0 = x^0(t)$,

$$S[g_{ij}, \phi, \tilde{N}] = \int dx^0 \frac{1}{\tilde{N}} \left[\frac{1}{4} \left(\text{tr} \, (G^{-1}\dot{G})^2 - (\text{tr} \, G^{-1}\dot{G})^2 \right) + \dot{\phi}^2 \right] \quad (9)$$

where we have set

$$ds^2 = -(\tilde{N}\sqrt{g})^2(dx^0)^2 + g_{ij}dx^i dx^j \quad (10)$$

and $\dot{F} \equiv dF/dx^0$. We have adopted matrix notations in which G stands for the $d \times d$ matrix with matrix elements (g_{ij}). The use of the rescaled lapse $\tilde{N} = N/\sqrt{g}$ makes the subsequent formulas simpler. We assume that x^0 increases as one goes to the singularity, i.e., we take the minus sign in the relation $dt = -\tilde{N}\sqrt{g}dx^0$ relating the proper time to x^0.

The action (9) is the action for a free particle with coordinates (g_{ij}, ϕ) moving in a curved space with metric

$$d\sigma^2 = \frac{1}{4} \left[\text{tr} \, (G^{-1}dG)^2 - (\text{tr} \, G^{-1}dG)^2 \right] + d\phi^2 \quad (11)$$

Observe that we designate by $d\sigma^2$ the line element in superspace to distinguish it from the line element in physical space time, which we denote by ds^2. The first two terms in the right-hand side of (11) define what is known as the De Witt supermetric in the space of the g_{ij}'s, considered first in [D67]. Furthermore, because of time reparametrization invariance, the velocities are constrained by

$$\frac{1}{4} \left(\text{tr} \, (G^{-1}\dot{G})^2 - (\text{tr} \, G^{-1}\dot{G})^2 \right) + \dot{\phi}^2 = 0 \quad (12)$$

as follows from extremizing (9) with respect to $\tilde{N}$. Thus, the motion is given by a null geodesic of the metric (11), as announced.

For diagonal metrics, the supermetric (11) reduces to

$$d\sigma^2 = \text{tr} \, (d\beta)^2 - (\text{tr} \, d\beta)^2 + d\phi^2 \quad (13)$$

The action becomes therefore

$$S[\beta^\mu, \tilde{N}] = \int dx^0 \frac{1}{\tilde{N}} G_{\mu\nu} \dot{\beta}^\mu \dot{\beta}^\nu \quad (14)$$

where we have used the notation $\beta^0 \equiv \phi$ for the dilaton (if it is present) and defined $G_{\mu\nu}$ through $d\sigma^2 = G_{\mu\nu}d\beta^\mu d\beta^\nu$ ($\mu, \nu = 0, 1, \cdots, d$ with a dilaton, $\mu, \nu = 1, \cdots, d$ without dilaton). It is natural to collect together the scale factors and the dilaton, since the dilaton can be viewed as a scale factor in one extra dimension. Although the metric (11) has a non-vanishing curvature, the metric (13) induced in the space of the scale factors (including the dilaton if any) is flat. The diagonal Kasner solution is thus just a lightlike straight line in the space of the scale factors. The Hamiltonian form of the action is

$$
S[\beta^\mu, \pi_\mu, \tilde{N}] = \int dx^0 \left[\pi_\mu \dot{\beta}^\mu - \frac{\tilde{N}}{4} G^{\mu\nu} \pi_\mu \pi_\nu \right] \tag{15}
$$

with

$$
G^{\mu\nu} \pi_\mu \pi_\nu \equiv \sum_{i=1}^{d} \pi_i^2 - \frac{1}{d-1} \left(\sum_{i=1}^{d} \pi_i \right)^2 + \pi_\phi^2 \tag{16}
$$

where $(\pi_i, \pi_\phi) \equiv (\pi_\mu)$ are the momenta respectively conjugate to β^i and ϕ, i.e., $\pi_\mu = 2\tilde{N}^{-1} G_{\mu\nu} \dot{\beta}^\nu$.

The metric (13) has Minkowskian signature $(-, +, +, \cdots, +)$. The minus sign is due to the presence of gravity, and more specifically to the conformal factor in the metric; for the other fields, the kinetic term is always positive definite. This reflects the familiar fact that the gravitational action is not bounded from below (switching to Euclidean signature will not change the De Witt supermetric). Conformal transformations of the metric, in which the scale factors are all scaled in the same fashion, make $d\sigma^2$ negative. It is this characteristic feature of gravity which is responsible for the Lorentzian type of the Kac-Moody algebras which emerge in the analysis of the billiard symmetries [DH01]. The Lorentzian signature of the metric in the space of the scale factors enables one to define the light cone through any point. We define the time-orientation to be such that future-pointing vectors v^μ have $\sum_i v^i > 0$. Small volumes (small g) mean large positive values of $\sum_i \beta^i$. In contrast, large volumes (large g) mean large negative values of $\sum_i \beta^i$. We are interested in the small volume limit.

2.2.2 Hyperbolic space.
With the gauge choice $\tilde{N} = 1$, the solution of the dynamical equations read

$$
\beta^\mu = v^\mu x^0 + \beta_0^\mu \tag{17}
$$

while the constraint is

$$
G_{\mu\nu} v^\mu v^\nu = 0. \tag{18}
$$

It follows from (18) that $\sum_i v^i \neq 0^2$. In fact, since $g \sim \exp[-2(\sum_i v^i)x^0]$ decreases as x^0 increases, $\sum_i v^i > 0$. The Kasner exponents p^μ, which are normalized to $\sum_i p^i = 1$, are related to the v^μ through $p^\mu = v^\mu/(\sum_i v^i)$.

Consider the trajectories that get inside the future light cone of some point, say the origin[3]. One can radially project these trajectories on the hyperbolic space H_d (if a dilaton is present) or H_{d-1} (no dilaton), realized as the future sheet of the unit hyperboloid of vectors with norm squared equal to -1. This is most readily done by introducing hyperbolic coordinates,

$$\beta^\mu = \rho \gamma^\mu, \tag{19}$$

where γ^μ are coordinates on the future sheet of the unit hyperboloid,

$$\gamma^\mu \gamma_\mu = -1 \tag{20}$$

In terms of ρ and γ^μ, the metric becomes

$$d\sigma^2 = -d\rho^2 + \rho^2 d\Sigma^2 \tag{21}$$

where $d\Sigma^2$ is the metric on the unit hyperboloid. Reinstating the rescaled lapse $\tilde{N}$, the Hamiltonian is

$$H = \frac{\tilde{N}}{4}\left[-\pi_\rho^2 + \frac{1}{\rho^2}\pi_\gamma^2\right] \tag{22}$$

where π_γ are the momenta conjugate to the (constrained) γ^μ. Equivalently,

$$H = \frac{\tilde{N}}{4\rho^2}\left[-\pi_T^2 + \pi_\gamma^2\right] \tag{23}$$

with $T = \ln\rho$ and π_T its conjugate momentum. In terms of these variables, the motion is simple in the gauge $\tilde{N} = \rho^2$, since the Hamiltonian becomes then just $(1/4)[-(\pi_T)^2 + (\pi_\gamma)^2]$. One can view T as an intrinsic time variable in terms of which one can describe the dynamics. The evolution is a geodesic motion on Γ.

---

[2] The Kasner metric (1) corresponds to the particular choice of integration constants $\sum_i v^i = 1$, $\beta_0^\mu = 0$, which implies $t = \exp(-x^0)$, i.e., $x^0 = \tau$ and $\sqrt{g} = t$. These "initial conditions" can be reached by using the symmetries of the problem. These are (i) independent rescalings of the coordinates $x^i \to x'^i = k^i x^i$; (ii) redefinition of the time x^0, $x^0 \to x'^0 = Bx^0 + C$, where B and C are constants, with $B = k^1 k^2 \cdots k^d$ so as to preserve the gauge condition $dt = -\sqrt{g}dx^0$; and (iii) $\phi \to \phi' = \phi + A$, where A is a constant. However, if one considers many Kasner solutions simultaneously, as we shall do below when analysing collisions from one Kasner regime to another one, one cannot make these simple choices of integration constants for all the solutions at the same time - one can make it only, say, for the "first" one. For this reason, we shall keep generic β_0^μ in (17) and we shall not impose $\sum_i v^i = 1$.
[3] This occurs for x^0 big enough when $v_\mu \beta_0^\mu < 0$ since $\beta^\mu \beta_\mu = 2v_\mu \beta_0^\mu x^0 + \beta_0{}_\mu \beta_0^\mu$. The condition $v_\mu \beta_0^\mu < 0$ can always be assumed to hold by shifting the origin of the β^i's if necessary.

On the solutions, x^0 (in the gauge $\tilde{N} = 1$) $\sim \ln \sqrt{g}$, $\rho \sim |\ln \sqrt{g}|^{1/2}$ and $T \sim \ln|\ln \sqrt{g}| \sim \ln|\ln t|$ ($v_\mu \beta_0^\mu < 0$). Small volume is thus equivalent to large ρ or what is the same, large T, i.e., large intrinsic time. We shall call the limit $\rho \to \infty$ the BKL limit.

3.   Kasner solution – Non-diagonal case

3.1   Iwasawa decomposition

In the homogeneous, vacuum context, the metric remains diagonal if the initial data are so. This is in general not true when matter or inhomogeneities are included, in which case off-diagonal components generically appear even if there is none initially. For this reason, it is important to understand the rôle of off-diagonal terms already in the simpler homogeneous context, by examining the evolution of initial data that are not diagonal. This is most simply done by performing a constant linear transformation on the Kasner solution, which is a symmetry for Bianchi I models. If L is the linear transformation needed to diagonalize the initial data, it is easy to see that the solution is given by

$$G(t) = {}^t\! L\, G_K(t)\, L \qquad (1)$$

where ^{t} denotes transposition and $G_K(t)$ is the above diagonal Kasner solution. The dilaton, being a scalar, is still given by the same expression. Note that $\det G(t) = (\det L)^2 t^2$.

To understand the qualitative behaviour of (1), it is convenient to perform the Iwasawa decomposition of the metric,

$$G = {}^t\!\mathcal{N}\,\mathcal{A}^2\,\mathcal{N} \qquad (2)$$

where $\mathcal{N}$ is an upper triangular matrix with 1's on the diagonal ($\mathcal{N}_{ii} = 1$, $\mathcal{N}_{ij} = 0$ for $i > j$) and where $\mathcal{A}$ is a diagonal matrix with positive elements, which we parametrize as

$$\mathcal{A} = \exp(-\beta), \quad \beta = \mathrm{diag}\,(\beta^1, \beta^2, \cdots, \beta^d). \qquad (3)$$

One can view the Iwasawa decomposition[4] as the Gram-Schmidt orthogonalisation of the initial basis, which is indeed a triangular process,

$$ds^2 = g_{ij}dx^i dx^j = \sum_{k=1}^{d} e^{(-2\beta_k)}(\theta^k)^2 \qquad (4)$$

---

[4]The Iwasawa decomposition applies to general symmetric spaces (see e.g. [H78]). In our case the relevant symmetric space is the coset space $SL(d, \mathbb{R})/SO(d)$ since the space of positive definite symmetric matrices can be identified with $GL(d, \mathbb{R})/O(d)$, which is isomorphic to $SL(d, \mathbb{R})/SO(d) \times \mathbb{R}^+$.

with

$$\theta^k = \sum_i \mathcal{N}_{ki}\, dx^i. \tag{5}$$

Starting with $\theta^d = dx^d$, one successively constructs the next θ^j's by adding linear combinations of the dx^p's $(p > j)$ in such a way that θ^{d-1} is orthogonal to θ^d, θ^{d-2} is orthogonal to both θ^d and θ^{d-1}, etc. Explicitly,

$$\theta^d = dx^d \tag{6}$$
$$\theta^{d-1} = dx^{d-1} + \mathcal{N}_{d-1\,d}\, dx^d, \tag{7}$$
$$\theta^{d-2} = dx^{d-2} + \mathcal{N}_{d-2\,d-1}\, dx^{d-1} + \mathcal{N}_{d-2\,d}\, dx^d, \tag{8}$$
$$\cdots . \tag{9}$$

Note that the dual basis $\{e_j\}$ reads

$$e_j = \sum_m (\mathcal{N}^{-1})_{mj}\, \frac{\partial}{\partial x^m}. \tag{10}$$

The variables β^i of the Iwasawa decomposition give the scale factors in the new, orthogonal, basis, while the $\mathcal{N}_{ij}$ characterize the change of basis that diagonalizes the metric and hence, parametrize the off-diagonal components of the original g_{ij}.

In the diagonal case, $\mathcal{N} = 1$ and the β's behave linearly with τ,

$$\beta^i = p^i \tau \quad \text{(diagonal Kasner)}. \tag{11}$$

In the general case, $\mathcal{N} \neq 1$ and the evolution of the variables β^i, $\mathcal{N}_{ij}$ of the Iwasawa decomposition is more complicated. However, it becomes simple for $\tau \to \pm\infty$. Indeed, the β's become asymptotically linear functions of τ, as in the diagonal case, with coefficients that are given by a permutation of the p^i's, whereas the $\mathcal{N}_{ij}$ become constant. More precisely, let $q^1, q^2, \cdots, q^d$ be the ordered Kasner exponents, $q^1 \leq q^2 \leq \cdots \leq q^d$, $q^i = p^{\sigma(i)}$, with σ an appropriate permutation of $1, 2, \cdots, n$. Then one finds (for generic L's):

- as $\tau \to -\infty$ $(t \to \infty)$, $\beta^i \sim q^{d-i}\tau$ and $\mathcal{N}_{ij} \to$ const;

- as $\tau \to +\infty$ $(t \to 0^+)$, $\beta^i \sim q^i\tau$ and $\mathcal{N}_{ij} \to$ const.

In both limits, one has $\beta^1 \leq \beta^2 \leq \cdots \leq \beta^d$. The motion does not change the Kasner exponents p^i, but just redistributes them among the β's so as to achieve $\beta^1 \leq \beta^2 \leq \cdots \leq \beta^d$ for both $\tau \to \infty$ and $\tau \to -\infty$. [The fact that one gets $\beta^1 \leq \beta^2 \leq \cdots \leq \beta^d$ (in that order) follows from the choice for the matrix $\mathcal{N}$ made in the Iwasawa decomposition; had we taken $\mathcal{N}$ to be e.g. lower triangular instead of upper-triangular, the inequalities would have been all reversed.]

One can arrive at the above conclusions concerning the asymptotic form of the solutions either by direct computation from (1) and (2), or by using the Hamiltonian formulation of the theory. We shall derive here the Hamiltonian analysis and give the direct calculation (in the specific $3 + 1$ case) in the first appendix.

3.2   Symmetry walls

The simplest way to understand the asymptotic dynamics of the scale factors in the non-diagonal case is to use the geometric picture developed in subsection 2.2.

3.2.1   BKL limit.   Using the Iwasawa decomposition, the supermetric (11) becomes

$$
d\sigma^2 = \text{tr}\,(d\beta)^2 - (\text{tr}\,d\beta)^2 + \frac{1}{2}\text{tr}\left[\mathcal{A}^2\,d\mathcal{N}\,\mathcal{N}^{-1}\,\mathcal{A}^{-2}\,{}^t(\mathcal{N}^{-1})\,d({}^t\mathcal{N})\right]
$$
$$
+(d\phi)^2, \tag{12}
$$

i.e.,

$$
d\sigma^2 = \sum_{i=1}^{d}(d\beta^i)^2 - (\sum_{i=1}^{d} d\beta^i)^2
$$
$$
+\frac{1}{2}\sum_{i<j}e^{2(\beta^j-\beta^i)}\left((d\mathcal{N}\mathcal{N}^{-1})_{ij}\right)^2 + d\phi^2 \tag{13}
$$

For $d = 3$, this expression reduces to the one explicitly given in [HPT82].

The Hamiltonian governing the evolution is therefore

$$
H = \frac{\tilde{N}}{4}\left[\sum_{i=1}^{d}(\pi_i)^2 - \frac{1}{d-1}(\sum_{i=1}^{d}\pi_i)^2 + (\pi_\phi)^2\right]
$$
$$
+\frac{\tilde{N}}{2}\sum_{i<j}e^{-2(\beta^j-\beta^i)}\left(\sum_m P_{im}\mathcal{N}_{jm}\right)^2 \tag{14}
$$

where the P_{im} are the momenta conjugate to the $\mathcal{N}_{im}$, $P_{im} = \partial\mathcal{L}/\partial\dot{\mathcal{N}}_{im}$. Because the terms involving the off-diagonal parameters are positive, the constraint $H = 0$ implies that the velocity $\dot{\beta}^\mu$ is timelike or null.

We are interested in the BKL limit, i.e. small volume (large $\sum_i \beta^i$) or large ρ. Making the change of variables (19) yields, instead of (23)

$$
H = \frac{\tilde{N}}{\rho^2}\left[\frac{1}{4}(-\pi_T^2 + \pi_\gamma^2) + \frac{\rho^2}{2}\sum_{i<j}e^{-2\rho(\gamma^j-\gamma^i)}\left(\sum_m P_{im}\mathcal{N}_{jm}\right)^2\right] \tag{15}
$$

As ρ becomes large, the terms $\rho^2 \exp[-2\rho(\gamma^j - \gamma^i)]$ either go to zero if $\gamma^j - \gamma^i$ is positive, or explode to $+\infty$ if $\gamma^j - \gamma^i$ is negative. Thus one can replace the exponentials by $\Theta[-2(\gamma^j - \gamma^i)]$, where $\Theta(x) = 0$ for $x < 0$ and $+\infty$ for $x > 0$[5]. Of course, $\Theta(2x) = \Theta(x)$, but we keep the factor 2 here to recall that the arguments of the exponentials from which the Θ-functions originate come with a well-defined normalization. Since the functions $\gamma^j - \gamma^i$ and $\beta^j - \beta^i$ have the same sign ($\rho > 0$) and since $\Theta(x) = \lambda\Theta(x)$ for $\lambda > 0$, one can replace $\Theta[-2(\gamma^j - \gamma^i)]$ by $\Theta[-2(\beta^j - \beta^i)]$. The Hamiltonian governing the dynamics of the scale factors β^μ for type-I, non-diagonal models, reads, in the BKL limit,

$$
H \;=\; \frac{\tilde{N}}{4}\left[\sum_{i=1}^{d}\pi_i^2 - \frac{1}{d-1}\left(\sum_{i=1}^{d}\pi_i\right)^2 + \pi_\phi^2 \right.
$$

$$
\left. + \sum_{i<j}\Theta[-2(\beta^j - \beta^i)]\right] \tag{16}
$$

The terms $\Theta[-2(\beta^j - \beta^i)]$ are present for all (i, j) with $i < j$ provided all non-negative terms $(\sum_m P_{im}\mathcal{N}_{jm})^2$ are different from zero. This is the case generically, i.e., for all initial data except a set of measure zero. For this reason, we shall assume that it is fulfilled from now on. As a rule, we shall always make similar "genericity" assumptions below concerning the coefficients of the various potentials that enter the discussion.

3.2.2 First encounter with billiards.

Because of the simplicity of (16), the dynamics is easy to describe in the BKL limit. First, since the off-diagonal parameters $\mathcal{N}_{ij}$ and their conjugate drop out from the Hamiltonian, they are constant for large ρ. This is what we saw in subsection (3.1). The only effect of the off-diagonal components is to induce the Θ-function potentials for the scale factors. The asymptotic freezing of the off-diagonal terms is further discussed in appendix B. Second, the Θ-terms constrain the scale factors to be in the region

$$
w_{(ij)}^S(\beta) \equiv w_{(ij)\mu}^S \beta^\mu \equiv \beta^j - \beta^i \geq 0 \quad (i < j). \tag{17}
$$

In that region the motion is free, i.e., is given by a lightlike straight line (17) for some v^μ fulfilling (18), exactly as for diagonal Kasner metrics. When the trajectory hits a wall $\beta^j - \beta^i = 0$, it undergoes a reflection. This reflection permutes two v^μ's (see below). After a finite number of collisions, the v^μ's get ordered, $v^1 \leq v^2 \cdots \leq v^d$, and the motion then goes on freely forever. Indeed, the subsequent evolution of the β's, once the v^μ's are ordered, can never hit a

---

[5]One should more properly write $\Theta_\infty(x)$, but since this is the only step function encountered in this article, we use the simpler notation $\Theta(x)$.

wall (where two β's are equal) any more. This is what we announced in the analysis of the asymptotic behaviour of the solutions in subsection 3.1. We stress that this simple evolution arises only in the BKL limit. Before the asymptotic regime is reached, one cannot replace the exponentials by Θ-functions. The collisions that take place then are not strictly localized in time and therefore may not be clearly separated from each other. In addition, there is a slow drift of the off-diagonal parameters.

The region $w^S_{(ij)}(\beta) \geq 0$ $(i < j)$, $\sum_i \beta^i \geq 0$ is a convex region of the space of the scale factors. We shall call the hyperplanes $w^S_{(ij)}(\beta) = 0$ the "symmetry walls". These hyperplanes are timelike since

$$G^{\mu\nu} w^S_{(ij)\mu} w^S_{(ij)\nu} = 2 \qquad (18)$$

and intersect therefore the hyperboloid $G_{\mu\nu}\beta^\mu\beta^\nu = -1$, $\sum_i \beta^i \geq 0$. The symmetry billiard is defined to be the region of hyperbolic space determined by these inequalities.

Among the $d(d-1)/2$ symmetry walls, only $d-1$ are relevant. Indeed, the inequalities (17) all follow from $w^S_{(i\,i+1)} \geq 0$. The symmetry billiard is thus defined by the $d-1$ inequalities

$$\beta^2 - \beta^1 \geq 0, \ \beta^3 - \beta^2 \geq 0, \ \cdots, \beta^d - \beta^{d-1} \geq 0 \qquad (19)$$

and extend to infinity. It has infinite volume.

In order to analyse the collision against the wall $\beta^{i+1} - \beta^i = 0$, we make linear redefinitions so that $\beta^{i+1} - \beta^i$ is one of the new variables,

$$\beta^\mu \rightarrow \beta'^\mu \ (\mu \neq i, i+1), \ u = \beta^{i+1} + \beta^i, \ z = \beta^{i+1} - \beta^i \qquad (20)$$

In these new variables, the potential associated with the symmetry wall $\beta^{i+1} - \beta^i = 0$ is simply $\Theta(-2z)$. Furthermore, since z is orthogonal to the other variables, the sole effect of the collision is to change the sign of $\dot{z}$, leaving the other velocities fixed. In terms of the original variables, this is just the permutation of the velocities v^i and v^{i+1},

$$v^i \rightarrow v^{i+1}, \ v^{i+1} \rightarrow v^i \ (\text{other } v^k \text{ unchanged}). \qquad (21)$$

This transformation belongs to the orthochronous Lorentz group because it preserves both the Lorentzian metric and the arrow of time ($G_{\mu\nu} v'^\mu v'^\nu = G_{\mu\nu} v^\mu v^\nu$, $\sum_i v'^i = \sum_i v^i$). More generally, an arbitrary number of collisions against the symmetry walls will lead to an arbitrary permutation of the v^i's, since the transpositions $(i, i+1)$ generate the symmetry group S_d in d elements.

Remarks

    i **Large volume limit**

It follows from the above analysis that in the small volume limit, the Kasner exponents get re-ordered so that the first spatial direction carries the smallest Kasner exponent, the second spatial direction carries the next to smallest Kasner exponent etc. In the particular homogeneous case considered here, one can also investigate the large volume limit by the same techniques. Although this cannot be extended to the general inhomogeneous case discussed below, we briefly indicate how this is achieved in the present special case.

In the large volume limit, the β's go to the past in the space of the scale factors. So one sets

$$\beta^\mu = -\rho\gamma^\mu, \tag{22}$$

where γ^μ are still coordinates on the future sheet of the unit hyperboloid, $\gamma^\mu\gamma_\mu = -1$. Making the same transformations as above, one finds that all the symmetry walls come with the other sign, i.e., in the large volume limit, the potential reduces to $\sum_{i>j}\Theta[-2(\beta^j - \beta^i)]$. This forces the β's to be ordered as $\beta^1 \geq \beta^2 \geq \cdots \geq \beta^d$. Accordingly, the collisions against the walls reorders now the velocities p^i in decreasing order, as we stated in subsection 3.1.

ii Are the symmetry walls a gauge artifact?

Since the metric can diagonalized at all times by a time-independent coordinate transformation $x^a \to x'^a = L^a{}_b x^b$, one might fear that the symmetry walls, which are related to the off-diagonal components, are somewhat a gauge artifact with no true physical content. This conclusion, however, would be incorrect. First, the transformation needed to diagonalize the metric may not be a globally well-defined coordinate transformation if the spatial sections have non-trivial topology, e.g., are tori, since it would conflict in general with periodicity conditions. Second, even if the spatial sections are homeomorphic to R^d, the transformation $x^a \to x'^a = L^a{}_b x^b$, although a diffeomorphism, is not a proper gauge transformation in the sense that it is generated by a non-vanishing charge. One should regard as distinct (although related by a symmetry) two solutions that differ by the transformation $x^a \to x'^a = L^a{}_b x^b$. Initial conditions for which the metric is diagonal – and hence the symmetry walls absent – form a set of measure zero.

iii Alternative description of symmetry walls

We have just shown that the Iwasawa decomposition of the spatial metric leads to a projected description of the $GL(d,\mathbb{R})/O(d)$-geodesics as motions in the space of the scale factors with exponential ("Toda-like") potentials. An alternative description exists, which is based on the decomposition $G = {}^t R A R$ of the spatial metric, where $R \in SO(d)$ and A is diagonal [R72a, KM02]. One then gets "Calogero-like" $\sinh^{-2}(\beta^i - \beta^j)$-potentials. In the BKL limit, these potentials can be replaced by sharp wall potentials but whether the system lies to the left or to the right of the wall $\beta^i - \beta^j = 0$ depends in this alternative description on the initial conditions.

4.   Asymptotic dynamics in the general case - Gravitational billiards

We shall now show that the same ideas apply in the general case described by the action (1). In the vicinity of a spacelike singularity, the spatial points decouple and the billiard picture remains valid at each spatial point. Inhomogeneities and p-forms bring in only new walls.

4.1   Pseudo-Gaussian coordinates

We thus assume that there is a spacelike singularity at a finite distance in proper time. We adopt a spacetime slicing adapted to the singularity, which "occurs" on a slice of constant time. We build the slicing from the singularity by taking pseudo-Gaussian coordinates defined by $\tilde{N} = 1$ and $N^i = 0$, where $\tilde{N}$ is again the rescaled lapse $\tilde{N} = N/\sqrt{g}$ and N^i is the standard shift. More precisely, we assume that in some spacetime patch, the metric reads

$$ds^2 = -g(x^0, x^i)(dx^0)^2 + g_{ij}(x^0, x^i)dx^i dx^j, \quad g \equiv \det(g_{ij}), \qquad (1)$$

where the local volume g collapses at each spatial point as $x^0 \to +\infty$, in such a way that the proper time $dt = -\sqrt{g}dx^0$ remains finite (and tends conventionally to 0^+). We shall make the further assumption that since the local volume tends to zero, the variable ρ introduced above can be used everywhere in a given region of space (in a neighborhood of the singularity) as a well-defined ("intrinsic") time variable that goes all the way to $+\infty$. We can then investigate the BKL limit $\rho \to \infty$. Differently put, we study the general solution of the equations of motion in the regime met above where the scale factors β^μ go to infinity inside the future light cone in superspace, $\beta^\mu = \rho\gamma^\mu$, $G_{\mu\nu}\gamma^\mu\gamma^\nu = -1$, $\sum_i \gamma^i > 0$, $\rho \to \infty$ (at each spatial point in the local region under consideration). Some aspects of the consistency of this limit are discussed in the appendix D.

One of the motivations for studying spacelike singularities comes from the singularity theorems [HP70]. Of course, not all singularities are spacelike, and furthermore, there is no claim here that all spacelike singularities are necessarily of the BKL type. We are just interested in this regime, which, as we shall see, has a high degree of self-consistency and is quite general in the sense that it involves as many arbitrary functions of space as the general solution does (see subsection 4.6).

The only coordinate freedom left in the pseudo-Gaussian gauge (after having fixed the hypersurface $t = 0$) is that of making time-independent changes of spatial coordinates $x^k \rightarrow x'^k = f^k(x^m)$. Since the rescaled lapse is a spatial density of weight minus one, such changes of coordinates have the unusual feature of also changing the slicing (in some determined way - the scalar lapse N changes in a well-defined manner).

The fields ϕ and $A^{(p)}_{\mu_1 \cdots \mu_p}$ are also a priori functions of both space and time. No symmetry condition is imposed. Although the equations at each point will be asymptotically the same as those of homogeneous cosmological models, this does not follow from imposing extra dimensional reduction conditions but is rather a direct consequence of the general dynamical equations.

We shall partially fix the gauge in the p-form sector by imposing the temporal gauge condition $A^{(p)}_{0\mu_2 \cdots \mu_p} = 0$. This leaves the freedom of performing time-independent gauge transformations.

4.2   Hamiltonian action

To focus on the features relevant to the billiard picture, we assume first that there is no Chern-Simons terms or couplings of the exterior form gauge fields through a modification of the curvatures $F^{(p)}$, which are thus taken to be Abelian, $F^{(p)} = dA^{(p)}$. We verify in subsection 4.7 below that these interaction terms do not change the analysis. The Hamiltonian action in the pseudo-Gaussian gauge reads

$$S[g_{ij}, \pi^{ij}, \phi, \pi_\phi, A^{(p)}_{m_1 \cdots m_p}, \pi_{(p)}^{m_1 \cdots m_p}] =$$

$$\int dx^0 \left(\int d^d x (\pi^{ij} \dot{g}_{ij} + \pi_\phi \dot{\phi} + \sum_p \pi_{(p)}^{m_1 \cdots m_p} \dot{A}^{(p)}_{m_1 \cdots m_p}) - H \right) \quad (2)$$

where the Hamiltonian is

$$H = \int d^d x \, \mathcal{H} \quad (3)$$

$$\mathcal{H} = L + M \quad (4)$$

$$L = \pi^{ij} \pi_{ij} - \frac{1}{d-1}(\pi_i^i)^2 + \frac{1}{4}(\pi_\phi)^2 +$$

$$\sum_p \frac{p! \, e^{-\lambda^{(p)}\phi}}{2} \pi_{(p)}^{m_1 \cdots m_p} \pi_{(p) \, m_1 \cdots m_p} \quad (5)$$

$$M = -Rg + g^{ij} g \partial_i \phi \partial_j \phi + \sum_p \frac{e^{\lambda^{(p)}\phi}}{2\,(p+1)!} g \, F^{(p)}_{m_1 \cdots m_{p+1}} F^{(p) \, m_1 \cdots m_{p+1}} \quad (6)$$

The dynamical equations of motion are obtained by varying the above action w.r.t. the spatial metric components, the dilaton, the spatial p-form components and their conjugate momenta. In addition, there are constraints on the dynamical variables, which are

$$\mathcal{H} = 0 \quad \text{("Hamiltonian constraint")}, \tag{7}$$

$$\mathcal{H}_i = 0 \quad \text{("momentum constraint")}, \tag{8}$$

$$\varphi_{(p)}^{m_1 \cdots m_{p-1}} = 0 \quad \text{("Gauss law" for each } p\text{-form)} \ (p > 0) \tag{9}$$

Here, we have set

$$\mathcal{H}_i = -2\pi_i^j{}_{|j} + \pi_\phi \partial_i \phi + \sum_p \pi_{(p)}^{m_1 \cdots m_p} F_{i m_1 \cdots m_p}^{(p)} \tag{10}$$

$$\varphi_{(p)}^{m_1 \cdots m_{p-1}} = -p\, \pi_{(p)}^{m_1 \cdots m_{p-1} m_p}{}_{|m_p} \tag{11}$$

In order to understand the asymptotic behaviour of the fields, we perform, at each spatial point, the Iwasawa decomposition of the metric. This is a point canonical transformation, extended to the momenta in the standard way, i.e., $\pi^{ij}\dot{g}_{ij} = \pi_i \dot{\beta}^i + \sum_{i<j} P_{ij}\dot{\mathcal{N}}_{ij}$. We then split the Hamiltonian in two parts, one denoted by K, which is the kinetic term for the local scale factors β^μ, and the other denoted by V, which contains everything else. It is indeed natural to group the kinetic term for the off-diagonal variables with the original potential terms because they asymptotically play the role of a potential for the scale factors. As we shall see, the same feature holds for the p-form kinetic terms. In fact, this is not surprising, at least when $p=1$, since 1-forms can be viewed as off-diagonal components of the metric in one dimension higher. Thus, we write

$$\mathcal{H} = K + V \tag{12}$$

$$K = \frac{1}{4}\left[\sum_{i=1}^{d}\pi_i^2 - \frac{1}{d-1}\left(\sum_{i=1}^{d}\pi_i\right)^2 + \pi_\phi^2\right] \tag{13}$$

$$V = V_S + V_G + \sum_p V_p + V_\phi \tag{14}$$

$$V_S = \frac{1}{2}\sum_{i<j} e^{-2(\beta^j - \beta^i)}\left(\sum_m P_{im}\mathcal{N}_{jm}\right)^2 \tag{15}$$

$$V_G = -Rg \tag{16}$$

$$V_{(p)} = V_{(p)}^{el} + V_{(p)}^{magn} \tag{17}$$

$$V_{(p)}^{el} = \frac{p!\, e^{-\lambda^{(p)}\phi}}{2}\pi_{(p)}^{m_1 \cdots m_p}\pi_{(p)\, m_1 \cdots m_p} \tag{18}$$

$$V_{(p)}^{magn} = \frac{e^{\lambda^{(p)}\phi}}{2\,(p+1)!}\, g\, F_{m_1 \cdots m_{p+1}}^{(p)} F^{(p)\, m_1 \cdots m_{p+1}} \tag{19}$$

$$V_\phi = g^{ij} g \partial_i \phi \partial_j \phi. \tag{20}$$

We know that in the BKL limit, the symmetry potential V_S becomes a sum of sharp wall potentials,

$$V_S \simeq \sum_{i<j} \Theta[-2(\beta^j - \beta^j)] \qquad (\rho \to \infty) \tag{21}$$

The computation that leads to (21) was carried out above in the spatially homogeneous context but remains clearly valid here since V_S contains no spatial gradients. This potential is "ultralocal", i.e., its value at any spatial point involves only the scale factors at that same spatial point. It is quite remarkable that the same sharp wall behavior emerges asymptotically for the other potentials.

4.3   Curvature (gravitational) walls

We first establish this fact for the gravitational potential.

4.3.1   Computation of curvature.   To that end, one must explicitly express the spatial curvature in terms of the scale factors and the off-diagonal variables $\mathcal{N}_{ij}$. The calculation is most easily done in the frame $\theta^k = \sum_i \mathcal{N}_{ki} dx^i$ in which the spatial metric is diagonal,

$$ds^2 = \sum_{k=1}^{d} a_k^2 (\theta^k)^2, \quad a_k = e^{-\beta^k} \tag{22}$$

Let $C^i{}_{mn}(x)$ be the structure functions of the basis $\{\theta^k\}$,

$$d\theta^i = -\frac{1}{2} C^i{}_{mn} \theta^m \theta^n \tag{23}$$

Here, d is the spatial exterior differential. The functions $C^i{}_{mn}(x)$ depend clearly only on the off-diagonal variables $\mathcal{N}_{ij}$ and not on the scale factors. Using the Cartan formulas for the connection 1-form $\omega^i{}_j$,

$$d\theta^i + \sum_j \omega^i{}_j \wedge \theta^j = 0 \tag{24}$$

$$d\gamma_{ij} = \omega_{ij} + \omega_{ji} \tag{25}$$

where $\gamma_{ij} = \delta_{ij} a_i^2$ is the metric in the frame $\{\theta^k\}$, one finds

$$\omega^k{}_\ell = \sum_j \frac{1}{2} \left(C^j{}_{k\ell} \frac{a_j^2}{a_k^2} + C^\ell{}_{kj} \frac{a_\ell^2}{a_k^2} - C^k{}_{\ell j} \right) \theta^j$$

$$+ \sum_j \frac{1}{2a_k^2} \left(\delta_{k\ell}(a_k^2)_{,j} + \delta_{kj}(a_k^2)_{,\ell} - \delta_{\ell j}(a_\ell^2)_{,k} \right) \theta^j \tag{26}$$

One then gets the Riemann tensor $R^k_{\ \ell mn}$, the Ricci tensor $R_{\ell n}$ and the scalar curvature R through

$$\mathcal{R}^k_{\ \ell} = d\omega^k_{\ \ell} + \sum_j \omega^k_{\ j}\omega^j_{\ \ell} \qquad (27)$$

$$= \frac{1}{2}\sum_{m,n} R^k_{\ \ell mn}\theta^m \wedge \theta^n \qquad (28)$$

($\mathcal{R}^k_{\ \ell}$ is the curvature 2-form) and

$$R_{\ell n} = R^k_{\ \ell kn}, \quad R = \sum_\ell \frac{1}{a_\ell^2}R_{\ell\ell}. \qquad (29)$$

Direct, but somewhat cumbersome, computations yield

$$R = -\frac{1}{4}\sum_{j,k,\ell}\frac{a_j^2}{a_\ell^2 a_k^2}(C^j_{\ k\ell})^2 + \sum_j \frac{1}{a_j^2}F_j(\partial^2\beta, \partial\beta, \partial C, C) \qquad (30)$$

where F_j is some complicated function of its arguments whose explicit form will not be needed here. The only property of F_j that will be of importance is that it is a polynomial of degree two in the derivatives $\partial\beta$ and of degree one in $\partial^2\beta$. Thus, the exponential dependence on the β^i's, which determines the asymptotic behaviour in the BKL limit, occurs only through the a_j^2-terms written explicitly in (30).

Without loss of generality, we can assume $j \neq k$ and $j \neq \ell$ in the first sum on the right-hand side of (30), since the terms with either $j = k$ or $j = \ell$ can be absorbed through a redefinition of F_j. Also, one has clearly $k \neq \ell$ because the structure functions $C^j_{\ k\ell}$ are antisymmetric in k, ℓ. We can thus write the gravitational potential as

$$V_G \equiv -gR = \frac{1}{4}\sum_{i\neq j, i\neq k, j\neq k} e^{-2\alpha_{ijk}(\beta)}(C^i_{\ jk})^2 - \sum_j e^{-2\mu_j(\beta)}F_j \qquad (31)$$

where the linear forms $\alpha_{ijk}(\beta)$ and $\mu_j(\beta)$ are given by

$$\alpha_{ijk}(\beta) = 2\beta^i + \sum_{m,\,(m\neq i, m\neq j, m\neq k)} \beta^m, \quad (i \neq j, i \neq k, j \neq k) \qquad (32)$$

and

$$\mu_j(\beta) = \sum_{m,\,(m\neq j)} \beta^m. \qquad (33)$$

respectively.

4.3.2   BKL limit.   In the BKL limit where the scale factors β^μ go to infinity inside the future light cone, the gravitational potential becomes a sum of sharp wall potentials,

$$V_G \simeq \sum_{i\neq j, i\neq k, j\neq k} \Theta[-2\alpha_{ijk}(\beta)] + \sum_i (\pm\Theta[-2\mu_i(\beta)]) \qquad (34)$$

The terms in the second sum seem to pose a problem because they do not have a definite sign. They are, however, in fact always zero in the BKL limit because $\mu_i(\beta) > 0$. Indeed, each linear form $\mu_i(\beta)$ is lightlike and hence, each hyperplane $\mu_i(\beta) = 0$ is tangent to the light cone along some generatrix. This means that the future light cone is entirely on one side of the hyperplane $\mu_i(\beta) = 0$ (i.e., either $\mu_i(\beta) > 0$ for all points inside the future light cone or $\mu_i(\beta) < 0$). Now, the point $\beta^1 = \beta^2 = \cdots = \beta^d = 1$ is inside the future light cone and makes all the μ_i's positive. Hence $\mu_i(\beta) > 0$ inside the future light cone for each i and $\Theta[-2\mu_i(\beta)] = 0$: we can drop the second term in the gravitational potential, which reduces to

$$V_G \simeq \sum_{i\neq j, i\neq k, j\neq k} \Theta[-2\alpha_{ijk}(\beta)]. \qquad (35)$$

Note that in $D = 3$ spacetime dimensions, the gravitational walls $\alpha_{ijk}(\beta) = 0$ ($i \neq j, i \neq k, j \neq k$) are absent, since one cannot find three distinct spatial indices. The first term in (30) is of the same type as the second term: the only gravitational walls are then all of the subdominant type μ_j and thus, in the BKL limit,

$$V_G \simeq 0 \quad (D = 3). \qquad (36)$$

We thus see that the gravitational potential becomes, in the BKL limit, a positive sum of sharp wall potentials,

$$V_G \simeq \sum_{i\neq j, i\neq k, j\neq k} \Theta[-2\alpha_{ijk}(\beta)] \ (D > 3), \qquad V_G \simeq 0 \ (D = 3) \qquad (37)$$

This is remarkable for many reasons. First, the final form of the potential is rather simple, even though the curvature is a rather complicated function of the metric and its derivatives. Second, the limiting expression of the potential is positive, even though there are (subdominant) terms in V_G with indefinite sign. Third, it is ultralocal in the scale factors, i.e. involves only the scale factors but not their derivatives. It is this fact that accounts for the decoupling of the various spatial points.

It follows from this analysis that the scale factors are constrained by the conditions

$$\alpha_{ijk}(\beta) \geq 0 \quad (D > 3) \qquad (38)$$

besides the symmetry inequalities (19). The hyperplanes $\alpha_{ijk}(\beta) = 0$ are called the "curvature" or "gravitational" walls.

4.3.3   Remarks.

i In $D > 3$ dimensions, one can argue more directly that the first term in
   V_G asymptotically dominates the second one in the BKL limit $\rho \to \infty$ by
   observing that the μ^j's are all positive when the α_{ijk} are positive, since
   they can be written as linear combinations with positive coefficients of the
   α_{ijk} (the converse is not true, one may have $\mu^j > 0$ for all j's together
   with $\alpha_{ijk} < 0$ for some α_{ijk}'s)[6]. Thus, in the region $\alpha_{ijk} > 0$, the
   exponentials in V_G all go to zero as $\rho \to \infty$ and one can replace V_G
   by zero. Conversely, when one of the μ_j's is negative, there is always
   at least one α_{ijk} that is also negative. Actually, assuming without loss
   of generality that the β^i's are ordered, $\beta^1 \leq \beta^2 \leq \cdots \leq \beta^d$, the most
   negative μ_i is $\mu_d = \beta^1 + \beta^2 + \cdots + \beta^{d-1}$. But one has $\alpha_{1\,d\,d-1} =$
   $2\beta^1 + \beta^2 + \cdots + \beta^{d-2} \leq \mu^d$, in fact $\alpha_{1\,d\,d-1} \leq \alpha_{ijk}$ (for all i, j, k) and
   $\alpha_{1\,d\,d-1} \leq \mu_j$ (for all j's). Thus, as $\rho \to \infty$ (keeping the γ^μ fixed), the
   behaviour of V_G is controlled by the exponential $\exp(-2\alpha_{1\,d\,d-1})$, which
   blows up the fastest. Since it is multiplied by a positive coefficient, we
   conclude that $V_G \to +\infty$ (even though the subdominant term with F_j
   has no definite sign[7]). If the β^i were not ordered, it would be a different
   α_{ijk} that would take over, leading to the same conclusion.

ii The potential $V_\phi = g g^{ij} \partial_i \phi \partial_j \phi$ of the dilaton has the same form as the
    subdominant gravitational terms that we have just neglected, since the
    exponentials that control its asymptotic behaviour are also $\exp[-2\mu_j(\beta)]$.
    Consequently, we can drop V_ϕ in the BKL limit.

iii The above curvature computation involves only the Cartan formulas.
    It would also hold true if $\theta^k = \sum_i \mathcal{N}_{ki} f^i$, where $f^i = f^i{}_j(x) dx^j$ is
    an arbitrary fixed frame. The structure functions $C^i{}_{mn}(x)$ would get
    contributions from both $\mathcal{N}_{ki}$ (and its spatial derivatives) and f^i. We have
    taken above $f^i{}_j = \delta^i{}_j$. Then, $C^i{}_{mn} = C^i{}_{mn}(\mathcal{N})$. The other extreme case
    is $\mathcal{N}_{ki} = \delta_{ij}$, in which case $C^i{}_{mn} = C^i{}_{mn}(f)$. In fact, not all gravitational
    walls α_{ijk} appear if we make the choice $f^i = dx^i$ since it then follows
    from the formulas (9) that $C^d{}_{ij} = 0$ $(i, j$ arbitrary), $C^{d-1}{}_{ij} = 0$ (with
    both $i, j \neq d$) etc. Hence, $(C^d{}_{ij})^2 = 0$ and the gravitational walls α_{dij}

---

[6]In $D = 3$, there is no α_{ijk}, so this property does not hold. However, when there are p-forms present,
one can develop the same more direct alternative proof that the μ^j's can be dropped because one can then
express the μ^j as positive linear combinations of the p-form wall forms.

[7]Note two things: (i) the exponent μ_d is of the same order as $\alpha_{1\,d\,d-1}$ when $\beta^1 \simeq \beta^2 \simeq \beta^{d-1}$. This
corresponds to the case of "small oscillations" considered by BKL [BKL70, BKL82], for which they verify
(in 3 dimensions) that the evolution is indeed controlled by the α_{ijk}-terms even in that region. (ii) For the
Kasner solution, one has $\exp(-2\mu_d) \simeq t^{1-p^d} \to 0$ and so the Kasner solution never reaches the region
where μ_d is arbitrarily negatively large.

are absent. Similarly, only the gravitational walls $\alpha_{d-1\,id}$ are present among the walls $\alpha_{d-1\,ij}$. To get all the gravitational walls, one needs a non-holonomic frame f^i. The dominant gravitational wall $\alpha_{1\,d-1\,d}$ is, however, always present, and this is what matters for the billiard (when gravitational walls are relevant at all).

iv The coefficients of the dominant exponentials involve only the undifferentiated structure functions $C^i{}_{jk}$. Thus, one can mimic at each point x the gravitational potential in the context of spatially homogeneous cosmologies, where the $C^i{}_{jk}$ are constant, by considering homogeneity groups that are "sufficiently" non-Abelian so that none of the coefficients of the relevant exponentials vanishes (Bianchi types VIII and IX for $d = 3$, other homogeneity groups for $d > 3$ - see [DDRH88]).

4.4 p-form walls

4.4.1 Electric walls.
We now turn to the electric potential $V^{el}_{(p)}$, which we express in the basis $\{\theta^k\}$. One has

$$V^{el}_{(p)} = \frac{p!}{2} \sum_{i_1,i_2,\cdots,i_p} e^{-2e_{i_1\cdots i_p}(\beta)} (\mathcal{E}^{i_1\cdots i_p}_{(p)})^2 \tag{39}$$

where $\mathcal{E}^{i_1\cdots i_p}$ are the components of the electric field $\pi^{m_1\cdots m_p}_{(p)}$ in the basis $\{\theta^k\}$,

$$\mathcal{E}^{i_1\cdots i_p}_{(p)} = \sum_{m_1,\cdots,m_p} \mathcal{N}_{i_1 m_1}\mathcal{N}_{i_2 m_2}\cdots\mathcal{N}_{i_p m_p}\pi^{m_1\cdots m_p}_{(p)} \tag{40}$$

and where $e_{i_1\cdots i_p}(\beta)$ are the electric linear forms

$$e_{i_1\cdots i_p}(\beta) = \beta^{i_1} + \cdots + \beta^{i_p} + \frac{\lambda^{(p)}}{2}\phi \quad \text{(all } i_j\text{'s distinct)} \tag{41}$$

(the indices i_j's are all distinct because $\mathcal{E}^{i_1\cdots i_p}_{(p)}$ is completely antisymmetric). The variables $\mathcal{E}^{i_1\cdots i_p}$ do not depend on the β^μ. It is thus rather easy to take the BKL limit. The exponentials in (39) are multiplied by positive factors which are different from zero in the generic case. Thus, in the BKL limit, $V^{el}_{(p)}$ becomes

$$V^{el}_{(p)} \simeq \sum_{i_1<i_2<\cdots<i_p} \Theta[-2e_{i_1\cdots i_p}(\beta)]. \tag{42}$$

Note that the transformation from the variables $(\mathcal{N}_{ij}, P_{ij}, A^{(p)}_{m_1\cdots m_p}, \pi^{m_1\cdots m_p}_{(p)})$ to the variables $(\mathcal{N}_{ij}, P_{ij}, A^{(p)}_{m_1\cdots m_p}, \mathcal{E}^{i_1\cdots i_p}_{(p)})$ is a point canonical transformation whose explicit form is obtained from

$$\sum_{i<j} P_{ij}\dot{\mathcal{N}}_{ij} + \sum_{p} \pi_{(p)}^{m_1\cdots m_p} \dot{A}_{m_1\cdots m_p}^{(p)} = \sum_{i<j} P'_{ij}\dot{\mathcal{N}}_{ij} + \sum_{p} \mathcal{E}_{(p)}^{i_1\cdots i_p} \dot{\mathcal{A}}_{m_1\cdots m_p}^{(p)}$$

(43)

(The momenta P_{ij} conjugate to $\mathcal{N}_{ij}$ get redefined by terms involving $\mathcal{E}$, $\mathcal{N}$ and $\mathcal{A}$ since the components $A_{m_1\cdots m_p}^{(p)}$ of the p-forms in the basis $\{\theta^k\}$ involve the $\mathcal{N}$'s,

$$\mathcal{A}_{i_1\cdots i_p}^{(p)} = \sum_{m_1,\cdots,m_p} (\mathcal{N}^{-1})_{m_1 i_1} \cdots (\mathcal{N}^{-1})_{m_p i_p} A_{(p)m_1\cdots m_p}.$$

This does not affect the symmetry walls in the BKL limit.)

4.4.2   Magnetic walls.

The magnetic potential is dealt with similarly. Expressing it in the $\{\theta^k\}$-frame, one obtains

$$V_{(p)}^{magn} = \frac{1}{2\,(p+1)!} \sum_{i_1,i_2,\cdots,i_{p+1}} e^{-2m_{i_1\cdots i_{p+1}}(\beta)} (\mathcal{F}_{(p)\,i_1\cdots i_{p+1}})^2$$

(44)

where $\mathcal{F}_{(p)\,i_1\cdots i_{p+1}}$ are the components of the magnetic field $F_{(p)m_1\cdots m_{p+1}}$ in the basis $\{\theta^k\}$,

$$\mathcal{F}_{(p)\,i_1\cdots i_{p+1}} = \sum_{m_1,\cdots,m_{p+1}} (\mathcal{N}^{-1})_{m_1 i_1} \cdots (\mathcal{N}^{-1})_{m_{p+1} i_{p+1}} F_{(p)m_1\cdots m_{p+1}}$$

(45)

and where $m_{i_1\cdots i_{p+1}}(\beta)$ are the magnetic linear forms

$$m_{i_1\cdots i_{p+1}}(\beta) = \sum_{j\notin\{i_1,i_2,\cdots i_{p+1}\}} \beta^j - \frac{\lambda^{(p)}}{2}\phi$$

(46)

(all i_j's are distinct). One sometimes rewrites $m_{i_1\cdots i_{p+1}}(\beta)$ as $b_{i_{p+2}\cdots i_d}$, where $\{i_{p+2}, i_{p+3}, \cdots, i_d\}$ is the set complementary to $\{i_1, i_2, \cdots i_{p+1}\}$; e.g.,

$$b_{1\,2\,\cdots\,d-p-1} = \beta^1 + \cdots + \beta^{d-p-1} - \frac{\lambda^{(p)}}{2}\phi = m_{d-p\cdots d}$$

(47)

The exterior derivative $\mathcal{F}$ of $\mathcal{A}$ in the non-holonomic frame $\{\theta^k\}$ involves of course the structure coefficients, i.e., of frame $\mathcal{F}_{(p)\,i_1\cdots i_{p+1}} = \partial_{[i_1} \mathcal{A}_{i_2\cdots i_{p+1}]} + $ "$\mathcal{C}\mathcal{A}$"-terms where $\partial_{i_1} \equiv \sum_{m_1}(\mathcal{N}^{-1})_{m_1 i_1}(\partial/\partial x^{m_1})$ is here the frame derivative.

Again, the BKL limit is quite simple and yields (assuming generic magnetic fields)

$$V_{(p)}^{magn} \simeq \sum_{i_1<\cdots<i_{d-p-1}} \Theta[-2b_{i_1\cdots i_{d-p-1}}(\beta)].$$

(48)

Note that, just as the off-diagonal variables, the electric and magnetic fields get frozen in the BKL limit since the Hamiltonian no longer depends on the p-form variables. These drop out because one can rescale the coefficient of any Θ-function to be one (when it is not zero), absorbing thereby the dependence on the p-form variables.

The scale factors are constrained by the further conditions

$$e_{i_1 \cdots i_p}(\beta) \geq 0, \quad b_{i_1 \cdots i_{d-p-1}}(\beta) \geq 0. \tag{49}$$

The boundary hyperplanes $e_{i_1 \cdots i_p}(\beta) = 0$ and $b_{i_1 \cdots i_{d-p-1}}(\beta) = 0$ are called the "electric" and "magnetic" walls, respectively.

4.5   Billiards

At this point, we see that the interesting dynamics as one goes towards the spacelike singularity is carried by the scale factors since the off-diagonal variables (including the p-form variables) asymptotically drop out. Furthermore, the evolution of the scale factors at each spatial point is, in the BKL limit, a broken null straight line of the metric $G_{\mu\nu}d\beta^\mu d\beta^\nu$ interrupted by collisions again the sharp walls

$$w_A(\beta) \equiv w_{A\mu}\beta^\mu = 0 \tag{50}$$

defined by the symmetry, gravitational and p-form potentials through $V = \sum_A \Theta(-2w_A(\beta))$. All these walls are timelike hyperplanes. Indeed, we have found above that the gradients to the symmetry walls have squared norm equal to 2, see Eq. (18). Similarly, the gradients to the gravitational walls $\alpha_{ijk} = 0$ are spacelike and, in fact, have the same squared norm $+2$. Finally, the gradients to the electric wall $e_{i_1 \cdots i_p}$ have also positive squared norm

$$\frac{p(d - p - 1)}{d - 1} + \left(\frac{\lambda^{(p)}}{2}\right)^2, \tag{51}$$

which is equal to the squared norm of the corresponding magnetic walls. Incidentally, this shows that the norm of the p-form walls is invariant under electric-magnetic duality.

Because the walls are timelike, the velocity undergoes under a collision a geometric reflection in the hyperplane, which is an element of the orthochronous Lorentz group. The reflection preserves the norm and the time-orientation (hence, the velocity vector remains null and future-oriented).

The billiard is obtained by radial projection onto hyperbolic space. The billiard ball is constrained to be in the region $w_A(\beta) \geq 0$. Not all the walls are relevant since some of the inequalities $w_A(\beta) \geq 0$ are consequences of the others [DH01]. Only the dominant wall forms, in terms of which all the other wall forms can be expressed as linear combinations with non-negative coefficients, are relevant for determining the billiard. Usually, these are the symmetry walls and some of the p-form walls. The billiard is in general non-compact because some walls meet at infinity. However, even when it is non-compact, it can have finite volume.

In fact, the geodesic motion in a billiard in hyperbolic space has been much studied. It is known that this motion is chaotic or non-chaotic according to whether the billiard has finite or infinite volume [Ma69, HM79, Z84, EMcM93].

Notes:

i Because of reparametrization invariance – in particular, time redefinitions –, some indicators of chaos must be used with care in general relativity, see [CL96, IM01] for a discussion of the original mixmaster model.

ii The hyperbolic billiard description of the (3+1)-dimensional mixmaster system was first worked out by Chitre [C72] and Misner [M94]. Our derivation of the asymptotic expression for the potential follows [K93, IKM94]. The extension to higher dimensions with perfect fluid sources was considered in [KM94], without symmetry walls. Exterior form sources were investigated in [IM95, IM99] for special classes of metric and p-form configurations.

4.6   Constraints

We have just seen that in the BKL limit, the evolution equations become ordinary differential equations with respect to time. Although the spatial points are decoupled in the evolution equations, they are, however, still coupled in the constraints. These constraints just restrict the initial data and need only be imposed at one time, since they are preserved by the dynamical equations of motion. Indeed, one easily finds that

$$\dot{\mathcal{H}} = 0 \tag{52}$$

since $[\mathcal{H}(x), \mathcal{H}(x')] = 0$ in the ultralocal limit. This corresponds simply to the fact that the collisions preserve the lightlike character of the velocity vector. Furthermore, the gauge constraints (9) are also preserved in time since the Hamiltonian constraint is gauge-invariant. Finally, the momentum constraint fulfills

$$\dot{\mathcal{H}}_k(x) = \partial_k \mathcal{H} \approx 0 \tag{53}$$

It is important to observe that the restrictions on the initial data do not constrain the coefficients of the walls in the sense that these may assume non-zero values. For instance, it is well known that it is consistent with Gauss law to take non-vanishing electric and magnetic energy densities; thus the coefficients of the electric and magnetic walls are indeed generically non-vanishing. In fact, the constraints are conditions on the spatial gradients of the variables entering the wall coefficients, not on these variables themselves. In some non-generic contexts, however, the constraints could force some of the wall coefficients to be zero; the corresponding walls would thus be absent. [E.g., for vacuum gravity in four dimensions, the momentum constraints for some Bianchi homogeneous models force some symmetry wall coefficients to vanish. But this is peculiar to the homogeneous case.]

It is easy to see that the number of arbitrary physical functions involved in the solution of the BKL equations of motion is the same as in the general solution of the complete Einstein-matter equations. Indeed, the number of constraints on the initial data and the residual gauge freedom are the same in both cases. Further discussion of the constraints in the BKL context may be found in [AR01, DHRW02].

4.7   Chern-Simons or Yang-Mills couplings, Chapline-Manton terms

The addition of Chern-Simons terms, Yang-Mills or Chapline-Manton couplings does not modify the billiard picture and, furthermore, does not bring in new walls. The only change in the asymptotic dynamics is a modification of the constraints.

Yang-Mills couplings

We start with the Yang-Mills coupling terms. The contribution to the energy density from the Yang-Mills field takes the same form as for a collection of abelian 1-forms, with the replacement of the momenta $\mathcal{E}^i$ by the Yang-Mills momenta $\mathcal{E}^i_a$, $a = 1, \cdots, N$ (with N the dimension of the internal Lie algebra) and of the magnetic fields by the corresponding non-abelian field strengths. As their abelian counterparts, these do not involve the scale factors β^μ. Because of this key property, the same analysis goes through. Each electric and magnetic 1-form wall is simply repeated a number of times equal to the dimension of the Lie algebra. Gauss law is, however, modified and reads now:

$$\mathcal{D}_i \mathcal{E}^i_a \equiv \nabla_i \mathcal{E}^i_a + f^b{}_{ac} \mathcal{E}^i_b A^c_i = 0. \tag{54}$$

Here, ∇_i is the standard metric covariant derivative. Similarly, the momentum constraints are modified and involves the non-abelian field strengths.

Chapline-Manton couplings and Chern-Simons terms

The discussion of Chapline-Manton couplings or Chern-Simons terms proceeds in the same way. The energy-density of the p-forms has the same dependence on the scale factors as in the absence of couplings, i.e., provides the same exponentials. The only difference is that the wall coefficients are different functions of the p-form canonical variables; but this difference is again washed out in the sharp wall limit, where the coefficients can be replaced by one (provided they are non zero). The momentum and Gauss constraints are genuinely different and impose different conditions on the initial data.

Note: although Chern-Simons terms do not generically change the features of the billiard (shape, volume), they may play a more significant rôle in peculiar contexts where only specialized field configurations are considered. This occurs for instance in [IM99], where it is shown that the 11-dimensional supergravity Chern-Simons term, in the case of spatially homogeneous metrics and magnetic fields, may constrain some electromagnetic walls to "accidentally" disappear. This makes the otherwise finite-volume billiard to be of infinite volume.

5. Velocity-dominance - Strong coupling/small tension limit

It is sometimes useful to separate in the Hamiltonian the time derivatives (conjugate momenta) from the space derivatives. This yields

$$\mathcal{H} = K' + \varepsilon V' \tag{55}$$

where $\varepsilon = \pm 1$ according to whether the spacetime signature is Lorentzian ($\varepsilon = 1$) or Euclidean ($\varepsilon = -1$). Here,

$$K' = K + V_S + V_{(p)}^{el} \tag{56}$$

and

$$V' = V_G + V_{(p)}^{magn} + V_\phi. \tag{57}$$

The reason that this splitting is useful is that for some models, the asymptotic dynamics is entirely controlled by K', i.e., by the limit $\varepsilon = 0$. This occurs whenever the billiard that emerges in the BKL limit is defined by the symmetry and electric walls, as it happens for instance for eleven-dimensional supergravity [DH00a], or the pure Einstein-Maxwell system in spacetime dimensions $D \geq 5$ [DH00b, KS00]. Curvature and magnetic walls are then subdominant, i.e., spatial gradients become negligible as one goes toward the singularity.

If the curvature and magnetic walls can indeed be neglected, the evolution equations are exactly the same as the equations of motion obtained by performing a direct torus reduction to $1 + 0$ dimensions. We stress, however, that no homogeneity assumption has ever been made. *This effective torus dimensional reduction follows from the dynamics and is not imposed by hand.*

The limit $\varepsilon = 0$ is known as the "zero signature limit" [T78] and lies half-way between spacetimes of Minkowskian or Lorentzian signature. It corresponds to a vanishing velocity of light (or vanishing "medium tension"); the underlying geometry is built on the Carroll contraction of the Lorentz group [H79]. [The terminology "strong coupling" is also used [I76] and stems from the fact that with appropriate redefinitions, $\mathcal{H}$ can formally be rewritten as $\mathcal{H} = GK' + (1/G)V'$

where G is Newton's constant.] A revival of interest in this ultrarelativistic limit has arisen recently [D98, LS01, A02].

When the billiard has infinite volume, the dynamics in the vicinity of the singularity is even simpler because there is only a finite number of collisions with the walls. The system generically settles after a finite time in a Kasner-like motion that lasts all the way to the singularity. The asymptotic dynamics is controlled solely by the kinetic energy K of the scale factors (after all collisions have taken place). This case, where both spatial gradients and matter (here p-form) terms can be neglected, has been called "velocity-dominated" in [E72] and enables a rigorous analysis of its asymptotic dynamics by means of "Fuchsian" techniques [AR01, DHRW02]. By contrast, rigorous results concerning the finite-volume case are rare (see, however, the recent analytic advances in [R01]). Besides the existing rigorous results, one should also mention the good wealth of numerical support of the BKL ideas [BGIMW98, B02].

6. Miscellany and Conclusions

In this paper, we have shown that theories involving gravity admits a remarkable asymptotic description in the vicinity of a spacelike singularity in terms of billiards in hyperbolic space. Depending on whether the actual billiard has finite or infinite volume, the dynamical evolution of the local scale factors is chaotic and of the mixmaster type, or monotone and Kasner-like. The billiard, and in particular its volume, is a fundamental characteristic of the theory, in the sense that it is determined solely by the field content and the parameters in the Lagrangian, and not by the initial conditions (in the generic case; i.e., there may be initial conditions for which some walls are absent – and the billiard is changed –, but these are exceptional). Although we have not investigated the physical implications of this property for cosmological scenarios (in particular, string-inspired cosmologies [GV92, BDV98, LWC00, W02, GV02]), nor its quantum analog, we believe that this result is already interesting in its own right because it uncovers an intrinsic feature of gravitational theories. Furthermore, the regularity properties of the billiards appear to give a powerful access to hidden symmetries, as we now briefly discuss.

6.1 Kac-Moody billiards

The billiard description holds for all systems governed by the action (1). In general, the billiard has no notable regularity property. In particular, the dihedral angles between the faces, which depend on the (continuous) dilaton couplings (when there is a dilaton), need not be integer submultiples of π. In some instances, however, the billiard can be identified with the fundamental Weyl chamber of a symmetrizable Kac-Moody algebra of indefinite type, with Lorentzian signature metric [DH01]. One then says that the billiard is a "Kac-

Moody billiard". [See [Kac90, MP95] for information on Kac-Moody (KM) algebras.] In [DH01], superstring models were considered and the rank 10 KM algebras E_{10} and BE_{10} were shown to emerge, in line with earlier conjectures made in [J80, J85][8]. This was then further extended to pure gravity in any number of spacetime dimensions, for which the relevant KM algebra is AE_d, and it was understood that chaos (finite volume of the billiard) is equivalent to hyperbolicity of the underlying KM algebra [DHJN01]. The original case of $D = 4$ pure gravity corresponds to the hyperbolic algebra AE_3 investigated in [FF83, N92][9]. Further examples of emergence of Lorentzian KM algebras, based on the models of [BMG88, CJLP99], are given in [DdBHS02].

The walls that determine the billiards are the dominant walls. For KM billiards, they correspond to the simple roots of the KM algebra. The sub-dominant walls also have an algebraic interpretation in terms of higher height positive roots [DHN02]. This enables one to go beyond the BKL limit and to see the beginning of a possible identification of the dynamics of the scale factors *and* of all the off-diagonal variables (including the p-form variables) with that of a non-linear sigma model obtained by formally taking the coset of the KM group with its maximal compact subgroup [DHN02].

6.2   Chaos versus non-chaos

We close our paper by indicating when the models described by the action (1) exhibit mixmaster behaviour (finite volume billiard) or Kasner-like behaviour (infinite volume billiard) (for generic initial conditions).

- Pure gravity billiards have finite volume for spacetime dimension $D \leq 10$ and infinite volume for spacetime dimension $D \geq 11$ [DHS85]. This can be understood in terms of the underlying Kac-Moody algebra [DHJN01].

- Gravity + dilaton has always an infinite volume billiard [BK73].

- Gravity + p-forms (with $p \neq 0$ and $p < D - 2$) and *no* dilaton has a finite volume billiard [DH00b]. In particular, 11-dimensional supergravity exhibits mixmaster behaviour, while vacuum gravity in 11 dimensions does not. The 3-form is crucial for closing the billiard. Similarly, the Einstein-Maxwell system in 4 (in fact any number of) dimensions has an finite-volume billiard (see [J86, L97, W99] for a discussion of four-dimensional homogeneous models with Maxwell field exhibiting mixmaster behaviour).

---

[8]Note that the Weyl groups of the E-family have been discussed in a similar vein in the context of U-duality [LPS96, OPR98].

[9]Note that in the original analysis of [BKL70, BKL82, C72, M94], the symmetry walls are not included; the KM algebra that arises has a 3×3 Cartan matrix given by $A_{ii} = 2$, $A_{ij} = -2$ $(i \neq j)$ and its fundamental Weyl chamber (radially projected on H_2) is the ideal equilateral triangle having its 3 vertices at infinity.

- The volume of the mixed Einstein-dilaton-p-form system depends on the dilaton couplings. For a given spacetime dimension D and a given menu of p-forms there exists a "subcritical" domain $\mathcal{D}$ in the space of of the dilaton couplings (an open neighbourhood of the origin $\lambda^{(p)} = 0$) such that: (i) when the dilaton couplings $\lambda^{(p)}$ belong to $\mathcal{D}$ the general behaviour is Kasner-like, but (ii) when the $\lambda^{(p)}$ do not belong to $\mathcal{D}$ the behaviour is oscillatory [DH00a, DHRW02]. For all the superstring models, the dilaton couplings do not belong to the critical domain and the billiard has finite volume.

Acknowledgments

M.H. is grateful to the organizers of the "School on Quantum Gravity" held at CECS Valdivia, Chile (January 2002), where some of the material contained in this article was presented. He is also grateful to Claudio Teitelboim for numerous discussions on the zero-velocity-of-light limit, which arose his interest in the BKL analysis. The work of M.H. is supported in part by the "Actions de Recherche Concertées" of the "Direction de la Recherche Scientifique - Communauté Française de Belgique", by a "Pôle d'Attraction Interuniversitaire" (Belgium), by IISN-Belgium (convention 4.4505.86), by Proyectos FONDE-CYT 1970151 and 7960001 (Chile) and by the European Commission RTN programme HPRN-CT-00131, in which he is associated to K. U. Leuven. H.N. is partially supported by the EU contract HPRN-CT-2000-00122. Both M.H. and H.N. would like to thank I.H.E.S. for hospitality while this work was carried out.

Appendix: A. Iwasawa decomposition and asymptotics of non-diagonal $3d$ Kasner metric

The Iwasawa decomposition for three-dimensional metrics has been explicitly analyzed in [HPT82]. Setting

$$\mathcal{N} = \begin{pmatrix} 1 & n_1 & n_2 \\ 0 & 1 & n_3 \\ 0 & 0 & 1 \end{pmatrix} \tag{A.1}$$

together with

$$\mathcal{A} = \begin{pmatrix} \exp(-\beta^1) & 0 & 0 \\ 0 & \exp(-\beta^2) & 0 \\ 0 & 0 & \exp(-\beta^3) \end{pmatrix}, \tag{A.2}$$

one finds

$$
\begin{aligned}
g_{11} &= e^{-2\beta^1}, \quad g_{12} = n_1 e^{-2\beta^1}, \quad g_{13} = n_2 e^{-2\beta^1}, & \text{(A.3)} \\
g_{22} &= (n_1)^2 e^{-2\beta^1} + e^{-2\beta^2}, \quad g_{23} = n_1 n_2 e^{-2\beta^1} + n_3 e^{-2\beta^2}, & \text{(A.4)} \\
g_{33} &= (n_2)^2 e^{-2\beta^1} + (n_3)^2 e^{-2\beta^2} + e^{-2\beta^3} & \text{(A.5)}
\end{aligned}
$$

from which one gets

$$\beta^1 = -\frac{1}{2}\ln g_{11}, \quad \beta^2 = -\frac{1}{2}\ln\left[\frac{g_{11}g_{22} - g_{12}^2}{g_{11}}\right], \tag{A.6}$$

$$\beta^3 = -\frac{1}{2}\ln\left[\frac{g}{g_{11}g_{22} - g_{12}^2}\right], \quad n_1 = \frac{g_{12}}{g_{11}}, \tag{A.7}$$

$$n_2 = \frac{g_{13}}{g_{11}}, \quad n_3 = \frac{g_{23}g_{11} - g_{12}g_{13}}{g_{11}g_{22} - g_{12}^2}. \tag{A.8}$$

On the other hand, (1) yields

$$g_{ij} = t^{2p^1} l_i l_j + t^{2p^2} m_i m_j + t^{2p^3} r_i r_j \tag{A.9}$$

with

$$L = \begin{pmatrix} l_1 & l_2 & l_3 \\ m_1 & m_2 & m_3 \\ r_1 & r_2 & r_3 \end{pmatrix}. \tag{A.10}$$

Combining these relations, one obtains the time dependence of the Iwasawa variables

$$\beta^1 = -\frac{1}{2}\ln X, \quad \beta^2 = -\frac{1}{2}\ln\left[\frac{Y}{X}\right], \tag{A.11}$$

$$\beta^3 = -\frac{1}{2}\ln\left[\frac{t^{2(p^1+p^2+p^3)}(\det L)^2}{Y}\right], \tag{A.12}$$

$$n_1 = \frac{t^{2p^1} l_1 l_2 + t^{2p^2} m_1 m_2 + t^{2p^3} r_1 r_2}{X}, \tag{A.13}$$

$$n_2 = \frac{t^{2p^1} l_1 l_3 + t^{2p^2} m_1 m_3 + t^{2p^3} r_1 r_3}{X}, \quad n_3 = \frac{Z}{Y} \tag{A.14}$$

with

$$X = t^{2p^1}(l_1)^2 + t^{2p^2}(m_1)^2 + t^{2p^3}(r_1)^2, \tag{A.15}$$

$$Y = t^{2p^1+2p^2}(l_1 m_2 - l_2 m_1)^2 + t^{2p^1+2p^3}(l_1 r_2 - l_2 r_1)^2$$
$$+ t^{2p^2+2p^3}(m_1 r_2 - m_2 r_1)^2, \tag{A.16}$$

$$Z = t^{2p^1+2p^2}(l_1 m_2 - l_2 m_1)(l_1 m_3 - l_3 m_1)$$
$$+ t^{2p^1+2p^3}(l_1 r_2 - l_2 r_1)(l_1 r_3 - l_3 r_1)$$
$$+ t^{2p^2+2p^3}(m_1 r_2 - m_2 r_1)(m_1 r_3 - m_3 r_1). \tag{A.17}$$

Without loss of generality, one can assume $p^1 \leq p^2 \leq p^3$. If necessary, this can be achieved by multiplying L by an appropriate permutation matrix. We shall in fact consider the case $p^1 < p^2 < p^3$, leaving the discussion of the limiting situations to the reader. One then finds, for generic L's ($l_1 \neq 0$, $r_1 \neq 0$, $l_1 m_2 - l_2 m_1 \neq 0$, $m_1 r_2 - m_2 r_1 \neq 0$), the following asymptotic behaviour:

$$\tau \to -\infty:$$

$$\beta^1 \sim p^3\tau, \quad \beta^2 \sim p^2\tau, \quad \beta^3 \sim p^1\tau,$$

$$n_1 \to \frac{r_2}{r_1}, \quad n_2 \to \frac{r_3}{r_1}, \quad n_3 \to \frac{m_1 r_3 - m_3 r_1}{m_1 r_2 - m_2 r_1} \tag{A.18}$$

$$\tau \to \infty :$$

$$\beta^1 \sim p^1\tau, \ \beta^2 \sim p^2\tau, \ \beta^3 \sim p^3\tau,$$

$$n_1 \to \frac{l_2}{l_1}, \ n_2 \to \frac{l_3}{l_1}, \ n_3 \to \frac{l_1 m_3 - l_3 m_1}{l_1 m_2 - l_2 m_1}. \qquad (A.19)$$

It is rather clear that the n_i's should asymptotically tend to constants since they are homogeneous functions of degree zero in the g_{ij}'s - in fact, ratios of polynomials of degree one or two in the t^{2p^i}. It is a bit more subtle that the scale factors $\exp(-2\beta^i)$, which are homogeneous of degree one in the g_{ij}'s, are not all driven by the fastest growing term (t^{2p^3} for $t \to \infty$ or t^{2p^1} for $t \to 0^+$). This is actually true only for the first scale factor $\exp(-2\beta^1)$. The second scale factor $\exp(-2\beta^2)$ feels the next subleading term t^{2p^2} because the fastest growing term drops from its numerator, equal to the minor $g_{11}g_{22} - (g_{12})^2$. Similar cancellations occur for the last scale factor $\exp(-2\beta^3)$, which feels only the smallest term t^{2p^1} ($t \to \infty$) or t^{2p^3} ($t \to 0^+$).

We come thus to the conclusion that the asymptotic behaviour of the Iwasawa variables is indeed simple: the scale factors asymptotically behave as the scale factors of the diagonal Kasner solutions, while the parameters n_i parametrizing the off-diagonal components approach constants. The "out"-values of the Kasner exponents differ from their "in"-values by a permutation such that the inequalities $\beta^1 \leq \beta^2 \leq \beta^3$ hold both for $\tau \to +\infty$ and $\tau \to -\infty$.

The frame $\{l, m, r\}$ where the spatial metric and the extrinsic curvature are simultaneously diagonal is called the "Kasner frame". Given that the time slicing has been fixed, this geometric frame is unique – up to individual normalization of each basis vector – when the eigenvalues of the extrinsic curvature are distinct, if one prescribes in addition some definite ordering of the eigenvalues (i.e. $p_1 < p_2 < p_3$ as above). For asymptotic values of τ, the Iwasawa frames $\{\theta^1, \theta^2, \theta^3\}$ (which are by contrast not unique since one can redefine the coordinates x^i) become approximately time-independent and hence, the extrinsic curvature become approximately diagonal. Yet, the Iwasawa frames do not tend to align with the Kasner frame. To understand this point, we assume for definiteness that $\tau \to +\infty$. In that case, one gets from (A.19) that

$$\theta^1 \to \lambda l, \quad \theta^2 \to \mu l + \nu m, \quad \theta^3 \to \varphi l + \chi m + \psi r \qquad (A.20)$$

where $\lambda = l_1^{-1}$, $\mu = -m_1/(l_1 m_2 - l_2 m_1)$, $\nu = l_1/(l_1 m_2 - l_2 m_1)$, and φ, χ, ψ are some constants. This implies that the misalignment of the Iwasawa frames with respect to the Kasner frame is a small effect, in the sense that it induces a small change in the metric – even though the coefficients $\lambda, \mu, \nu, \varphi, \chi$ and ψ are of order unity. Indeed, the change of frame $\{l, m, r\} \to \{l, m + \alpha l, r + \beta l + \gamma m\}$, with α, β, γ constants[10], induces changes Δg_{ab} which fulfill the smallness condition $\Delta g_{ab} << \sqrt{g_{aa}}\sqrt{g_{bb}}$ (where g_{ab} is the metric in the frame $\{l, m, r\}$). For instance, $\Delta g_{11} \sim t^{2p_2} << t^{2p_1}$. Thus, it is perfectly consistent to find that the Iwasawa frames become asymptotically constant as $\tau \to +\infty$ without aligning with the principal axes of the extrinsic curvature. [Note that the extrinsic curvature becomes asymptotically diagonal in the Iwasawa frame precisely in the sense that the off-diagonal components fulfill $|K_{ij}| << \sqrt{|K_{ii}|}\sqrt{|K_{jj}|}$ ($i \neq j$) for large τ. The mixed components $K^i{}_j$ tend to a non-diagonal matrix, however.]

---

[10] In order to preserve diagonality of the metric, there are of course unwritten additional small correction terms to the new vectors; for instance, the new vector l contains a term proportional to m with coefficient $t^{-2(p_1 - p_2)}$.

Appendix: B. Freezing the off-diagonal variables: a toy model

We have seen in the text that the off-diagonal variables and the p-form variables get frozen to constant values in the BKL limit. We provide here a more detailed understanding of this property by discussing a simpler model which shares the same features.

Consider a system with two canonically conjugate pairs (q, p), (Q, P) and time-dependent Hamiltonian

$$H = \frac{1}{2}p^2 + \frac{1}{2}P^2 \rho^k e^{-\rho q} \qquad (k \text{ positive integer}) \tag{B.1}$$

where ρ is $\rho \equiv \exp(T)$, with T the time. One can think of (q, p) as mimicking the scale factors, while (Q, P) mimicks the off-diagonal components or the p-form variables. In (B.1), there is only one potential wall for q (namely, the second term). We shall consider later the case with several walls.

One has $\dot{P} = 0$, hence $P = P_0$ where P_0 is a constant which we assume to be different from zero. In the limit of large times, the motion in q is a free motion interrupted by a collision against the potential wall,

$$q = |T - T_0| + q_0 \tag{B.2}$$

where T_0 is the time of the collision and q_0 the turning point. We take a unit initial velocity. The location of the turning point is determined by

$$P_0^2 \rho_0^k e^{-\rho_0 q_0} = 1. \tag{B.3}$$

The time length ΔT of the collision is roughly of the order $1/\rho_0$: the later the collision, the sharper the wall. Let us evaluate the change in Q in the collision. To that end, we need to compute

$$\Delta Q = P_0 \int_{-\infty}^{\infty} dT \rho^k e^{-\rho(|T - T_0| + q_0)} \tag{B.4}$$

since $\dot{Q} = P\rho^k \exp(-\rho q)$. The integrand is maximum at $T = T_0$. We can approximate the integral by the value at the maximum times the time length of the collision. Using (B.3), one gets

$$\Delta Q \approx \frac{1}{P_0 \rho_0}. \tag{B.5}$$

Hence, the variable Q receives a kick during the collision (which can be of order one at early times), but the later the collision, the smaller the kick.

Assume now that there is another wall with the same time dependence, say at $q = d$, so that q bounces between these two walls,

$$V_{additional} = \frac{1}{2}P^2 \rho^k e^{-\rho(d-q)}.$$

Because the speed of q remains constant (in the large T limit), the collisions are equally spaced in T. At each collision, Q receives a kick of order $1/\rho_0$. The total change in Q is obtained by summing all the individual changes, which yields

$$(\Delta Q)_{Total} \sim \sum_n e^{nd} \tag{B.6}$$

(the time interval between two collisions is d since we assumed unit velocity). This sum converges. After a while, one can neglect the further change in Q, i.e., assume $\dot{Q} = 0$. The Hamiltonian describing the large time limit is obtained by taking the sharp wall limit in the above H, and reads therefore

$$H = \frac{1}{2}p^2 + \Theta(-q) + \Theta(q - d). \tag{B.7}$$

The pair (Q, P) drops out because it is asymptotically frozen. Our analysis justifies taking the sharp wall limit directly in H for this system, which is the procedure we followed in the text to get the gravitational billiards.

Appendix: C. Kasner frame versus Iwasawa frames

In the original analysis of BKL [BKL70, BKL82], the description of the evolution of the fields is not carried out in the Iwasawa frames defined algebraically using the Iwasawa decomposition, but rather in the geometrically defined Kasner frame where the spatial metric and the extrinsic curvature are simultaneously diagonal and the Kasner exponents are ordered (during the Kasner "epoch" under study). Belinskii, Khalatnikov and Lifshitz found that the Kasner axes undergo, under collisions with the gravitational walls, changes of order unity with respect to time-independent spatial frames (frames having zero Lie bracket with $\partial/\partial t$), no matter how close one gets to the singularity [BKL72]; therefore, the Kasner axes generically never come to rest if there is an infinite number of collisions. The purpose of this appendix is to reconcile this result with the above conclusion that the off-diagonal components $\mathcal{N}_{ij}$ tend to constants as one approaches the singularity. The key point is, of course, that Kasner frames and Iwasawa frames do not coincide in general. We shall treat explicitly the $3 + 1$-case.

In the Kasner frame, the metric takes the form of Eq. (A.9) with t^{2p_1}, t^{2p_2} and t^{2p_3} replaced respectively by a^2, b^2 and c^2. Far from the gravitational walls, the functions a^2, b^2 and c^2 are given by t^{2p_1}, t^{2p_2} and t^{2p_3} where p_1, p_2 and p_3 depend only on the spatial coordinates, and the frame components l_i, m_i and n_i are also time-independent. Neither of these properties hold in the collision region. The collision against the gravitational wall induces the familiar transition between the Kasner exponents

$$p_1' = \frac{-p_1}{1 + 2p_1}, \quad p_2' = \frac{p_2 + 2p_1}{1 + 2p_1}, \quad p_3' = \frac{p_3 + 2p_1}{1 + 2p_1} \tag{C.1}$$

and also a change in the frame components given by [BKL72]

$$l_i' = l_i, \quad m_i' = m_i + \sigma_m l_i, \quad r_i' = r_i + \sigma_r l_i \tag{C.2}$$

where σ_m and σ_r are of order unity. This formula holds for pure gravitational collisions (far from the symmetry walls), i.e., under the assumption that both b^2 and c^2 are very small compared with a^2 [BKL72] so that only one term in the potential is non-negligible, namely, the curvature term proportional to a^4 (the only case we shall consider explicitly). To avoid "interference" with the symmetry wall $b^2 = c^2$, we impose also the condition $c^2 << b^2$, although this is actually not necessary for showing that the Iwasawa parameters n_i are constant. Now, under the assumption $b^2 << a^2$ and $c^2 << b^2$, it is easy to see, using the formulas (A.11) through (A.17) with t^{2p_1}, t^{2p_2} and t^{2p_3} replaced by a^2, b^2 and c^2, that n_1, n_2 and n_3 are respectively given by

$$n_1 = \frac{l_2}{l_1}, \quad n_2 = \frac{l_3}{l_1}, \quad n_3 = \frac{l_1 m_3 - l_3 m_1}{l_1 m_2 - l_2 m_1} \tag{C.3}$$

(just like in (A.19)). It is clear that if we substitute in these formulas l_i', m_i' and r_i' for l_i, m_i and r_i according to (C.2), we get no change in the off-diagonal variables n_1, n_2 and n_3, as we wanted to show. There is thus no contradiction between the change of Kasner axes and the freezing of the off-diagonal Iwasawa variables. The same conclusion holds for collisions against the other types of walls, where the Kasner axes "rotate" as in (C.2).

Appendix: D. Hamiltonian reduction

We provide in this appendix a derivation of the BKL limit through a (partial) Hamiltonian reduction of the dynamics, along the lines discussed in subsection 2.2 for homogeneous models of Bianchi type I. ["Partial" because we only take care of the Hamiltonian constraint; the other constraints must still be imposed.] Our considerations are of a purely local nature.

We proceed as in section 4 but, to begin with, we impose only the orthogonal gauge condition $N^i = 0$, without fixing the lapse. The metric reads thus

$$ds^2 = -(\tilde{N}\sqrt{g})^2(dx^0)^2 + g_{ij}(x^0, x^i)dx^i dx^j. \tag{D.1}$$

The action is, in terms of the Iwasawa variables,

$$S[\beta^\mu, \mathcal{N}_{ij}, \pi_\mu, P_{ij}, A^{(p)}_{m_1\cdots m_p}, \pi^{m_1\cdots m_p}_{(p)}] =$$

$$\int dx^0 \left(\int d^d x (\pi_\mu \dot{\beta}^\mu + \sum_{i<j} P_{ij}\dot{\mathcal{N}}_{ij} + \sum_p \pi^{m_1\cdots m_p}_{(p)} \dot{A}^{(p)}_{m_1\cdots m_p}) - H \right) \tag{D.2}$$

where the Hamiltonian is

$$H = \int d^d x\, \tilde{N}\mathcal{H} \tag{D.3}$$

(with $\mathcal{H}$ given by (12)).

As explained in the text, our main assumption is that $\beta^\mu \beta_\mu$ is < 0 for large x^0 and that $\rho^2 = -\beta^\mu \beta_\mu$ monotonously tends to $+\infty$ as $x^0 \to +\infty$ (once x^0 is big enough). In terms of $T \equiv \ln \rho$ and the hyperboloid variables γ^μ (constrained by $\gamma^\mu \gamma_\mu = -1$), the action is

$$S[T, \gamma^\mu, \mathcal{N}_{ij}, \pi_T, \tilde{\pi}_\mu, P_{ij}, A^{(p)}_{m_1\cdots m_p}, \pi^{m_1\cdots m_p}_{(p)}] =$$

$$\int dx^0 \left(\int d^d x (\pi_T \dot{T} + \tilde{\pi}_\mu \dot{\gamma}^\mu + \sum_{i<j} P_{ij}\dot{\mathcal{N}}_{ij} + \sum_p \pi^{m_1\cdots m_p}_{(p)} \dot{A}^{(p)}_{m_1\cdots m_p}) - H \right)$$

$$\tag{D.4}$$

with

$$H = \int d^d x \frac{\tilde{N}}{4\rho^2} \tilde{\mathcal{H}} \tag{D.5}$$

$$\tilde{\mathcal{H}} = -\pi_T^2 + (\tilde{\pi}_\mu)^2 + 4\rho^2(V_S + V_G + V_\phi + \sum_p V_{(p)}) \tag{D.6}$$

Here, $\tilde{\pi}_\mu$ are the (constrained) momenta conjugate to the hyperboloid variables. The coordinate T is clearly a timelike variable in the space of the scale factors; large T is the same as large $\sum_i \beta^i$ or small g, exactly as in the homogeneous case (the formulas are the same since the supermetric is ultralocal).

Since T is assumed to monotonously increase to $+\infty$ as $x^0 \to +\infty$, we can use it as a time coordinate, i.e., impose the gauge condition $T = x^0$. The reduced action in that gauge is

$$S[\gamma^\mu, \mathcal{N}_{ij}, \tilde{\pi}_\mu, P_{ij}, A^{(p)}_{m_1\cdots m_p}, \pi^{m_1\cdots m_p}_{(p)}] =$$

$$\int dT \left[\int d^d x (\tilde{\pi}_\mu \frac{d\gamma^\mu}{dT} + \sum_{i<j} P_{ij} \frac{d\mathcal{N}_{ij}}{dT} \right.$$

$$\left. + \sum_p \pi^{m_1\cdots m_p}_{(p)} \frac{dA^{(p)}_{m_1\cdots m_p}}{dT}) - H_T \right] \tag{D.7}$$

where $H_T = -p_T$ is the Hamiltonian in the gauge $T = x^0$ (for which $\dot{T} = 1$),

$$H_T = \sqrt{(\tilde{\pi}_\mu)^2 + \rho^2(V_S + V_G + V_\phi + \sum_p V_{(p)})}. \tag{D.8}$$

H_T is explicitly time-dependent (through $\rho = \exp(T)$).

So far, no approximation has been made. We now investigate the large time (large T) limit, which can be taken since T is assumed to monotonously increase to $+\infty$. This is similar to investigating the large time dynamics of a Hamiltonian system with time-dependent Hamiltonian $H(q, p, t)$ by taking the $t \to \infty$ limit directly in the Hamiltonian (if it exists). As explained in the text, the potential becomes in that limit a sum of sharp wall potentials, so that one can replace (D.8) by

$$H_T = \sqrt{(\tilde{\pi}_\mu)^2 + \sum_A \Theta(-2w_A(\gamma))},$$
(D.9)

which is time-independent. Because the walls are timelike and the free motion is lightlike, the asymptotic motion of the scale factors is a succession of future-oriented lightlike straight line segments and hence, it is indeed timelike. This provides a self-consistency check of the assumption that ρ increases and tends to infinity.

Of course, the replacement of (D.8) by (D.9) is permissible only if the coefficients of the exponentials do not grow too fast, so that, as $T \to +\infty$, the variables $\mathcal{N}_{ij}$, P_{ij}, $A_{m_1 \cdots m_p}^{(p)}$ and $\pi_{(p)}^{m_1 \cdots m_p}$ as well as the spatial derivatives of the scale factors do not outgrow the exponentials. This other consistency check is also verified since we have shown that $\mathcal{N}_{ij}$, P_{ij}, $A_{m_1 \cdots m_p}^{(p)}$ and $\pi_{(p)}^{m_1 \cdots m_p}$ get asymptotically frozen. Similarly, from (17), one sees that, between collisions, $\partial \beta \sim \ln t$ and $\partial^2 \beta \sim \ln t$ (with a coefficient of order one which changes in each collision) so that the terms $(\partial \beta)^2$ or $\partial^2 \beta$ that multiply the subdominant gravitational walls do not outgrow the exponentials. [This is actually a bit trickier because it is not entirely clear that the coefficients of $\ln t$ remain of order one during collisions. This is because the evolution is independent at each spatial point, so that the β's might not remain differentiable and the spatial gradients might become more singular. This has been argued to lead to a kind of turbulent gravitational behaviour in which energy is pumped into shorter and shorter length scales [KK87, B92].]

Since the Hamiltonan does not depend explicitly on time, $\pi_T \equiv -H_T$ is constant in time. On the other hand, $\pi_T = \pi_\mu \beta^\mu$ transforms non-trivially under spatial coordinate transformations, so one can achieve, locally at least, $\pi_T = 1$ by a spatial diffeomorhism. The rescaled lapse is determined by the gauge condition $T = x^0$ and the equation for T,

$$1 = \dot{T} = \frac{\tilde{N}}{2\rho^2} \pi_T = \frac{\tilde{N}}{2\rho^2}$$
(D.10)

This shows that in the gauge $T = x^0$, the rescaled lapse $\tilde{N}$ reduces to $2\rho^2$ and depends accordingly (asymptotically) only on time, which implies that the equal-time slices in the gauge $T = x^0$ are the same as the equal-time slices of the pseudo-Gaussian gauge ($\tilde{N} = 1$) considered in the text. The difference between the two cooordinate systems is a mere space-independent relabeling of the time coordinate. [If one imposes the further spatial coordinate condition that $\sqrt{g}$ does not depend on the spatial coordinates, these slices are also the slices of a Gaussian coordinate system.]

There are thus two equivalent descriptions of the asymptotic evolution:

- the reduced description, in which the motion is at each spatial point a unit velocity "relativistic" billiard motion in hyperbolic space γ^μ with Hamiltonian (D.9);

- the unreduced description, in which the motion is a lightlike motion in the space of all the scale factors β^μ, interrupted by collisions against the hyperplanes $w_A(\beta) = 0$ ($\Leftrightarrow w_A(\gamma) = 0$).

In the second description, the Minkowskian time ($\sim \sum \beta^\mu$) between two collisions grows and there is, at each collision, a redshift of the momentum because the walls are receding. In the first, projected description, the walls are fixed, so that the (average) time between two collisions is constant, as well as the average change in the momenta conjugate to the reduced variables. To be more precise, the momenta $\tilde{\pi}^\mu$ conjugate to the γ^μ's remain of order unity, while the momenta $\pi^\mu \sim \rho^{-1}\tilde{\pi}^\mu$ conjugate to the β^μ's go to zero.

Notes

i We have stressed that the symmetry, dominant gravitational and p-form walls are all timelike. This provides an important consistency check of the BKL picture. The only spacelike wall that we know of is the cosmological constant term $\Lambda g \sim \exp[-2(\sum_i \beta^i)]$. Depending on the initial conditions – which do indeed set the scales –, this wall either prevents the system to reach the BKL small volume regime (there could be a bounce like in the de Sitter solution) or does not prevent the collapse, in which case it does not affect the BKL picture since the cosmological potential Λg goes to zero as g goes to zero.

ii **Number of collisions** Since the the Hamiltonian is asymptotically T-independent in the reduced description, the number of collisions per unit time T is asymptotically constant. Hence, T is a measure of the number of collisions. One has $g \sim t^2$ and $T \sim \ln \ln t$ so the number of collisions goes like $\ln \ln t$.

References

[A02] E. Anderson, "Strong-coupled relativity without relativity," arXiv:gr-qc/0205118.

[AR01] L. Andersson and A.D. Rendall, "Quiescent cosmological singularities," Commun. Math. Phys. **218**, 479-511 (2001) [arXiv:gr-qc/0001047].

[B92] V. A. Belinsky, JETP Letters **56**, 422 (1992).

[BK73] V.A. Belinskii and I.M. Khalatnikov, "Effect of scalar and vector fields on the nature of the cosmological singularity," Sov. Phys. JETP **36**, 591-597 (1973).

[BKL70] V.A. Belinskii, I.M. Khalatnikov and E.M. Lifshitz, "Oscillatory approach to a singular point in the relativistic cosmology," Adv. Phys. **19**, 525 (1970);

[BKL72] V.A. Belinskii, I.M. Khalatnikov and E.M. Lifshitz, "Construction of a general cosmological solution of the Einstein equation with a time singularity", Sov. Phys. JETP **35**, 838-841 (1972).

[BKL82] V.A. Belinskii, I.M. Khalatnikov and E.M. Lifshitz, "A general solution of the Einstein equations with a time singularity," Adv. Phys. **31**, 639 (1982).

[B02] B. K. Berger, "Numerical Approaches to Spacetime Singularities," arXiv:gr-qc/0201056.

[BGIMW98] B. K. Berger, D. Garfinkle, J. Isenberg, V. Moncrief and M. Weaver, "The singularity in generic gravitational collapse is spacelike, local, and oscillatory," Mod. Phys. Lett. A **13**, 1565 (1998) [arXiv:gr-qc/9805063].

[BRWN82] E. Bergshoeff, M. de Roo, B. de Wit and P. van Nieuwenhuizen, "Ten-Dimensional Maxwell-Einstein Supergravity, Its Currents, And The Issue Of Its Auxiliary Fields," Nucl. Phys. B **195**, 97 (1982).

[BMG88] P. Breitenlohner, D. Maison and G. W. Gibbons, "Four-Dimensional Black Holes From Kaluza-Klein Theories," Commun. Math. Phys. **120**, 295 (1988).

[BDV98] A. Buonanno, T. Damour and G. Veneziano, "Pre-big bang bubbles from the gravitational instability of generic string vacua," Nucl. Phys. B **543**, 275 (1999) [arXiv:hep-th/9806230].

[CM83] G. F. Chapline and N. S. Manton, "Unification Of Yang-Mills Theory And Supergravity In Ten-Dimensions," Phys. Lett. B **120**, 105 (1983).

[CB83] D. F. Chernoff and J. D. Barrow, Phys. Rev. Lett. **50**, 134 (1983).

[C72] D.M. Chitre, Ph. D. thesis, University of Maryland, 1972.

[CL96] N. J. Cornish and J. J. Levin, "The mixmaster universe is chaotic," Phys. Rev. Lett. **78**, 998 (1997) [arXiv:gr-qc/9605029]; "The mixmaster universe: A chaotic Farey tale," Phys. Rev. D **55**, 7489 (1997) [arXiv:gr-qc/9612066].

[CJLP99] E. Cremmer, B. Julia, H. Lu and C. N. Pope, arXiv:hep-th/9909099.

[CJS78] E. Cremmer, B. Julia and J. Scherk, "Supergravity Theory In 11 Dimensions," Phys. Lett. B **76**, 409 (1978).

[DdBHS02] T. Damour, S. de Buyl, M. Henneaux and C. Schomblond, "Einstein billiards and overextensions of finite-dimensional simple Lie algebras," arXiv:hep-th/0206125.

[DH00a] T. Damour and M. Henneaux, "Chaos in superstring cosmology," Phys. Rev. Lett. **85**, 920 (2000) [arXiv:hep-th/0003139]; [See also short version in Gen. Rel. Grav. **32**, 2339 (2000).]

[DH00b] T. Damour and M. Henneaux, "Oscillatory behaviour in homogeneous string cosmology models," Phys. Lett. B **488**, 108 (2000) [arXiv:hep-th/0006171].

[DH01] T. Damour and M. Henneaux, "E(10), BE(10) and arithmetical chaos in superstring cosmology," Phys. Rev. Lett. **86**, 4749 (2001) [arXiv:hep-th/0012172].

[DHJN01] T. Damour, M. Henneaux, B. Julia and H. Nicolai, "Hyperbolic Kac-Moody algebras and chaos in Kaluza-Klein models," Phys. Lett. B **509**, 323 (2001) [arXiv:hep-th/0103094].

[DHN02] T. Damour, M. Henneaux and H. Nicolai, "E(10) and a 'small tension expansion' of M theory," arXiv:hep-th/0207267.

[DHRW02] T. Damour, M. Henneaux, A. D. Rendall and M. Weaver, "Kasner-like behaviour for subcritical Einstein-matter systems," arXiv:gr-qc/0202069, to appear in Ann. Inst. H. Poincaré.

[D98] G. Dautcourt, "On the ultrarelativistic limit of general relativity," Acta Phys. Polon. B **29**, 1047 (1998) [arXiv:gr-qc/9801093].

[DDRH88] J. Demaret, Y. De Rop and M. Henneaux, "Chaos In Nondiagonal Spatially Homogeneous Cosmological Models In Space-Time Dimensions > 10," Phys. Lett. B **211**, 37 (1988).

[DHHST86] J. Demaret, J.L. Hanquin, M. Henneaux, P. Spindel and A. Taormina, "The Fate Of The Mixmaster Behavior In Vacuum Inhomogeneous Kaluza-Klein Cosmological Models," Phys. Lett. B **175**, 129 (1986).

[DHS85] J. Demaret, M. Henneaux and P. Spindel, "Nonoscillatory Behavior In Vacuum Kaluza-Klein Cosmologies," Phys. Lett. **164B**, 27 (1985).

[D67] B. S. Dewitt, "Quantum Theory Of Gravity. 1. The Canonical Theory," Phys. Rev. **160**, 1113 (1967).

[E72] D. Eardley, E. Liang and R. Sachs, "Velocity-Dominated Singularities In Irrotational Dust Cosmologies," J. Math. Phys. **13**, 99 (1972).

[EMcM93] A. Eskin and C. McMullen, "Mixing, counting and equidistribution in Lie groups", Duke Math. J. **71**, 181-209 (1993).

[FF83] A.J. Feingold and I.B. Frenkel, Math. Ann. **263**, 87 (1983)

[GV92] M. Gasperini and G. Veneziano, "Pre - big bang in string cosmology," Astropart. Phys. **1**, 317 (1993) [arXiv:hep-th/9211021].

[GV02] M. Gasperini and G. Veneziano, "The pre-big bang scenario in string cosmology," arXiv:hep-th/0207130.

[HP70] S. W. Hawking and R. Penrose, "The Singularities Of Gravitational Collapse And Cosmology," Proc. Roy. Soc. Lond. A **314**, 529 (1970).

[H78] S. Helgason, "Differential Geometry, Lie Groups, and Symmetric Scpaces", Graduate Studies in Mathematics vol. 34, American Mathematical Society, Providence 2001

[H79] M. Henneaux, "Geometry Of Zero Signature Space-Times," Print-79-0606 (Princeton), published in Bull. Soc. Math. Belg. **31**, 47 (1979) (note the misprints in the published version, absent in the preprint version).

[HPT82] M. Henneaux, M. Pilati and C. Teitelboim, "Explicit Solution For The Zero Signature (Strong Coupling) Limit Of The Propagation Amplitude In Quantum Gravity," Phys. Lett. B **110**, 123 (1982).

[HM79] R. E. Howe and C. C. Moore, "Asymptotic properties of unitary representations," J. Functional Analysis **32**, 72-96 (1979)

[IM01] G. Imponente and G. Montani, "On the Covariance of the Mixmaster Chaoticity," Phys. Rev. D **63**, 103501 (2001) [arXiv:astro-ph/0102067].

[I76] C. J. Isham, Proc. Roy. Soc. Lond. A **351**, 209 (1976).

[IKM94] V. D. Ivashchuk, A. A. Kirillov and V. N. Melnikov, JETP Lett. **60**, N4 (1994).

[IM95] V.D. Ivashchuk and V.N. Melnikov, "Billiard Representation For Multidimensional Cosmology With Multicomponent Perfect Fluid Near The Singularity," Class. Quantum Grav. **12**, 809 (1995).

[IM99] V. D. Ivashchuk and V. N. Melnikov, "Billiard representation for multidimensional cosmology with intersecting p-branes near the singularity," J. Math. Phys. **41**, 6341 (2000) [arXiv:hep-th/9904077].

[J86] R.T. Jantzen, "Finite-dimensional Einstein-Maxwell-scalar-field system," Phys. Rev. D **33**, 2121-2135 (1986).

[J01] R. T. Jantzen, "Spatially homogeneous dynamics: A unified picture," arXiv:gr-qc/0102035.

[J80] B. Julia, LPTENS 80/16, Invited paper presented at Nuffield Gravity Workshop, Cambridge, Eng., Jun 22 - Jul 12, 1980.

[J85] B. Julia, in *Lectures in Applied Mathematics*, AMS-SIAM, vol 21 (1985), p.355.

[Kac90] V.G. Kac, Infinite Dimensional Lie Algebras, 3rd edn., Cambridge University Press, 1990

[KM02] A. M. Khvedelidze and D. M. Mladenov, "Bianchi I cosmology and Euler-Calogero-Sutherland model," arXiv:gr-qc/0208037.

[K93] A. A. Kirillov, Sov. Phys. JETP **76**, 355 (1993).

[KK87] A. A. Kirillov and A. A. Kochnev, JETP Letters **46**, 436 (1987).

[KM94] A. A. Kirillov and V. N. Melnikov, "Dynamics Of Inhomogeneities Of Metric In The Vicinity Of A Singularity In Multidimensional Cosmology," Phys. Rev. D **52**, 723 (1995) [gr-qc/9408004];

[KS00] A. A. Kirillov and G. V. Serebryakov, "Origin of a classical space in quantum cosmologies," Grav. Cosmol. **7**, 211 (2001) [arXiv:hep-th/0012245].

[L97] V.G. LeBlanc, "Asymptotic states of magnetic Bianchi I cosmologies," Class. Quantum Grav. **14**, 2281-2301 (1997); "Bianchi II magnetic cosmologies," Class. Quantum Grav. **15**, 1607-1626 (1998).

[LWC00] J. E. Lidsey, D. Wands and E. J. Copeland, "Superstring cosmology," Phys. Rept. **337**, 343 (2000) [arXiv:hep-th/9909061].

[LLK71] E. M. Lifshitz, I. M. Lifshitz and I. M. Khalatnikov, Sov. Phys. JETP **32**, 173 (1971).

[LS01] U. Lindstrom and H. G. Svendsen, "A pedestrian approach to high energy limits of branes and other gravitational systems," Int. J. Mod. Phys. A **16**, 1347 (2001) [arXiv:hep-th/0007101].

[LPS96] H. Lu, C. N. Pope and K. S. Stelle, "Weyl Group Invariance and p-brane Multiplets," Nucl. Phys. B **476**, 89 (1996) [arXiv:hep-th/9602140].

[Ma69] G. A. Margulis, "Applications of ergodic theory to the investigation of manifolds of negative curvature," Funct. Anal. Appl. **4**, 335 (1969)

[M69a] C.W. Misner, "Mixmaster universe," Phys. Rev. Lett. **22**, 1071-1074 (1969).

[M69b] C. W. Misner, "Quantum Cosmology. 1," Phys. Rev. **186**, 1319 (1969); "Minisuper-space," *In *J R Klauder, Magic Without Magic*, San Francisco 1972, 441-473.*

[M94] C.W. Misner, in: D. Hobill et al. (Eds), Deterministic chaos in general relativity, Plenum, 1994, pp. 317-328 [gr-qc/9405068].

[MP95] R.V. Moody and A. Pianzola, Lie Algebras with Triangular Decomposition, Wiley, New York, 1995

[N92] H. Nicolai, "A Hyperbolic Lie Algebra From Supergravity," Phys. Lett. B **276**, 333 (1992).

[OPR98] N. A. Obers, B. Pioline and E. Rabinovici, "M-theory and U-duality on T**d with gauge backgrounds," Nucl. Phys. B **525**, 163 (1998) [arXiv:hep-th/9712084].

[R01] H. Ringström, "The Bianchi IX attractor," Ann. H. Poincaré **2**, 405-500 (2001) [arXiv:gr-qc/0006035].

[R72a] M. P. Ryan, "The oscillatory regime near the singularity in Bianchi-type IX universes," Ann. Phys. (N.Y.) **70**, 301 (1972).

[R72b] M. P. Ryan, "Hamiltonian cosmology," *Springer-Verlag, Heidelberg (1972).*

[RS75] M. P. Ryan and L. C. Shepley, "Homogeneous Relativistic Cosmologies," *Princeton, USA: Univ. Pr. (1975) 320 P. (Princeton Series In Physics).*

[T78] C. Teitelboim, "The Hamiltonian Structure Of Space-Time," PRINT-78-0682 (Princeton), in: General Relativity and Gravitation, vol 1, A. Held ed., Plenum Press, 1980.

[W02] D. Wands, "String-inspired cosmology," Class. Quant. Grav. **19**, 3403 (2002) [arXiv:hep-th/0203107].

[W99] M. Weaver, "Dynamics of magnetic Bianchi VI_0 cosmologies," Class. Quantum Grav. **17**, 421-434 (2000) [arXiv:gr-qc/9909043].

[Z84] R. Zimmer, *Ergodic Theory and Semisimple Groups*, (Birkhauser, Boston, 1984)

TALL TALES FROM DE SITTER SPACE

Robert C. Myers
Perimeter Institute for Theoretical Physics, Waterloo, Ontario N2J 2W9, Canda
Department of Physics, University of Waterloo, Waterloo, Ontario N2L 3G1, Canada
Department of Physics, McGill University, Montréal, Québec H3A 2T8l, Canada
rmyers@perimeterinstitute.ca

Abstract This is a short summary of lectures on some of the recent ideas emerging in the discussion of quantum gravity with a positive cosmological constant. The lectures given at the School on Quantum Gravity, at Centro de Estudios Científicos, Valdivia, Chile in January 2002. The following summary is largely based on material appearing in refs. [1, 2, 3].

1. Prologue

Recent observations suggest that our universe is proceeding toward a phase where its evolution will be dominated by a small positive cosmological constant — see, *e.g.,* [4]. These results have increased the urgency with which theoretical physicists are addressing the question of understanding the physics of de Sitter-like universes — see, *e.g.,* [5, 6, 7, 8, 9, 10, 11, 12, 13, 14, 15, 16, 17]. While de Sitter (dS) space does not itself represent a phenomenologically interesting cosmology, it does present a simple framework within which we may investigate the physics of quantum gravity with a positive cosmological constant. In particular, the cosmological horizon of de Sitter space is an interesting and oft-discussed feature which appears in any spacetime that approaches dS space asymptotically, as might our own universe. Such a cosmological horizon naturally has a Bekenstein-Hawking entropy and a temperature [18]. Given the recent successes in understanding the analogous horizon entropy for black holes in terms of microscopic degrees of freedom [20], one might ask if a similar interpretation arises for de Sitter entropy. There have been a number of dramatic ideas proposed to answer this question. These lectures will focus on two of these: the Λ-N correspondence [7, 8] and the dS/CFT duality [5]. While both of these approaches must still be regarded as bold conjectures, each represents a radical shift of the framework in which we attempt to understand quantum gravity with a positive cosmological constant. Hence they have the potential of providing deep new insights into unresolved issues surrounding the latter. Aspects of these developments are the topic of the present lecture summary.

In all, I gave four lectures in Valdiva on widely separated topics. In the first, I described some of the recent calculations of black hole entropy using techniques involving D-branes. In particular, I focussed on the original calculations of Strominger and Vafa [19]. These were the first calculations of any sort which successfully determined the Bekenstein-Hawking entropy with a statistical mechanical model in terms of some underlying microphysical states. There are already several extensive reviews of the D-brane description of black hole microphysics [20]. I would also highly recommend Juan Maldacena's Ph.D. thesis [21] as a well-written and pedagogical introduction to this topic. The interested reader may also enjoy the discussion in ref. [22]. Of course, Clifford Johnson's lectures at this school give a very good introduction to required D-brane physics, as does his new book [23].

The second lecture focussed on the non-Abelian action describing the physics on nearly coincident D-branes, as well as some of the interesting physical effects that are revealed by studying this action. Of course, the same material already appears in various conference proceedings [24], but I also still recommend reading the original paper on the dielectric effect for D-branes [25].

The last two lectures had the common theme of recent attempts to understand quantum gravity with a positive cosmological constant. In particular, I described aspects of the Λ-N correspondence [7, 8] and the dS/CFT duality [5], and these topics are the only ones represented in the following summary of my lectures. The interested reader may find more details in the original work appearing in refs. [1, 2, 3].

Now in order to fulfill the mission of this school to provide a cultural as well as a scientific exchange between scientists and students from both North and South America, I must digress to elucidate my title. Tall tales are a form of story-telling that originated amongst European settlers in early expansion into the United States and Canada. These stories have a number of distinctive features such as [26]:

a) A larger-than-life or superhuman character with a specific job.

b) A problem that is solved in a funny way.

c) Exaggerated details that describe things as greater than they are.

In particular, the key to a successful tall tale is *exaggeration*. Now while it may be amusing to try to determine ways in which these features may apply in the following scientific discussion, the irony of the title is, of course, that none of them apply there. Rather 'tall' is a technical term that we apply in the description of a class of asymptotically de Sitter spacetimes. Given this brief cultural discourse, we now continue with the scientific discussion.

2. De Sitter space basics

The simplest construction of the $(n+1)$-dimensional de Sitter spacetime is through an embedding in Minkowski space in $n+2$ dimensions, where it may be defined as the hyperboloid

$$\eta_{AB} X^A X^B = \ell^2. \tag{1}$$

The resulting surface is maximally symmetric, *i.e.*,

$$R_{ijkl} = \frac{1}{\ell^2} \left(g_{ik}\, g_{jl} - g_{il}\, g_{jk} \right), \tag{2}$$

which also ensures that the geometry is locally conformally flat. Hence dS space solves Einstein's equations with a positive cosmological constant,

$$R_{ij} = \frac{2\Lambda}{n-1} g_{ij} \quad \text{with} \quad \Lambda = \frac{n(n-1)}{2\ell^2}. \tag{3}$$

The topology of the space is $R \times S^n$, that is spatial slices are n-spheres which evolve in time. From this embedding (1), we see that the time evolution proceeds as follows: In the distant past, the n-sphere is large — the radius diverges as $X^0 \to -\infty$. These spatial slices shrink as time evolves forward, reaching the minimum radius of ℓ at $X^0 = 0$. This is followed by an expanding phase where the size of the n-sphere again diverges as $X^0 \to +\infty$.

One may present the metric on dS space in many different coordinate systems, the choice of which will depend on the application under consideration. One particularly useful choice is conformal coordinates, which allows us to understand the causal structure of dS space:

$$ds^2 = \frac{\ell^2}{\sin^2 \tau} \left[-d\tau^2 + d\theta^2 + \sin^2\theta\, d\Omega_{n-1}^2 \right], \tag{4}$$

where $d\Omega_{n-1}^2$ denotes the unit metric on an $(n{-}1)$-sphere. Here the conformal time covers the entire cosmological evolution in a finite interval, $\tau \in [0, \pi]$, with the conformal factor diverging at the endpoints of this interval. Note that the polar angle on the n-sphere, θ, runs over the same range between the north ($\theta = 0$) and south ($\theta = \pi$) poles. Stripping off the overall conformal factor leaves the metric on the corresponding Penrose diagram. Hence, the latter is usually represented by a square [27], as illustrated in figure 1. Any horizontal cross section of the square is an n-sphere, so that any point in the interior of the diagram represents an $(n-1)$-sphere. At the right and left edges, the points correspond to the north and south poles of the n-sphere, respectively. The infinite future boundary I^+ and past boundary I^- correspond to the slices $\tau = 0$ and π, respectively. Note that diagram is just 'tall' enough that a light

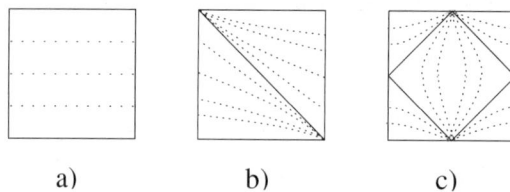

a)   b)   c)

Figure 1.   Constant t slices in (a) the spherical slicing, (b) the flat slicing, and (c) the hyperbolic slicing of de Sitter space.

cone emerging from a point on I^- expands into the space and then reconverges at precisely the antipodal point on the n-sphere at I^+.

We are naturally lead to introduce three other coordinate systems by interpreting dS space as a cosmological evolution. These three choices come from foliating the embedding space above with flat hypersurfaces, $n_A X^A = constant$. The three distinct choices correspond to the cases where the normal vector n_A is time-like, null or space-like. With these distinct choices, a given hypersurface intersects the hyperboloid (1) on a spatial section which has a spherical, flat or hyperbolic geometry, respectively. Following the standard notation for Friedmann-Robertson-Walker (FRW) cosmologies, we denote these three cases as $k = +1, 0$ and -1, respectively. Then the three corresponding metrics on dS space can be written in a unified way as follows:

$$ds^2 = -dt^2 + a_k^2(t)d\Sigma_{k,n}^2 , \tag{5}$$

where the n-dimensional Euclidean metric $d\Sigma_{k,n}^2$ is

$$d\Sigma_{k,n}^2 = \begin{cases} \ell^2 d\Omega_n^2 & \text{for } k = +1 \\ \sum_{i=1}^n dx_i^2 & \text{for } k = 0 \\ \ell^2 d\Xi_n^2 & \text{for } k = -1 , \end{cases} \tag{6}$$

where following the previous notation, $d\Omega_n^2$ is the unit metric on S^n. The 'unit metric' $d\Xi_n^2$ is the n–dimensional hyperbolic space (H^n) which can be obtained by analytic continuation of that on S^n. For $k = \pm 1$ we assume that $n \geq 2$.

The scale factor in each of these cases would be given by

$$a_k(t) = \begin{cases} \cosh(t/\ell) & \text{for } k = +1 \\ \exp(t/\ell) & \text{for } k = 0 \\ \sinh(t/\ell) & \text{for } k = -1 . \end{cases} \tag{7}$$

Note that, as is standard for FRW metrics, the time coordinate in eq. (5) corresponds to the proper time in a particular comoving frame. The $k = +1$ coordinates correspond to standard global coordinates which cover the entire spacetime. In this case, the proper time t is related to the conformal time τ in eq. (4) by $\cosh t/\ell = 1/\sin\tau$. The choice $k = 0$ corresponds to the standard inflationary coordinates, where the flat spatial slices experience an exponential expansion. In this case, $t = -\infty$ corresponds to a horizon (*i.e.*, the boundary of the causal future) for a comoving observer emerging from I^-. Hence these coordinates only cover half of the full dS space but, of course, substituting a minus sign in the exponential of eq. (7) yields a metric which naturally covers the lower half. The choice $k = -1$ yields a perhaps less familiar coordinate choice where the spatial sections have constant negative curvature. In this case, $t = 0$ represents a horizon. However, this horizon is the future null cone of an actual point inside dS space. Figure 1 illustrates slices of constant t on a conformal diagram of dS space. Note that exponential expansion dominates the late time evolution of all three slicings, independent of the spatial curvature, *i.e.*, $a_k(t) \sim \exp(t/\ell)$ as $t \to \infty$ for all k. The interested reader is referred to ref. [2] for further discussion of these metrics.

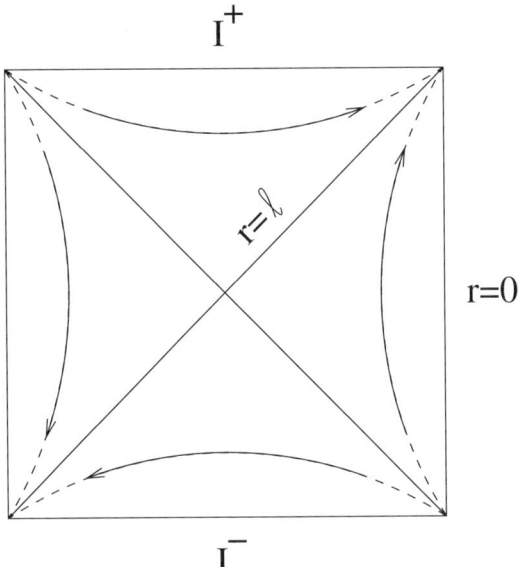

Figure 2. Constant r slices and motion generated by ∂_t in static patch coordinates.

Finally we consider the 'static patch' coordinates for dS_{n+1}:

$$ds^2 = -V(r)\,dt^2 + \frac{1}{V(r)}dr^2 + r^2 d\Omega_{n-1}^2 , \qquad (8)$$

where

$$V(r) = 1 - \frac{r^2}{\ell^2} . \qquad (9)$$

This coordinate system is adapted to discussing the physics of an inertial observer in dS space. The observer's worldline corresponds to $r = 0$. The other distinguished radius is $r = \ell$, which corresponds to a cosmological horizon. As illustrated in figure 2, this horizon corresponds to the boundary of the intersection of the observer's causal future and her causal past. In the static patch, the time t is a Killing coordinate which is simply related to the original construction (1) described above. The motions generated by the Killing vector ∂_t correspond to boosts in the embedding space, along the direction joining the origin to the observer's worldline. Extending the coordinates past the horizon at $r = \ell$, we see that ∂_t is spacelike in the quadrants neighboring the static patch, and is past directed in the quadrant around the antipodal point to the observer.

The cosmological horizon in dS space differs in two important respects from the event horizon of a black hole. First of all, the horizon is observer dependent. That is, different dS horizons are associated with inertial observers emerging from different points on I^- and reaching different points on I^+. However, one may show that the trajectories of any two such inertial observers are related by an isometry of the dS space. Hence the same isometry will map one observer's horizon on to the other's, and the geometric properties of the horizons must be the same. Secondly, the area of, say, a Schwarzschild black hole is determined by its mass, a variable parameter of the solution. In contrast, the area of the dS horizon is set by the cosmological constant, which is a fixed fundamental parameter of the theory. However, in both cases, ordinary entropy is lost to the observer when matter crosses behind the horizon. Hence just as with a black hole horizon, in order to maintain a generalized second law of thermodynamics [28], the dS horizon must be assigned a Bekenstein-Hawking entropy equal to its area divided by $4G_N$ [18]. For empty dS space in $n{+}1$ dimensions, the result is

$$S = \frac{A}{4G_N} = \frac{\Omega_{n-1}\ell^{n-1}}{4G_N} \tag{10}$$

where $\Omega_{n-1} = 2\pi^{n/2}/\Gamma(n/2)$ is the area of a unit $(n{-}1)$-sphere. The usual considerations of quantum fields in curved spacetime — see the lectures by T. Jacobson in this volume — lead one to conclude that there is also a Hawking temperature associated with the horizon [18] given by

$$T = \frac{1}{2\pi\ell} \tag{11}$$

in any number of dimensions.

Hence we find that the de Sitter spacetime, which may have a natural interpretation in terms of the vacuum solution, is actually a thermal system. Given the recent successes in understanding the analogous Bekenstein-Hawking entropy for black holes in terms of microscopic degrees of freedom [20], one might ask the question: Does de Sitter entropy have a microphysical interpretation?

i dS entropy is a formal analogy, which should not be interpreted as a real statistical mechanical quantity. This seems unlikely given that it stands on an equal footing with black hole entropy in, for example, considerations of the generalized second law.

ii dS entropy arises from quantum entanglement with degrees of freedom hidden behind the horizon. This point of view was recently pursued in the context of braneworld scenarios [29].

iii dS entropy arises by quantizing horizon degrees of freedom, in analogy to Carlip's treatment of black holes in 2+1 dimensions [30]. A directly analogous calculation has been performed for 2+1 dimensions [31], but more recently higher dimensions were considered in ref. [32].

iv dS entropy counts the number of microscopic configurations that are macroscopically de Sitter space, in analogy to the recent work on stringy black holes[20]. One recent attempt in this direction is the dS/CFT duality [5, 11].

v dS entropy gives the finite dimension of the Hilbert space describing quantum gravity in de Sitter space. A conjecture along these lines was recently formulated as the Λ-N correspondence [7, 8].

The focus of the remainder of this paper will be on aspects of responses 4 and 5.

3.   The Hunt for Λ-N

We begin with response 5. Here the general framework for the discussion is in attempting to determine properties of de Sitter quantum gravity by applying the general consequences which we may expect from a complete quantum theory, namely semi-classical gravity and the holographic principle [33, 34, 35]. A more complete description of the results in this section appears in ref. [1].

One recent line of thought in this direction was initiated by Banks and Fischler [8, 7]. They focus on the semi-classical result [18] that empty dS space has a finite entropy, which is inversely proportional to the cosmological constant:

$$S_0 = \frac{\Omega_{d-2}}{4G_N} \left[\frac{(d-1)(d-2)}{2\Lambda} \right]^{\frac{d-2}{2}} \tag{12}$$

in d dimensions.[1] As above, Ω_{d-2} is the area of a unit $(d-2)$-sphere. Further in the presence of a positive cosmological constant, any matter structures

---

[1]Recall that $d = n + 1$ in the previous section.

tend to be inflated away so that generically the spacetime approaches dS space asymptotically. In such an asymptotically dS spacetime, any physicist is again surrounded by a horizon, the boundary of her causal past forms which possesses precisely the same entropy (12). It follows immediately from the generalized second law that at any stage in the preceding evolution the combined entropy of any matter and the cosmological horizon is less than that of empty dS space [36]:

$$S_{\text{matter}} + S_{\text{hor}} \leq S_0 \ . \tag{13}$$

Hence S_0 is the largest entropy (counting both horizons and matter) observable in any asymptotically dS spacetimes.

>From this conclusion, one may deduce a lower bound for the underlying quantum theory. Namely, the physics of these cosmological evolutions should be described by a theory with no less than e^{S_0} independent quantum states. Motivated by discussions of black hole complementarity [33, 37], Banks and Fischler reason that any additional states would be superfluous, as they would render the theory more complex than the phenomena it describes — an uneconomical and arbitrary excess. Hence, the class of universes with an asymptotic dS future (dS$^+$) should be fully described by a quantum gravity theory with a Hilbert space $\mathcal{H}$ of finite dimension [8, 7]

$$\dim \mathcal{H} = e^{S_0}. \tag{14}$$

For any quantum system, we define N, the *number of degrees of freedom*, to be the logarithm of the dimension of the Hilbert space. (With this definition, degrees of freedom are spin-like, rather than fields or harmonic oscillators.) Then we may restate the first conclusion as follows: The quantum description of the dS$^+$ universes requires only a finite number of degrees of freedom,

$$N = S_0 \ . \tag{15}$$

The further conjecture is that the correct quantum theory underlying dS$^+$ universes contains precisely N degrees of freedom. In its strongest form, this reasoning leads to a new perspective on the origin of vacuum energy [7], known as the "Λ-N correspondence" [38]. The cosmological constant Λ should be understood as a direct consequence of the finite number of states e^N in the Hilbert space describing the world. Λ effectively provides a cutoff on observable entropy, ensuring that the theory need never describe phenomena requiring a larger number of states; the smaller the Hilbert space, the larger the cosmological constant. This conjecture would point at a new class of theories, distinct from those describing asymptotically flat or AdS spacetimes. The finiteness of N may then be a crucial qualitative feature underlying a successful description of dS$^+$ spacetimes, which may well include our own universe.

The above observations are specifically made in the context of dS$^+$ spaces, *i.e.,* spacetimes that asymptote to dS space in the future [8, 7]. However, for reasons first noted in ref. [38] (see also ref. [1]), it is unnatural for any particular theory to describe *only* spacetimes of this type. A positive cosmological constant certainly does not guarantee the presence of asymptotic dS regions; worse, their existence can be affected by small deformations of Cauchy data. Hence it seems that the conjectured quantum gravity theories with finite Hilbert spaces must be describing a larger class of semiclassical spacetimes. In view of the fact that the Λ-N correspondence attempts to make a statement about the fundamental theory rather than about a particular class of spacetimes, the simplest proposal is to consider the set of all solutions of Einstein's equations coupled to a positive cosmological constant $\Lambda(N)$, irrespective of asymptotic conditions and types of matter present [38].

Bousso then proposed a precise test of the Λ-N correspondence. Namely, the latter should imply that physicists should be limited in their observations in any spacetime solving the relevant low energy equations of motion. Hence Bousso's "N-bound" is [38]:

> In any universe with a positive cosmological constant Λ (as well as arbitrary additional matter that may well dominate at all times), the observable entropy S is bounded by $N=S_0$, the entropy of empty dS space.

For the special case of central observers in spherically symmetric spacetimes with $\Lambda(N)$, the N-bound was proven [38] using the covariant entropy bound [34, 35]. This proof relies on general geometric properties of spacetimes with positive cosmological constant, and on a bound on matter entropy in de Sitter space [36]. While the original analysis was made with $d = 4$, it is easily extended to all $d \geq 4$.

However, as shown in ref. [1] and as will be described below, there are counterexamples to the N-bound for $d > 4$. They are not counterexamples to the covariant entropy bound, but they evade the proof of ref. [38] by violating the assumption of spherical symmetry. Their key novel ingredient is flux, which is used to stabilize a product metric with one large or non-compact factor. In this way, these spacetimes contradict the original intuition that the presence of a positive cosmological constant should result in any matter structures being inflated away so that the solution asymptotes to an empty dS space.

This shows, in particular, that the mere specification of a positive cosmological constant does not suffice to guarantee a finite observable entropy. Λ cannot be in correspondence with N unless some additional conditions hold that exclude our counterexamples. Note that our results do not emphatically rule out finite Hilbert space theories as being relevant for quantum gravity.

3.1   Product spacetimes with flux

We consider solutions to the following action in $d = p + q$ dimensions:

$$I = \frac{1}{16\pi G_N} \int d^{p+q}x \sqrt{-g} \left(R - 2\Lambda - \frac{1}{2\,q!}F_q^2 \right). \tag{16}$$

This describes Einstein gravity with a cosmological constant Λ, taken to be positive, coupled to a q-form field strength, $F_q = dA_{q-1}$. The equations of motion may be written as

$$R_{MN} = \frac{1}{2(q-1)!} F_{MP_2 \cdots P_q} F_N{}^{P_2 \cdots P_q} - \frac{(q-1)}{2(d-2)\,q!} g_{MN} F_q^2 + \frac{2}{d-2} g_{MN} \Lambda \,,$$

$$d * F_q = 0 \,. \tag{17}$$

The most symmetric solution is d-dimensional dS space, with $F = 0$ and the metric as given in, *e.g.*, eq. (8). Recall from eq. (3) that the radius of curvature is related to the cosmological constant by

$$\ell^2 = \frac{(d-1)(d-2)}{2\Lambda} \,. \tag{18}$$

Next, we find product solutions of the form $K_p \times M_q$, where K_p is Lorentzian with coordinates x^μ, M_q is Riemannian with coordinates y^α, and both factors are Einstein:

$$R_{\mu\nu} = \frac{p-1}{L^2} g_{\mu\nu} \,, \qquad R_{\alpha\beta} = \frac{q-1}{R^2} g_{\alpha\beta} \,. \tag{19}$$

Additionally, we take the field strength F_q to be proportional to the volume form on M_q:

$$F_q = c \, \mathrm{vol}_{M_q} \,, \tag{20}$$

where vol_{S^q} is normalized so that $\oint \mathrm{vol}_{S^q} = R^q \, \Omega_q$. This field strength automatically satisfies the Maxwell equation (17) and the Bianchi identity $dF = 0$. Einstein's equations now permit a family of solutions, parametrized by the dimensionless flux

$$\mathcal{F} \equiv \frac{c^2}{4\Lambda} \,. \tag{21}$$

The curvature radii L, R satisfy

$$\frac{p-1}{L^2} = \frac{2\Lambda}{d-2} \left[1 - (q-1)\mathcal{F} \right] \,, \qquad \frac{q-1}{R^2} = \frac{2\Lambda}{d-2} \left[1 + (p-1)\mathcal{F} \right] \,. \tag{22}$$

Since $\mathcal{F} > 0$ by definition and we assume $\Lambda > 0$, equation (22) requires $R^2 > 0$ as well. Hence, M_q must have positive curvature. We will generally take $M_q = S^q$, the q-dimensional sphere. From stability considerations, one finds that this is the most interesting choice [1].

On the other hand, we see that L^2 has indefinite sign. For small $\mathcal{F}$, one finds $L^2 > 0$. This means that K_p is positively curved and can be taken to be dS$_p$. At the value

$$\mathcal{F} = \mathcal{F}_m \equiv \frac{1}{q-1} \,, \tag{23}$$

the curvature radius diverges. In this case K_p is flat; it can be taken to be p-dimensional Minkowski space, for example. For $\mathcal{F} > \mathcal{F}_m$, L^2 becomes negative. This corresponds to a change of sign of the Ricci scalar. The Lorentzian factor will be negatively curved, and it is useful to define $\widetilde{L}^2 \equiv -L^2$. We can take K_p to be p-dimensional anti-de Sitter space (with real curvature radius $\widetilde{L}$) in this case. Note that $\widetilde{L}$ satisfies

$$\frac{p-1}{\widetilde{L}^2} = \frac{2\Lambda}{d-2}\left[(q-1)\mathcal{F}-1\right].$$

(24)

We observe that as $\Lambda \to 0$ (with c fixed), R and $\widetilde{L}$ remain finite and our solutions reproduce the usual Freund-Rubin compactification [39] with geometry $\mathrm{AdS}_p \times S^q$.

Independently of the sign of L^2, all metrics we consider for K_p may be written in the form given in eq. (8), with $n = d - 1$ replaced by $p - 1$. Solving Einstein's equations does not require K_p to be maximally symmetric, rather it must simply satisfy (19). Hence, for $p > 2$, K_p can be taken to be a p-dimensional Schwarzschild-(anti)-de Sitter solution with:

$$V(r) = 1 - \frac{\mu}{r^{p-3}} - \frac{r^2}{L^2}.$$

(25)

This introduces an additional parameter, the "mass" μ, into the space of solutions. We will ignore this freedom in the $L^2 > 0$ case, where we set $\mu = 0$ because empty $\mathrm{dS}_p \times S^q$ has the largest horizon area in that family. However, in the $L^2 < 0$ case, we will find that black holes offer a convenient way of adding unlimited entropy without affecting the stability of an asymptotically $\mathrm{AdS}_p \times S^q$ solution.

3.2 N-bound fails!

At first sight, it seems obvious that the solutions in the preceding section spell disaster for the conjectured N-bound [38]. If we consider the $\mathrm{dS}_p \times S^q$ solutions, *i.e.*, those with $\mathcal{F} < \mathcal{F}_m$, the entropy is determined by the area of the cosmological horizon in dS_p times the volume of the S^q:

$$S(\Lambda, \mathcal{F}) = \frac{1}{4G_N}\Omega_{p-2}\,\Omega_q\left(\frac{d-2}{2\Lambda}\right)^{\frac{d-2}{2}}\left(\frac{p-1}{1-(q-1)\mathcal{F}}\right)^{\frac{p-2}{2}}\left(\frac{q-1}{1+(p-1)\mathcal{F}}\right)^{\frac{q}{2}}.$$

(26)

Hence the ratio between (26) and the entropy of the dS_d horizon (12) is

$$\frac{S(\Lambda, \mathcal{F})}{S_0(\Lambda)} = \frac{\Omega_{p-2}\,\Omega_q}{(d-1)^{\frac{d-2}{2}}\Omega_{d-2}}\left(\frac{p-1}{1-(q-1)\mathcal{F}}\right)^{\frac{p-2}{2}}\left(\frac{q-1}{1+(p-1)\mathcal{F}}\right)^{\frac{q}{2}}.$$

(27)

If this ratio is greater than unity, the N-bound (15) appears to be violated. For fixed Λ, the ratio begins less than unity for $\mathcal{F} = 0$, and as $\mathcal{F}$ increases, the ratio further decreases until

$$\mathcal{F} = \mathcal{F}_s \equiv \frac{1}{(p-1)(q-1)} \, . \tag{28}$$

Above $\mathcal{F}_s$ the ratio begins to increase. It crosses unity at some value $\mathcal{F}_{\text{crit}}$ and actually diverges at $\mathcal{F} = \mathcal{F}_m$. Hence for a range of values, the ratio is actually larger than one.

Rather than immediately declaring the demise of the N-bound, we observe that the stability of these solutions should first be examined. The solutions of the preceding section contain no entropy in the form of ordinary matter systems — hence all potential contributions to entropy come from the Bekenstein-Hawking entropy of event horizons. Event horizons are determined by the global structure of a spacetime, not by its shape at an instant of time. In an unstable product solution, the thermal fluctuations associated with the horizon itself would destabilize the spacetime, and the far future (including any event horizon) would differ from the solution assumed unstable. Such a scenario does not present an immediate inconsistency, *i.e.*, it could be that the true evolution contains a horizon with an even larger area than that originally anticipated for the static product solution. However, to be confident that we have properly determined the entropy and, where applicable, that we have correctly identified violations of the N-bound, we limit our attention to stable solutions.

The stability analysis of the solutions in section 3.1 is an interesting and subtle project in its own right. However, due to space constraints, we present only a brief summary of the results here and refer the interested reader to [1] for the complete details.

Beginning with the $dS_p \times S^q$ solutions, one finds that for $q = 2$ and 3, all of the solutions are stable regardless of the value of p. That is, any solution with $\mathcal{F} < \mathcal{F}_m$ corresponds to a stable spacetime when $q = 2$ or 3. For $q = 4$, there is a window of stability with

$$\frac{1}{3(p-1)} \leq \mathcal{F} \leq \frac{1}{2(p-1)} \, . \tag{29}$$

Finally for $q \geq 5$, all of the $dS_p \times S^q$ solutions are unstable. In any event, given the stability of the solutions with $q = 2$ and 3, it is clear from eqs. (26) and (27) that there are solutions for which the entropy can be made arbitrarily large and, hence, which present a clear violation of the N-bound!

In the case of the $AdS_p \times S^q$ solutions, an intricate pattern of instabilities emerges once more. For $q = 2$ and 3, the spacetimes are stable for all values of $\mathcal{F} \geq \mathcal{F}_m$ regardless of p. For $q \geq 4$, a perturbative tachyon always exists for q odd, but for even q there is a window of stability for $\mathcal{F}$ sufficiently large. Further one may argue that the stability analysis of these solutions extends to the case where a black hole is introduced in the AdS$_p$ spacetime, as in eq. (25). This point is important as there is no entropy associated with an empty AdS$_p \times S^q$ spacetime. However, of course, the presence of a black hole implies the solution has a Bekenstein-Hawking entropy associated with the event horizon. Further, at fixed $\mathcal{F}$, the horizon area of the stable Schwarzschild-AdS$_p \times S^q$ solutions (with $p > 2$) can be made arbitrarily large (*i.e.,* with $\mu \to \infty$). Hence the stable Schwarzschild-AdS$_p \times S^q$ solutions present another violation of the N-bound!

3.3   Discussion

Although some progress has been reported (see, e.g., Refs. [40, 41, 42, 43]), there is presently no fully satisfactory embedding of de Sitter space into string theory. It is important to understand whether this is only a technical problem, or whether significant new developments (comparable, e.g., to the discovery of D-branes [23]) will be required for progress. Interpreting the Λ-N correspondence as a statement that the fundamental theory describing dS space has only a finite number of degrees of freedom, one is faced with a clear challenge to string theory, *i.e.,* it would rule out string theory as a viable description.

However, we have found [1] that some spacetimes with positive cosmological constant $\Lambda(N)$ contain observable entropy greater than N. Hence they cannot be described by a theory with a Hilbert space of finite dimension e^N. Further the entropy was unbounded for the solutions describing a black hole in the stable AdS$_p \times S^q$ backgrounds, and so specifying a positive cosmological constant does not even guarantee a finite observable entropy. While our analysis shows that the N-bound as conjectured by Bousso [38] is incorrect, we have only illustrated that, if correct, the conjectured Λ-N correspondence [7] must be more subtle than first thought. In particular, it cannot be broadly interpreted as a statement about the fundamental theory which at low energies gives rise to Einstein gravity coupled to a positive cosmological constant. Rather at best, it will have an interpretation as a statement about a certain class of solutions within this low energy theory.

Experience with string theory has taught us that the same low energy Lagrangian (in fact, the same fundamental theory) can have different 'superselection' sectors described by different dual theories. For example, in the AdS/CFT correspondence, the rank of the gauge group of the CFT dual to a particular Freund-Rubin compactification (with $\Lambda = 0$) depends on the flux. In the present context of gravity with positive Λ, it could be that a particular dual theory with a specific number of degrees of freedom N will only capture the spacetime physics of a certain sector of a given low energy theory. It follows from the above results that specifying the (positive) value of Λ alone is not sufficient to characterize the dual theory. That is, we found that there cannot be a straightforward correspondence between Λ and the number of degrees of freedom, N (a Λ-N correspondence).

Similarly additional data is required to specify the class of spacetimes dual to a finite Hilbert space theory. Certainly a natural approach to defining the corresponding class would be to include fluxes along with the cosmological constant[2] among the parameters whose specification is necessary in order to ensure that the observable entropy can not exceed N. This appears to produce rather complicated conditions whose sufficiency is not obvious [1].

Of course, this interpretation of the Λ-N correspondence leaves open the possibility that the full theory of quantum gravity which describes, *e.g.,* dS^+ universes will make use of an infinite-dimensional Hilbert space. Hence it cannot be regarded as presenting a fundamental obstacle to a description of dS space by string theory. It is still remarkable, however, that a finite Hilbert space is in principle sufficient to describe such spacetimes. By contrast, the Hilbert space describing asymptotically flat or AdS spacetimes must necessarily contain an infinite number of states, to accommodate arbitrarily large entropy (for example, in the form of large black holes).

Finally, we should remark that the above discussion following from the demise of the N-bound only applies for $d > 4$. One finds that $p = 2$ is a special case in the analysis of sections 3.1 and 3.2. We refer the interested reader to ref. [1] for the details, but the essential comment is that there is no obvious way to violate the N-bound in this case. Further the sphere component of the product solutions must have $q \geq 2$ in order to satisfy eq. (22) with positive Λ. Combining these two results means that we have no counterexamples to the N-bound for $d = 4$ [1]. The significance of this exception is not clear.

4.   Introduction to the dS/CFT correspondence

Recently Strominger [5] conjectured that quantum gravity in asymptotically dS universes has a dual description in terms of a Euclidean conformal field theory on the future boundary (I^+) and/or the past boundary (I^-). Much of the motivation for this dS/CFT duality comes from our understanding of the AdS/CFT duality [46]. Of course, the nature of dS space is quite different from its AdS counterpart. In particular, the conformal boundaries, which one expects to play a central role in any dS/CFT correspondence, are hypersurfaces of Euclidean signature. As a result, one expects the dual field theory to be a Euclidean field theory. Further in de Sitter space, there are two such hypersurfaces: the future boundary, I^+, and the past boundary, I^-. Hence one must ask whether the proposed duality will involve a single field theory [5, 6] or two

---

[2]The specification of flux will identify an isolated sector only if flux-changing instantons [44, 45] are completely suppressed. This is the case for the $\Lambda = 0$ product solutions studied in the AdS/CFT context, but one might expect an onset of non-perturbative instabilities if $\Lambda > 0$.

[47]. Unfortunately at present, the most striking difference from the AdS/CFT duality is the fact that we have no rigorous realizations of the dS/CFT duality — some progress has been made with pure three-dimensional de Sitter gravity [15]. However, the idea of a dS/CFT correspondence is a powerful and suggestive one that could have fundamental implications for the physics of our universe. A present difficulty is that rather little is known about the Euclidean field theory which is to be dual to physics in the bulk. Hence the goal in the discussion below will be to explore further the requirements for a candidate dual field theory, under the assumption that the bulk physics must reproduce standard background quantum field theory in the low energy limit. In particular, much of the following will focus on the interpretation of scalar field theory in a de Sitter background in the context of the dS/CFT correspondence. As the AdS/CFT correspondence [46] motivates many of the calculations and their interpretation, we begin with a brief review of the latter duality (section 4.1) and then present the most relevant aspects of dS/CFT (section 4.2). The reader interested in the mathematical details can consult the appendices to which we will refer in the following.

4.1 Brief AdS/CFT review

Consider probing anti-de Sitter space with a massive scalar field. We consider the following metric on $(n+1)$-dimensional AdS space,[3]

$$ds^2 = dr^2 + e^{2r/\tilde{\ell}}\eta_{\mu\nu}dx^\mu dx^\nu \, , \tag{30}$$

and the standard equation of motion for the scalar,

$$\left[\Box - M^2\right]\phi = 0 \, . \tag{31}$$

Then to leading order in the asymptotic region $r \to \infty$, the two independent solutions take the form [48]

$$\phi_\pm \simeq e^{-\Delta_\pm r/\tilde{\ell}}\phi_{0\pm}(x^\mu) \qquad \text{where} \qquad \Delta_\pm = \frac{n}{2} \pm \sqrt{\frac{n^2}{4} + M^2\tilde{\ell}^2} \, . \tag{32}$$

Now the interpretation of these results depends on the value of the mass, and there are three regimes of interest:

$$\text{(i) } M^2 > 0 \, , \quad \text{(ii) } 0 > M^2 > -\frac{n^2}{4\tilde{\ell}^2} \quad \text{and} \quad \text{(iii) } M^2 < -\frac{n^2}{4\tilde{\ell}^2} \, . \tag{33}$$

---

[3]The essential feature for the following analysis is the exponential expansion of the radial slices with proper distance r. While we have chosen to consider pure AdS space in Poincaré coordinates for specificity, this expansion, of course, arises quite generally in the asymptotic large-radius region for any choice of boundary metric and for any asymptotically AdS spacetime.

In case (i), Δ_- is negative and so the corresponding "perturbation" is actually divergent in the asymptotic regime. Hence in constructing a quantum field theory on AdS space, only the ϕ_+ modes would be useful for the construction of an orthogonal basis of normalizable mode functions [49]. In particular, the bulk scalar wave operator is essentially self-adjoint and picks out the boundary condition that the ϕ_- modes are not excited dynamically. In the context of the AdS/CFT then, the ϕ_{0-} functions are associated with source currents (of dimension Δ_-). These may then be used to generate correlation functions of the dual CFT operator of dimension Δ_+ through the equivalence [48, 50]

$$Z_{\mathrm{AdS}}(\phi) = \int D\phi \, e^{iI_{\mathrm{AdS}}(\phi_-,\phi_+)} = \left\langle e^{i \int \phi_{0-} \mathcal{O}_-} \right\rangle_{\mathrm{CFT}} . \tag{34}$$

On the other hand, the boundary functions ϕ_{0+} are associated with the expectation value for states where the dual operator has been excited [49].

In case (ii), the lower limit corresponds precisely to the Breitenlohner-Freedman bound [51, 52]. While the scalar appears tachyonic, it is not truly unstable and it is still possible to construct a unitary quantum field theory on AdS space. Further, in this regime, both sets of solutions (32) are well-behaved in the asymptotic region. However, together they would form an over-complete set of modes. The theory must therefore be supplemented with a boundary condition at AdS infinity which selects out one set of modes to define a self-adjoint extension of the scalar wave operator (and thus the time evolution operator). For $0 > M^2\tilde{\ell}^2 > 1 - n^2/4$, there is a unique boundary condition which produces an AdS invariant quantization [51]. However, for

$$1 - n^2/4 > M^2\tilde{\ell}^2 > -n^2/4 , \tag{35}$$

the boundary condition is ambiguous. The AdS/CFT interpretation is essentially the same as above. That is, the ϕ_{0+} and ϕ_{0-} functions may be associated with expectation values and source currents of the dual CFT operator, respectively. For the ambiguous regime (35), there is a freedom in this equivalence associated with a Legendre transformation of the generating functional [53].

Finally in case (iii), the mass exceeds the Breitenlohner-Freedman bound [51, 52] and the scalar field is actually unstable; no sensible quantization is possible. However, if one were to attempt an AdS/CFT interpretation analogous to those above, the dimension Δ_+ of the dual CFT operator would be complex, which might be interpreted as indicating that the corresponding theory is not unitary. Hence one still seems to have agreement on both sides of the correspondence as to the unsuitability of the regime $M^2\tilde{\ell}^2 < -n^2/4$.

For any of these regimes, an important part of the AdS/CFT duality is the UV/IR correspondence [54]. The main lesson here is that asymptotic regions (near the boundary) of AdS space (connected with long wavelength or IR behavior in the bulk) are associated with short distance (UV) physics in the field theory dual. Similarly, regions deeper inside AdS (connected with short wavelengths or UV behavior in the bulk) are associated with long distance (IR) physics in the dual theory. This relation essentially follows from the action of the symmetry group SO(n,2) in each theory, and can be visualized by the standard association of the field theory with the boundary of AdS. In the Poincaré coordinates above, the scaling transformation $(r, x^\mu) \to (r - \tilde{\ell} \log \lambda, \lambda x^\mu)$ for $\lambda \in \mathbb{R}^+$ constitutes an AdS translation. This symmetry makes it clear that motion toward the boundary (large r) of AdS is associated with smaller and smaller scales in the field theory, which lives on the space labelled by the x^μ. In this context, we emphasize that the CFT does not live on the boundary of the AdS space. Usually one has chosen a particular foliation of AdS [55], and the bulk space calculations are naturally compared to those for the field theory living on the geometry of the surfaces comprising this foliation. Via the UV/IR correspondence, each surface in the bulk foliation is naturally associated with degrees of freedom in the CFT at a particular energy scale [56].

As remarked above, changing the boundary conditions of the scalar fields through the addition of ϕ_- modes corresponds to a deformation of the CFT, which, in general, breaks the conformal symmetry. However, if the deformation corresponds to a relevant (or marginal) operator,[4] the theory remains nearly conformal in the UV. In turn then, one expects the gravity dual to remain asymptotically anti-de Sitter. Indeed, one finds that the associated bulk perturbations remain small near the boundary. However, the field theory behavior at intermediate scales and in the IR can be quite nontrivial. The corresponding bulk perturbation becomes large as one proceeds inward from the boundary and it is natural to seek a gravitating dual by solving the exact classical equations of motion of the gravitating theory — see, for example, [58]. Thus, it is natural to relate the 'radial evolution' of the gravitating spacetime with a field theory renormalization group flow [59]. One of the interesting results to emerge from this discussion is a gravitational 'c-theorem'. That is, one seems to be able to define a c-function as a local geometric quantity in the asymptotically AdS space, that gives a measure of the number of degrees of freedom relevant for physics of the dual field theory at different energy scales [59]. In particular, one is able to show that the bulk equations of motion dictate that this c-function must decrease in evolving from the UV into the IR regions.

4.2   Some dS/CFT basics

Given the brief overview of the AdS/CFT correspondence, we now turn to asymptotically de Sitter spaces, where one would like to study the possibility of a similar duality between quantum gravity in the bulk and a Euclidean CFT

---

[4]Ref. [57] discusses certain irrelevant deformations in the AdS/CFT.

[5]. As in the previous review, we focus the present discussion on the case of a pure de Sitter space background:

$$ds^2 = -dt^2 + \cosh^2 t/\ell \, d\Omega_n^2 \,, \tag{36}$$

where $d\Omega_n^2$ is the standard round metric on an n-sphere. This metric solves Einstein's equations, $R_{ij} = 2\Lambda/(n-1) \, g_{ij}$, in $n+1$ dimensions. The curvature scale ℓ is related to the cosmological constant by $\ell^2 = n(n-1)/(2\Lambda)$. Again, the important feature of this geometry is the exponential expansion in the spatial metric in the asymptotic regions, *i.e.*, $t \to \pm\infty$. Much of the following discussion carries over to spacetimes that only resemble dS asymptotically[5] and indeed, if the proposed dS/CFT duality is to be useful, it must extend to such spacetimes. We will explore certain aspects of the dS/CFT for such backgrounds in section 7.

Consider a free scalar field propagating on the above background (36), which we wish to treat in a perturbative regime where the self-gravity is small. Hence, the equation of motion is

$$\left[\Box - M^2\right] \phi = 0 \,. \tag{37}$$

In general, the effective mass may receive a contribution from a nonminimal coupling to the gravitational field [60]. Therefore we write

$$M^2 = m^2 + \xi R \,, \tag{38}$$

where m^2 is the mass squared of the field in the flat space limit and ξ is the dimensionless constant determining the scalar field's coupling to the Ricci scalar, R. In the dS background (36), we have $R = n(n+1)/\ell^2$. A case of particular interest in the following section will be that of the conformally coupled massless scalar field, for which $m^2 = 0$, $\xi = (n-1)/4n$ and hence $M^2 = (n^2-1)/4\ell^2$. With these parameters, the solutions of eq. (37) transform in a simple way under local conformal scalings of the background metric [60].

In parallel with the AdS case, scalar fields propagating in de Sitter space can have two possible behaviors near the boundaries. Let us for the moment think of defining these boundary conditions at past infinity (I^-). Equation (37) above is readily solved [5] near I^- to yield two independent solutions with the asymptotic form $\phi \sim e^{h_\pm t/\ell}$, where

$$h_\pm = \frac{n}{2} \pm \sqrt{\frac{n^2}{4} - M^2\ell^2} \,. \tag{39}$$

Note that this asymptotic time dependence is independent of the details of the spatial mode. In the pure dS background (36), the same exponents also govern the behavior of the fields at future infinity — see the appendix for details.

---

[5]Many explicit examples of such backgrounds may be found in ref. [2].

The fact that the boundaries are spacelike in de Sitter space means that the 'boundary conditions' have a different conceptual status than in the AdS setting. In particular, requiring that the bulk evolution is well-defined in dS space will not impose any restrictions on past or future boundary conditions. So in contrast to the AdS/CFT correspondence, in the dS/CFT correspondence, both the ϕ_+ and ϕ_- modes appear on an equal footing. Certainly, a complete description of physics in the bulk must include both sets of modes as dynamical quantum fields. Following the analogy with the AdS/CFT correspondence and in accord with the preceding discussion, it is natural then to associate both modes $\phi_\pm$ with source currents for dual field theory operators $\mathcal{O}_\pm$, with conformal dimensions $h_\mp$ [5]. As we will discuss shortly, this matching of modes with dual operators is further supported by a bulk construction of a generating functional for correlation functions in the CFT.

As in the AdS case, one can classify the scalars displaying distinct types of boundary behavior in three different regimes:

$$ \text{(i) } M^2 > \frac{n^2}{4\ell^2}, \quad \text{(ii) } \frac{n^2}{4\ell^2} > M^2 > 0 \quad \text{and} \quad \text{(iii) } M^2 < 0. \tag{40} $$

These three regimes also appear in discussions in the mathematics literature — see, *e.g.,* [61, 62]. There the scalar field is classified according to M^2 regarded as its $SO(n+1, 1)$ Casimir. A common nomenclature for the three possibilities delineated above is the (i) principal, (ii) complementary (or supplementary) and (iii) discrete series of representations of $SO(n+1, 1)$. As is evident from eq. (39), the distinguishing feature of scalar fields in the principal series is that they are oscillatory near past (or future) infinity. In contrast, the exponents for fields in the complementary series are real and positive, and so their asymptotic behavior is a purely exponential damping near both boundaries.

Let us consider case (iii) $M^2 < 0$ in detail. While $h_\pm$ are both real, the modes $\phi_- \sim e^{h_- t}$ diverge as one approaches I^-, since $h_- < 0$. One finds similar divergent behavior for one of the modes at the future boundary I^+. The discrete series then corresponds to special values of the mass in this range where a subset of the modes display the convergent h_+ behavior at both $I^\pm$ — see the appendix and refs. [61, 62]. However, we emphasize that even in these special cases, the full space of solutions still includes modes diverging at both asymptotic boundaries. In a physical situation then, the uncertainty principle would not allow us to simply set the amplitude of the divergent modes to zero. Hence the formal mathematical analysis of these fields is only of limited physical interest and we will not consider them further in the following. Of course, the divergence of the generic field configuration is simply an indication that treating the tachyonic fields as linearized perturbations is inappropriate. Nonlinear field theories with potentials including unstable critical points may play an important role in the dS/CFT correspondence, *e.g.,* in constructing models of inflationary cosmology. The essential point though is that one must study the full nonlinear evolution of such fields, including their backreaction on the spacetime geometry.

Considering the principal series, (i) $M^2 > n^2/4\ell^2$, in more detail, one expects to find a pair of dual operators $\mathcal{O}_\pm$ with *complex* conformal dimensions $h_\mp$. Having operators with a complex conformal weight suggests that the dual CFT is nonunitary [5]. We add the brief observation that, since, in the quantum field theory, the two sets of independent modes $\phi_\pm$ correspond roughly to creation and annihilation operators (positive and negative frequency modes) in the bulk (see, *e.g.*, [11, 63, 64, 65]), the corresponding operators in the dual theory should have nontrivial commutation relations.

Finally we consider the complementary series, (ii) $n^2/4\ell^2 > M^2 \geq 0$. In this case, the time dependence for both sets of modes is purely a real exponential decay near both $I^\pm$. In the bulk, the two linearly independent solutions may be chosen to be real, as is readily verified by explicit computations — see the appendix. Because the $\phi_\pm$ solutions are real, they each have zero norm in the usual Klein-Gordon inner product, while a nonvanishing inner product arises from (ϕ_+, ϕ_-). It follows that, upon quantization, the corresponding operator coefficients are analogous to position and momentum operators, rather than creation and annihilation operators. That is, these degrees of freedom are canonically conjugate. In any event, both types of modes are again required to describe standard quantum field theory in the bulk.

As before, the dual CFT should contain a pair of operators $\mathcal{O}_\pm$ dual to the $h_\mp$ modes. In this case, the operators have real conformal weights and must be distinct, as their weights are different. One can readily see that both $\mathcal{O}_\pm$ will have local correlation functions: One simply notes that the corresponding source currents are obtained from the bulk scalar field through

$$J_-(\Omega) \equiv \lim_{t \to -\infty} e^{-h_- t/\ell} \phi(\Omega, t) , \tag{41}$$

$$J_+(\Omega) \equiv \lim_{t \to -\infty} e^{-h_+ t/\ell} [\phi(\Omega, t) - e^{h_- t/\ell} J_-(\Omega)] ,$$

where Ω denotes a point on the n-sphere. As these constructions are local in position, their two-point functions will also be local. Note that the above discussion of inner products indicates that the operators $\mathcal{O}_\pm$ should have nontrivial commutation relations with each other but vanishing commutators amongst themselves.

Next we consider the generator of correlation functions in the dual field theory. A natural construction proposed in [5] for a free bulk field theory is

$$\mathcal{F} = \lim_{t,t' \to -\infty} \int d\Sigma^\mu d\Sigma'^\nu \phi(x) \overleftrightarrow{\partial}_\mu G(x, x') \overleftrightarrow{\partial}_\nu \phi(x') . \tag{42}$$

In the original proposal of [5], $G(x, x')$ was chosen as the Hadamard two-point function

$$G(x, x') = \langle 0 | \{ \phi(x), \phi(x') \} | 0 \rangle \tag{43}$$

in the Euclidean vacuum. This two-point function is symmetric in its arguments. Generalizing this construction to other two-point functions was considered in [11, 66]. These alternatives all provide essentially the same short distance singularities discussed below.

One proceeds by evaluating the generating functional $\mathcal{F}$. First, the boundary conditions (39) at I^- yield

$$\lim_{t \to -\infty} \phi(\Omega, t) \simeq \phi_{0+}(\Omega)\, e^{h_+ t} + \phi_{0-}(\Omega)\, e^{h_- t} , \tag{44}$$

where we imagine that $M^2 > 0$ so that the above shows no divergent behavior. Now the dS-invariant two-point function may also be expanded in the limit that $t, t' \to -\infty$, the result being

$$G(x, x') \simeq c_+ \frac{e^{-h_+(t+t')}}{(w^i w'^i - 1)^{h_+}} + c_- \frac{e^{-h_-(t+t')}}{(w^i w'^i - 1)^{h_-}} , \tag{45}$$

where c_+ and c_- are constants and w^i denote direction cosines on S^n. Using the notation of [66], one has $w^1 = \cos\theta_1$, $w^2 = \sin\theta_1 \cos\theta_2$, ..., $w^d = \sin\theta_1 \ldots \sin\theta_{n-1} \sin\theta_n$. Note in particular that, with this choice of coordinates, when the points on the sphere coincide, one has $w^i w'^i = 1$, while for antipodal points, one has $w^i w'^i = -1$. Taking into account the measure factors, the final result for the generating functional reduces to

$$\mathcal{F} = -\frac{(h_+ - h_-)^2}{2^{2n}} \int d\Omega d\Omega' \left[c_+ \frac{\phi_0_-(\Omega)\,\phi_0_-(\Omega')}{(w^i w'^i - 1)^{h_+}} + c_- \frac{\phi_0_+(\Omega)\,\phi_0_+(\Omega')}{(w^i w'^i - 1)^{h_-}} \right] . \tag{46}$$

Note that the Klein-Gordon inner product has eliminated the cross-terms (which were potentially divergent). Further, the coincidence singularities in eq. (46) are proportional to the Euclidean two-point function on a n-sphere, *i.e.*,

$$\Delta_{h_\pm} \simeq \frac{1}{(w^i w'^i - 1)^{h_\pm}} , \tag{47}$$

for operators with conformal weights $h_\pm$. Hence $\mathcal{F}$ appears to be a generating functional for CFT correlation functions, with $\phi_{0\pm}$ acting as source currents for operators with conformal dimensions $h_\mp$. The above relies on having a free field theory in the bulk dS space, but extending the construction to an interacting field theory was considered in [66].

5.   A generalized de Sitter c-theorem

The discussion of the dS/CFT correspondence has also been extended beyond pure dS space to more general backgrounds with asymptotically dS regions. Indeed, if the proposed dS/CFT duality is to be useful, it must extend to such spacetimes. In analogy to the AdS/CFT duality, such backgrounds might have an interpretation in terms of 'renormalization group flows' in the dual field theory [9, 10, 2, rgflows]. In the AdS context, one of the interesting features is the UV/IR correspondence [54]. That is, physics at large (small) radii in the AdS space is dual to local, ultraviolet (nonlocal, infrared) physics in the dual CFT. As was extensively studied in gauged supergravity — see, *e.g.,* [58] — 'domain wall' solutions which evolve from one phase near the AdS boundary to another in the interior can be interpreted as renormalization group flows of the CFT when perturbed by certain operators. In analogy to Zamolodchikov's results for two-dimensional CFT's [67], it was found that a c-theorem could be established for such flows [59] using Einstein's equations. The c-function defined in terms of the gravity theory then seems to give a local geometric measure of the number of degrees of freedom relevant for physics at different energy scales in the dual field theory.

In the dS/CFT duality, there is again a natural correspondence between UV (IR) physics in the CFT and phenomena occurring near the boundary (deep in the interior) of dS space. Therefore, in the context of more general solutions that are asymptotically dS, one has an interpretation in terms of renormalization group flows, which should naturally be subject to a c-theorem [9, 10]. The original investigations [9, 10] considered only solutions with flat spatial sections ($k = 0$), and we generalize these results in the following to include spherical and hyperbolic sections ($k = \pm 1$). We also consider the flows involving anisotropic scalings of the boundary geometry, but our results are less conclusive in these cases.

5.1   The c-function

The foliations of spacetimes of the form given in eq. (5) are privileged in that time translations (a) act as a scaling on the spatial metric, and thus in the field theory dual, and (b) preserve the foliation and merely move one slice to another. In the context of the dS/CFT correspondence, these properties naturally lead to the idea that time evolution in these spaces should be interpreted as a renormalization group flow [9, 10]. Certainly, the same properties apply for time evolution independent of the curvature of the spatial sections. Hence if a c-theorem applies for the $k = 0$ solutions [9, 10], one might expect that it should extend to $k = \pm 1$ and perhaps other cases if properly generalized.

For $k = 0$, the proposed c-function [9, 10], when generalized to $n+1$ dimensions, is

$$c \simeq \frac{1}{G_N \left| \frac{\dot{a}}{a} \right|^{n-1}} . \tag{48}$$

The Einstein equations ensure that $\partial_t \left(\dot{a}/a \right) < 0$, provided that any matter in the spacetime satisfies the null energy condition [27]. This result then guarantees that c will always decrease in a contracting phase of the evolution and increase in an expanding phase.

For our general study, we wish to define a c-function which can be evaluated on each slice of some foliation of the spacetime. Of course, our function should satisfy a 'c-theorem', *e.g.*, our function should monotonically decrease as the surfaces contract in the spacetime evolution. Further, it should be a geometric function built from the intrinsic and extrinsic curvatures of a slice. Toward this end, we begin with the idea that the c-function is known for any slice of de Sitter space, and note that, in this case, eq. (48) takes the form

$$c \sim \frac{1}{G_N \Lambda^{(n-1)/2}} . \tag{49}$$

Thus, if our slice can be embedded in some de Sitter space (as is the case for any isotropic homogeneous slice — see eq. (5)), the value of the c-function should be given by eq. (49). In other words, we can associate an effective cosmological constant Λ_{eff} to any slice that can be embedded in de Sitter space and we can then use this Λ_{eff} to define our c-function.

It is useful to think a bit about this embedding in order to express Λ_{eff} directly in terms of the intrinsic and extrinsic curvatures of our slice. The answer is readily apparent from the general form of the 'vacuum' Einstein equations with a positive cosmological constant: $G_{ij} = -\Lambda g_{ij}$. Contracting these equations twice along the unit normal n^i to the hypersurface gives the Hamiltonian constraint, which is indeed a function only of the intrinsic and extrinsic curvature of the slice.[6] The effective cosmological constant defined by such a local matching to de Sitter space is therefore given by

$$\Lambda_{eff} = G_{ij} n^i n^j . \tag{50}$$

For metrics of the general form (5), this becomes

$$\Lambda_{eff} = \frac{n(n-1)}{2} \left[\left(\frac{\dot{a}}{a} \right)^2 + \frac{k}{a^2} \right] . \tag{51}$$

---

[6]The momentum constraints vanish in a homogeneous universe, and time derivatives of the extrinsic curvature only appear in the dynamical equations of motion.

Taking the c-function to be a function of this effective cosmological constant, dimensional analysis then fixes it to be

$$c \sim \frac{1}{G_N \Lambda_{eff}^{(n-1)/2}} = \frac{1}{G_N} \left(G_{ij} n^i n^j \right)^{-(n-1)/2} . \tag{52}$$

For the $k = 0$ isotropic case, it is clear that this reduces to the c-function (48) given previously in [9, 10]. For other isotropic cases, it is uniquely determined by the answer for the corresponding slices of de Sitter space. The same holds for an anisotropic slice that can be embedded in de Sitter (see, *e.g.,* [2] for examples). While the choice (52) is not uniquely determined by the constraints imposed thus far for any slice which cannot be so embedded, it does represent a natural generalization and, as we will see below, this definition allows a reasonable 'c-theorem' to be proven.

5.2   The c-theorem

For any of the homogeneous flows as considered in the previous section, it is straightforward to show that our c-function (52) always decreases (increases) in a contracting (expanding) phase of the evolution. However, we would like to give a more general discussion which in particular allows us to consider anisotropic geometries, as well as these isotropic cases.

To prove our theorem, we note that the Einstein equations relate our effective cosmological constant to the energy density ρ on the hypersurface,

$$\Lambda_{eff} = G_{ij} \, n^i n^j = T_{ij} \, n^i n^j = \rho. \tag{53}$$

Consider now the 'matter energy' $E = \rho V$ contained in the volume V of a small co-moving rectangular region on the homogeneous slice. That is, we take

$$V = \int_R \sqrt{g} \, d^n x \tag{54}$$

for some small co-moving region R of the form $R = \{x | x_a^i < x^i < x_b^i\}$ where x^i denote co-moving spatial coordinates. We also introduce $\delta x^i = x_b^i - x_a^i$, the co-moving size of R in the ith direction. Since R is small, each coordinate x^i can be associated with a scale factor $a^i(t)$ such that the corresponding physical linear size of R is $a^i(t) \delta x^i$.

Without loss of generality, let us assume that the coordinates x^i are aligned with the principle pressures P_i, which are the eigenvalues of the stress tensor on the hypersurface. Let us also introduce the corresponding area A_i of each face. Note that a net flow of energy into R from the neighboring region is forbidden by homogeneity. As a result, energy conservation implies that $dE = -P_i A_i d(a \delta x^i)$ as the slice evolves. However, clearly $dE = \rho dV + V d\rho$, so that we have

$$dΛ_{eff} = dρ = -\sum_i (ρ + P_i) \, d\ln a_i. \tag{55}$$

Now we will assume that any matter fields satisfy the weak energy condition so that $ρ + P_i ≥ 0$. Thus, if all of the scale factors are increasing, we find that the effective cosmological constant can only decrease in time.

This result provides a direct generalization of the results of [9, 10] to slicings that are not spatially flat. In particular, in the isotropic case (where all scale factors are equal, $a ≡ a_i = a_j$), it follows that $c(a)$ as given in eq. (51) is, as desired, a monotonically increasing function in any expanding phase of the universe.

Note, however, that the anisotropic case is not so simple to interpret. For example, it may be that the scale factors are expanding in some directions and contracting in others. In this case our effective cosmological constant may either increase or decrease, depending on the details of the solution.

5.3   Complete Flows versus Bouncing Universes

The general flows are further complicated by the fact that they may 'bounce', *i.e.*, the evolution of the scale factor(s) may reverse itself. The simplest example of this would be the $k = +1$ foliation of dS space in eq. (5). In this global coordinate system, the scale factor (7) begins contracting from $a(t = -∞) = ∞$ to $a(t = 0) = 1$, but then expands again toward the asymptotic region at $t = +∞$. In contrast, we refer to the $k=0$ and -1 foliations as 'complete'. By this we mean that within a given coordinate patch, the flow proceeds monotonically from $a = ∞$ in the asymptotic region to $a = 0$ at the boundary of the patch — the latter may be either simply a horizon (as in the case of pure dS space) or a true curvature singularity.

For any homogeneous flows, such as those illustrated in eq. (5), it is not hard to show that the $k=0$ and -1 flows are always complete and that only the $k = +1$ flows can bounce. The essential observation is that for $a(t)$ to bounce the Hubble parameter $\dot{a}/a$ must pass through zero. Now the (tt)-component of the Einstein equations yields

$$\left(\frac{\dot{a}}{a}\right)^2 = T_{tt} - \frac{k}{a^2} . \tag{56}$$

As long as the weak energy condition applies,[7] it is clear that the right-hand-side is always positive for $k = 0$ and -1 and so $\dot{a}/a$ will never reach zero. On the other hand, no such statement can be made for $k = +1$ and so it is only in this

---

[7]Note that if $k = 0$ and the energy density is identically zero, it follows that a is a constant. Hence in this case, we will not have an asymptotically dS geometry.

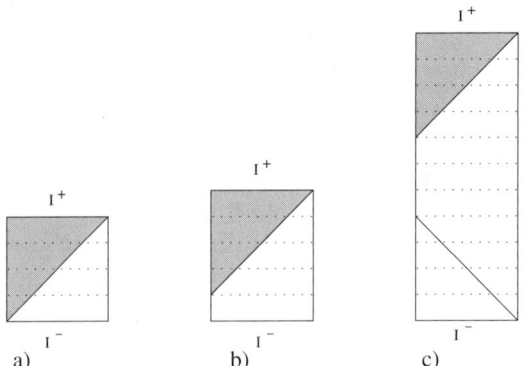

Figure 3.   Conformal diagrams of (a) de Sitter space, (b) perturbed de Sitter space, and (c) a very tall asymptotically de Sitter spacetime. The worldline of the 'central observer' is the right boundary of each diagram and various horizons related to her worldline are shown. Shaded regions cannot send signals to this observer.

case that bounces are possible. Further, one might observe that this analysis does not limit the number of bounces which such a solution might undergo. In certain cases with a simple matter content, *e.g.*, dust or radiation, one may show that only a single bounce is possible. However in (slightly) more complex models, multiple bounces are possible [2]. In the case of anisotropic solutions — see, *e.g.*, [2] — the characterization of the flows as complete or otherwise is more complicated.

6.   The global perspective

One feature of dS space, which presents a puzzle for the dS/CFT duality, is the fact that there are two conformal boundaries, $I^{\pm}$. In particular, one might ask whether there is a dual CFT to be associated with each boundary or a single CFT for both. Early discussions of the role of these surfaces[5, 6] emphasized the causal connection between points on the two boundaries. In particular, a light cone emerging from a point on I^- expands into the space and reconverges at the antipodal point on the sphere at I^+. As a consequence, the singularity structure of certain boundary correlation functions is left invariant when, *e.g.*, a local operator on I^- is replaced by a corresponding local operator at the antipodal point on I^+ [5]. Not only does this observation suggest that there is a single dual field theory, but further that dual operators associated with the two boundaries are simply related by the antipodal map on the sphere.

However, as will be discussed in section 7, this intuition does not seem to withstand closer scrutiny. In particular, one finds that the causal connection between the conformal boundaries of dS space is modified for generic perturbations. The relevant results follow from Corollary 1 of [68], which we paraphrase as follows:

> *Let the spacetime* (M, g_{ij}) *be null geodesically complete and satisfy the weak null energy condition and the null generic condition. Suppose in addition that* (M, g_{ij}) *is globally hyperbolic with a compact Cauchy surface* Σ. *Then there exist Cauchy surfaces* Σ_1 *and* Σ_2 *(of the same compact topology, and with* Σ_2 *in the future of* Σ_1*) such that if a point q lies in the future of* Σ_2*, then the entire Cauchy surface* Σ_1 *lies in the causal past of q.*

Hence the conformal diagrams, for universes that are asymptotically de Sitter in both the future and past, are 'tall', *i.e.*, they are taller than they are wide. Physically, an entire compact Cauchy surface will be visible to observers at some finite time, and, hence, perturbations of dS space may bring features that originally lay behind a horizon into an experimentally accessible region. This is shown in figure 3(b). Pushing this somewhat further, one can imagine that in certain circumstances asymptotically de Sitter spacetimes of the sort shown in figure 3(c) may arise. That is, in these spacetimes, a compact Cauchy surface lies in the intersection of the past *and* future of a generic worldline.

It follows that the relation between dual operators on the two boundaries must be manifestly nonlocal when we consider such 'tall' backgrounds. In a tall spacetime, the light rays emerging from a point on I^- reconverge,[8] but this occurs at a finite time, long before they reach I^+. After passing through the focal point, the rays diverge again to enclose a finite region on I^+. This observation refutes any intuition that the dual operators associated with the two boundaries could be related by a simple local map (*e.g.*, the antipodal map) on the sphere.

We will return to this discussion in section 7. In the following, we will discuss certain unusual features of the flows, in the sense of section 5, which become apparent from the global structure of the spacetimes. We assume that the slices are isotropic, and take each of the three possible cases (spheres, flat slices, and hyperbolic slices) in turn.

6.1   Flat slices ($k = 0$)

We now wish to construct the conformal diagram for flows with flat spatial sections. In order to draw useful two-dimensional diagrams, we shall use the common trick of studying rotationally symmetric spacetimes and drawing conformal diagrams associated with the 'r-t plane,' *i.e.*, associated with a hypersurface orthogonal to the spheres of symmetry.

---

[8]For simplicity, our description is restricted to spherically symmetric foliations [2]. Generically the converging light rays would not be focussed to a single point.

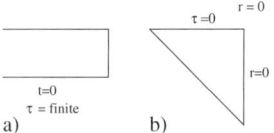

Figure 4.   In the distant past τ is (a) finite and (b) infinite.

For later use, we begin with an arbitrary $(n+1)$-dimensional spatially homo-geneous and spherically symmetric metric in proper time gauge:

$$ds^2 = -dt^2 + a^2(t)\left(dr^2 + \hat{R}^2(r)d\Omega_{n-1}^2\right),\tag{57}$$

where $d\Omega_{n-1}^2$ is the metric on the unit $n-1$ sphere and the form of $\hat{R}(r)$ depends on the spatial geometry: $\hat{R}(r) = \sin(r), r, \sinh(r)$ for spherical, flat, and hyperbolic geometries respectively. In fact, the function $\hat{R}(r)$ will not play a role below as our diagrams will depict the conformal structure only of the $(1+1)$-dimensional metric $ds_{1+1}^2 = -dt^2 + a^2(t)dr^2$. However, it will be important to note that r takes values only in $[0, \pi]$ for the spherical geometry but takes values in $[0, \infty]$ for the flat and hyperbolic cases. The usual change of coordinates to conformal time $\tau(t)$ defined by $d\tau = \frac{dt}{a}$ leads to the conformally Minkowski metric

$$ds_{1+1}^2 = a^2(t)(-d\tau^2 + dr^2).\tag{58}$$

Let us assume that our foliation represents an expanding phase that is asymp-totically de Sitter in the far future. That is, for $t \to +\infty$ the scale factor a di-verges exponentially. There are now two possibilities. Suppose first that $a = 0$ at some finite t. If a^{-1} diverges as a small enough power of t then τ will only reach a finite value in the past and the spacetime is conformal to a half-strip in Minkowski space.

In contrast, if a vanishes more quickly, or if it vanishes only asymptotically, then τ can be chosen to take values in $[-\infty, 0]$. From (58) we see that the region covered by our foliation is then conformal to a quadrant of Minkowski space. We take this quadrant to be the lower left one so that we may draw the conformal diagram as in figure 4(b).

We now wish to ask whether the region shown in figure 4(b) is 'complete' in some physical sense. In particular, we may wish to know whether light rays can reach the null 'boundary' in finite affine parameter. A short calculation shows that the affine parameter λ of a radial null ray is related to the original time coordinate t by $d\lambda = adt$. The affine parameter is clearly finite if a vanishes at finite t. In the remaining case, we have seen that ρ is bounded below. As a result, a must vanish at least exponentially and the affine parameter is again finite.

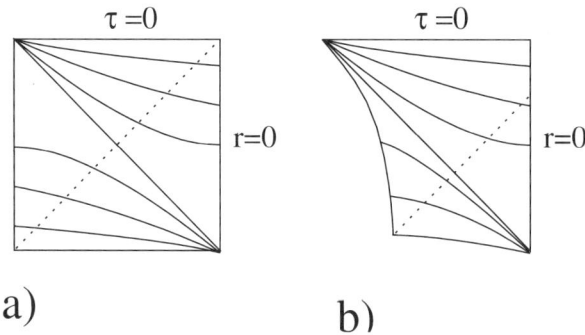

Figure 5. (a) In a square diagram, a light ray reaches the antipodal point only at I^+. (b) The generic conformal diagram for an asymptotically de Sitter space with flat surfaces of homogeneity. A light ray starting in the lower left corner reaches the antipodal point at a finite time.

Thus, this null surface represents either a singularity or a Cauchy horizon across which our spacetime should be continued. This statement is essentially a restricted version of the results of [69] (see also [70] for other interesting constraints on the 'beginning' of inflation). Note that there is no tension between our possible Cauchy horizon and the claim of a "singularity" in these references, as their use of the term singularity refers only to the geodesic incompleteness of the expanding phase.

>From (55) we see that unless $\rho + P_i$ vanishes as $a \to 0$, the energy density must diverge and a curvature singularity will indeed result. However, a proper tuning of the matter fields can achieve a finite ρ at $a = 0$. It is therefore interesting to consider solutions which are asymptotically de Sitter near $t = 0$, so that a vanishes exponentially. In this case, the $a = 0$ surface represents a Cauchy horizon across which we should continue our spacetime. We will focus exclusively on such cases below.

Since the boundary is a Cauchy horizon, there is clearly some arbitrariness in the choice of extension. We make the natural assumption here that the spacetime beyond the horizon is again foliated by flat hypersurfaces. Although at least one null hypersurface ($t = -\infty$) will be required, it can be shown that the surfaces of homogeneity must again become spacelike across the horizon if the spacetime is smooth. The key point here is that the signature can be deduced from the behavior of $a^2(t)$, which gives the norm $|\xi|^2$ of any Killing vector field ξ associated with the homogeneity. We impose a "past asymptotic de Sitter boundary condition" so that the behavior of this quantity near the Cauchy horizon must match that of some de Sitter spacetime. Consider in particular the behavior along some null geodesic crossing the Cauchy horizon and having affine parameter λ. It is straightforward to verify that $\lambda \sim a$, so that matching derivatives of $|\xi|^2$ across the horizon requires ξ to again become spacelike beyond the horizon.

It follows that the region beyond the Cauchy surface is just another region of flat spatial slices, but this time in the contracting phase. It is therefore conformal to the upper right quadrant of Minkowski space. However, having drawn the above diagram for our first region we have already used a certain amount of the available conformal freedom. Thus, it may not be the case that the region beyond the Cauchy surface can be drawn as an isosceles right triangle. The special case where this is possible is shown in figure 5(a). The exceptional nature of this case can be seen from the fact that it allows a spherical congruence of null geodesics to proceed from the upper right corner of I^+ (where it would have zero expansion) to the lower right corner of I^- (where it would also have zero expansion). Assuming, as usual, the weak energy condition, it follows that this congruence encountered no focusing anywhere along its path, *i.e.*, $\rho + P = 0$. Given the high degree of symmetry that we have already assumed, this can happen only in pure de Sitter space. The correct diagram for the general case is shown in figure 5(b) (see [2] for a complete derivation).

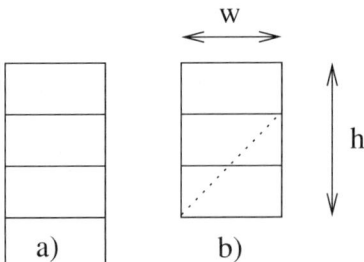

Figure 6.   Conformal diagrams for spherical surfaces of homogeneity (a) for the case where τ diverges in the past and (b) for the case where τ converges in the past.

6.2   Spherical slices ($k = +1$)

The conformal diagrams in this case are relatively simple. Since the radial coordinate now takes values only in an interval, we see from (58) that the conformal diagram is either a rectangle or a half-vertical strip, depending on whether or not τ is finite at the past boundary.

All such rectangles with the same ratio h/w (see figure 6(b)) can be mapped into each other via conformal transformations. For the case of pure de Sitter space we have h=w. On the other hand, for any spacetime satisfying the generic condition (so that null geodesics suffer some convergence along their trajectory), we know from [68] that the region to the past of any point p sufficiently close to I^+ must contain an entire Cauchy surface. Thus,[9] for such cases we have h > w.

6.3   Hyperbolic slices ($k = -1$)

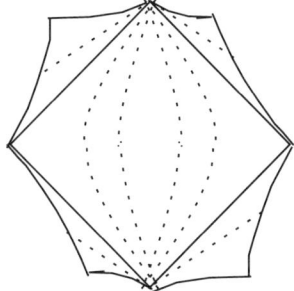

Figure 7.   The general conformal diagram for an appropriately complete asymptotically de Sitter spacetime with hyperbolic surfaces of homogeneity. A conformal frame has been chosen such that the diagram has a Z_2 reflection symmetry through the center.

Recall that the hyperbolic flows are complete, *i.e.*, a reaches 0 at finite t (say, $t = 0$), and vanishes at least as fast as t. The asymptotically de Sitter boundary conditions also require that a diverge exponentially as $t \to +\infty$. Note that since a vanishes quickly, τ will diverge at $t = 0$ and the region is again conformal to a quadrant of Minkowski space. As usual, this may or may not be singular, depending on the matter present.

Consider in particular the asymptotically de Sitter case where a vanishes linearly. One then finds that the affine parameter λ along a null ray near the horizon is asymptotically $\lambda \sim a^2$. The Killing vector field that implements

---

[9]This conclusion may also be reached by considering the sphere of null geodesics that begins in, say, the lower left corner and progresses toward the upper right and using the non-increase of the expansion θ implied by the weak null energy condition.

spatial translations in the direction along this null ray has a norm given by $a^2 \sim \lambda$ along it and so, if the spacetime is smooth, must become timelike beyond the horizon. Thus, the homogeneous surfaces must become timelike on the other side. It is not hard to show that an asymptotically de Sitter region foliated by timelike hyperbolic slices (i.e., copies of de Sitter space) has an 'r-t plane' that is conformal to a diamond in Minkowski space [2].

Assuming that no singularities are encountered within this diamond or on its boundaries, this provides three further Cauchy horizons across which we would like to extend our spacetime. A study of the norms of the Killing fields tells us that the foliation must again become spacelike beyond these horizons. Just as we saw for the flat foliations, we are therefore left with the task of attaching pieces conformal to various quadrants of Minkowski space. By the same reasoning as for $k = +1$ [2], the complete conformal diagram can be drawn as shown in figure 7.

7.   CFT on two boundaries

As remarked in the previous section, de Sitter space has two conformal boundaries and so one may ask the question as to whether the dS/CFT correspondence involves a single dual field theory or two. One simple argument in favor of one CFT is as follows [71]: The isometry group of $(n+1)$-dimensional dS space is $SO(n+1, 1)$, which agrees with the symmetries of a single Euclidean CFT in n dimensions. Further note that the global Killing vector fields corresponding to these isometries in dS space act nontrivially on both $I^{\pm}$. Hence there is a simple correlated action on source currents or dual operators identified with each of the boundaries. Hence, given the single symmetry group, it is natural to think that the dual description involves a single CFT.

Further, we recall our experience from the AdS/CFT correspondence. A central point in this context is that the CFT does not 'live' on the boundary of the AdS space. Usually one has chosen a particular foliation of AdS [55], and the bulk space calculations are naturally compared to those for the field theory living on the geometry of the surfaces comprising this foliation. Via the UV/IR correspondence, each surface in the bulk foliation is naturally associated with degrees of freedom in the CFT at a particular energy scale [56]. The boundary of AdS space plays a special role in calculations as this is a region of the geometry where the separation of operator insertions and expectation values is particularly simple. One notable exception where two CFT's seem to play a role is the eternal black hole [72, 73, 74]. In this case, however, the bulk geometry has two causally disconnected boundaries. In fact, one can show that for any solution of Einstein's equations with more than one asymptotically AdS boundary, the boundaries are all causally disconnected from each other [75]. In the case of dS space, the past and future boundaries are certainly causally connected and so it seems $I^\pm$ can be considered as two (special) slices in a certain foliation (36) of the spacetime. Hence this reasoning suggests that one should only consider a single CFT in the dual description.

7.1 Nonlocality in the boundary map

Next we turn to Strominger's observation [5] that the generating functional (42) can, in certain circumstances, be extended to incorporate sources on I^+. Certainly, the construction of the generating functional in the previous section produces essentially the same result if we replace both of the limits in eq. (42) with $t, t' \to +\infty$. This would produce an analogous generating functional with source currents defined by the asymptotic behavior of the scalar near I^+, *i.e.*,

$$\lim_{t \to +\infty} \phi(\Omega, t) \simeq \tilde{\varphi}_{0+}(\Omega) \, e^{-h_+ t} + \tilde{\varphi}_{0-}(\Omega) \, e^{-h_- t} \,. \tag{59}$$

However, it is also interesting to consider the case where only one of the limits in eq. (42) is replaced with one approaching I^+,

$$\widetilde{\mathcal{F}} = \lim_{t \to +\infty, t' \to -\infty} \int d\Sigma^\mu d\Sigma'^\nu \phi(x) \overleftrightarrow{\partial}_\mu G(x, x') \overleftrightarrow{\partial}_\nu \phi(x') \,. \tag{60}$$

Now an essential observation [5, 6] is the causal connection between points on the two boundaries $I^\pm$. In particular, a null geodesic emerging from a point on I^- expands out into the dS spacetime and refocuses precisely at the antipodal point on the n-sphere at I^+. Hence the two-point function in eq. (60) (or any dS-invariant Green's function) will introduce singularities when the point on I^+ approaches the antipode to the point on I^-, as the proper separation of these points vanishes. In fact, in certain circumstances (see the details below), evaluating the above expression yields the simple result:

$$\widetilde{\mathcal{F}} = -\frac{(h_+ - h_-)^2}{2^{2n}} \int d\Omega d\Omega' \left[\tilde{c}_+ \frac{\tilde{\varphi}_{0-}(\Omega) \, \phi_{0-}(\Omega')}{(w^i w'^i + 1)^{h_+}} + \tilde{c}_- \frac{\tilde{\varphi}_{0+}(\Omega) \, \phi_{0+}(\Omega')}{(w^i w'^i + 1)^{h_-}} \right] \,. \tag{61}$$

This expression incorporates the same Euclidean two-point function except that the singularities now arise as the sources $\tilde{\varphi}_{0\pm}(\Omega)$ approach antipodes on the n-sphere.

These results suggest that one need only consider a single copy of the CFT and that an operator on I^+ is identified with the same operator on I^- after an antipodal mapping. One finds further support for this interpretation by considering the isometries of dS space, for example, the isometry[10] which produces a dilatation around a point on I^-. On I^+, the same symmetry corresponds to a dilatation around the antipodal point on the n-sphere.

---

[10]This isometry corresponds to the action of a time translation ∂_t in the static patch coordinates [76].

However, this suggestion for identifying operators at I^+ and I^- is easily seen to require some revision as follows. As discussed in the introduction, bulk correlators are naturally related by time evolution. The key ingredient is simply the free field evolution of the scalar, which, given some configuration specified on a n-dimensional hypersurface, is characterized by the formula

$$\phi(x') = \int d\Sigma^\mu \, \phi(x) \, \overleftrightarrow{\partial}_\mu \, G_R(x, x') , \tag{62}$$

where $G_R(x, x')$ is a retarded Green's function, *i.e.,* it vanishes for $t > t'$. Now as an example, the integral appearing in the generating functional (42) is covariant and so should be invariant when evaluated on any time slices t and t'. The advantage of pushing these slices to I^- (or I^+) lies in the fact that one can easily separate the source currents according to their conformal weights.

We can explicitly consider the relation between currents on the past and future boundaries simply by following the classical evolution (62) of the fields from I^- to I^+. Unfortunately, it is clear that, generically, there is no simple local relation between the currents on I^- and those on I^+. This remark comes from the observation that, in general, the retarded Green's function will have support throughout the interior of the light cone. This intuition is readily confirmed by explicit calculations. Ref. [61] presents explicit Green's functions for generic masses in four-dimensional de Sitter space. So, for example, for scalar fields in the principal series, the retarded Green's function becomes, for large timelike proper separation,

$$G_R(t, \Omega; t', \Omega') \propto \frac{\sin^{n/2} \tau \, \sin^{n/2} \tau'}{(w^i w'^i - \cos \tau \, \cos \tau')^{n/2}} \, \theta(\tau' - \tau) , \tag{63}$$

where τ is the conformal time coordinate, $\sin \tau = 1/\cosh t/\ell$ — see eq. (4), below. Here the θ-function ensures the proper time-ordering of the points. In any event, eq. (63) illustrates how the field 'leaks' into the interior of the lightcone with the classical evolution. Generically this leads to a nonlocal mapping between the currents on I^- and I^+. This complication will only be avoided in certain exceptional cases, for example, if the retarded Green's function only has support precisely on the light cone — a point to which we return below.

The nonlocal relation between the currents on I^- and those on I^+ can be made more explicit through the mode expansion of the fields on dS space — see the appendix. A well-documented feature of cosmological spacetimes is mode-mixing or particle creation [60]. For the present case of dS space, this corresponds to the fact that a mode of the scalar field with a given boundary behavior on I^-, *e.g.,* having h_- scaling, will usually have a mixture of $h_\pm$ scaling components at I^+. The appendix provides a detailed discussion of the mode expansions on dS space as well as the Bogolubov transformation relating the modes with a simple time dependence (scaling behavior) near I^- to those near I^+. Using these results, we may discuss the mapping between the currents on the conformal boundaries. Following the notation of the appendix, we decompose the asymptotic fields in terms of spherical harmonics on the n-sphere

$$\phi_{0\pm}(\Omega) = \sum_{L,j} a_{\pm Lj} Y_{Lj}, \qquad \tilde{\varphi}_{0\pm}(\Omega) = \sum_{L,j} \tilde{a}_{\pm Lj} Y_{Lj}. \qquad (64)$$

Denoting the antipodal map on the n-sphere as $\Omega \to J\Omega$, one has[11] $Y_{Lj}(\Omega) = (-)^L Y_{Lj}(J\Omega)$. Now let us imagine that $\phi_{0\pm}$ and $\tilde{\varphi}_{0\pm}$ are related by the antipodal map, *i.e.*, $\phi_{0\pm}(\Omega) = z\,\tilde{\varphi}_{0\pm}(J\Omega)$ with some constant phase z. Then one must have

$$a_{\pm Lj} = z\,(-)^L\,\tilde{a}_{\pm Lj}, \qquad (65)$$

where, in particular, the constant z is independent of L.

However, in general, the Bogolubov transformation given in the appendix gives a more complex mapping. For example, from eq. (A.17) for the principal series, one finds

$$a_{\pm Lj} = C_-^-(\omega)e^{\pm 2i\theta_L}\tilde{a}_{\pm Lj} + C_-^+(\omega)\tilde{a}_{\mp Lj}. \qquad (66)$$

Now given eq. (A.18) for n odd, with both $C_-^-(\omega)$ and $C_-^+(\omega)$ nonvanishing, certainly eq. (65) is inapplicable. One comes closer to realizing the desired result with even n, for which $C_-^-(\omega) = 1$ and $C_-^+(\omega) = 0$. However, for either n odd or even, the phase θ_L always introduces a nontrivial L dependence (beyond the desired $(-)^L$) as shown in eq. (A.19). Thus while the mapping between I^- and I^+ may look relatively simple in this mode expansion, it will clearly be nonlocal when expressed in terms of the boundary data $\phi_{0\pm}(\Omega)$ and $\tilde{\varphi}_{0\pm}(\Omega)$.

The complementary series gives some more interesting possibilities with

$$
\begin{aligned}
a_{-Lj} &= \bar{C}_-^-(\mu)\,\tilde{a}_{-Lj} + \bar{C}_-^+(\mu)\,\tilde{a}_{+Lj}, \\
a_{+Lj} &= \bar{C}_+^-(\mu)\,\tilde{a}_{-Lj} + \bar{C}_+^+(\mu)\,\tilde{a}_{+Lj}.
\end{aligned} \qquad (67)
$$

In particular, for n odd and μ half-integer, one finds $\bar{C}_+^+(\mu) = 0 = \bar{C}_+^-(\mu)$ and $\bar{C}_-^-(\mu) = (-)^{\frac{n}{2}+\mu}(-)^L = \bar{C}_+^+(\mu)$. Note that these special cases include

---

[11]This result becomes clear when the n-sphere is embedded in R^{n+1} with $(x^1)^2+(x^2)^2+\cdots+(x^{n+1})^2 = 1$. In this case, the spherical harmonics Y_{Lj} may be represented in terms of symmetric traceless tensors, $Z_{i_1 i_2 \cdots i_L} x^{i_1} x^{i_2} \cdots x^{i_L}$, and hence it is clear that the antipodal map, which takes the form $J : x^i \to -x^i$, produces an overall factor of $(-)^L$.

$\mu = 1/2$, which corresponds to the conformally coupled massless scalar field to which we will return in the following section. Similarly, for n even and μ integer: $\bar{C}_-^- = \bar{C}_+^+ = (-1)^{\frac{n}{2}+\mu+1}(-1)^L$ and $\bar{C}_-^+ = 0 = \bar{C}_+^-$. Hence the coefficients for these special cases give a precise realization of eq. (65). Further for these cases then, the generating functional considered in eq. (60) will take the simple form given in eq. (61).

Hence, when considering the principal series or generic masses in the complementary series, it seems that nonlocality will be an unavoidable aspect of the relation between field theory operators associated with two conformal boundaries. The essential point is that the time evolution of the scalar generically introduces nonlocality in the mapping because the retarded Green's function smears a point-like source on I^- out over a finite region on I^+. However, note that one reproduces precisely the same boundary correlators, but after some nonlocal reorganization of the degrees of freedom within the dual field theory. It seems appropriate to refer to such relations as *nonlocal dualities* within the field theory. On the other hand, the complementary series does seem to provide some situations where the mapping of the boundary data between I^- and I^+ is local. In the absence of a working example of the proposed dS/CFT duality, one might interpret these results as a hint towards the specific types of fields that would appear in a successful realization of the dS/CFT. Unfortunately, however, this selection rule based on locality of the mapping between boundaries does not seem to survive in more interesting applications, as we will see in the following.

7.2   Nonlocal dualities in 'tall' spacetimes

It is of interest to extend the application of the dS/CFT correspondence from dS space to more general spacetimes with asymptotically dS regions. As a consequence of a theorem of Gao and Wald [68], such a (nonsingular) background will be 'tall' [2]. That is, the conformal diagram for such spacetimes must be taller in the timelike direction than it is wide in the spacelike direction. Of course, this feature has important implications for the causal connection between the past and future boundaries, and hence for the relation between the dual field theory operators defined at these surfaces. In particular, the latter relation becomes manifestly nonlocal.

We may explicitly illustrate the causal structure of the tall spacetimes by working in conformal coordinates. For asymptotically dS spacetimes which are homogeneous on spherical hypersurfaces, the metric may be written

$$ds^2 = C(\tau)\left[-d\tau^2 + d\theta^2 + \sin^2\theta\, d\Omega_{n-1}^2\right] . \tag{68}$$

Recall from eq. (4) that for pure dS space, $C(\tau) = \ell^2/\sin^2\tau$. For a tall spacetime, the conformal time above would run over an extended range $0 \leq$

$\tau \le \pi + \Delta$ where $\Delta > 0$. The assumption that the background is asymptotically dS means that the conformal factor has the following behavior near $I^{\pm}$:

$$\lim_{\tau \to 0} C(t) = \frac{\ell^2}{\sin^2 \tau}, \tag{69}$$

$$\lim_{\tau \to \pi + \Delta} C(t) = \frac{\tilde{\ell}^2}{\sin^2(\tau - \Delta)},$$

where we have allowed for the possibility that the cosmological 'constant' is different at I^+ than at I^-. This possibility may be realized in a model where a scalar field rolls from one critical point of its potential to another [2]. In any event, the corresponding conformal diagram will be a rectangle with height $\delta\tau = \pi + \Delta$ and width $\delta\theta = \pi$ (see figure 8).

This increase in the height of the conformal diagram modifies the causal connection between $I^{\pm}$ in an essential way. Consider the null rays emerging from the north pole ($\theta = 0$) at I^- ($\tau = 0$). This null cone expands out across the n-sphere reaching the equator ($\theta = \pi/2$) at $\tau = \pi/2$, and then begins to reconverge as it passes into the southern hemisphere. The null rays focus at the south pole ($\theta = \pi$) at $\tau = \pi$, however, in this tall spacetime, this event corresponds to a finite proper time for an observer at the south pole. Beyond this point, the null cone expands again and intersects I^+ ($\tau = \pi + \Delta$) on the finite-sized ($n-1$)-sphere at $\theta = \pi - \Delta$.

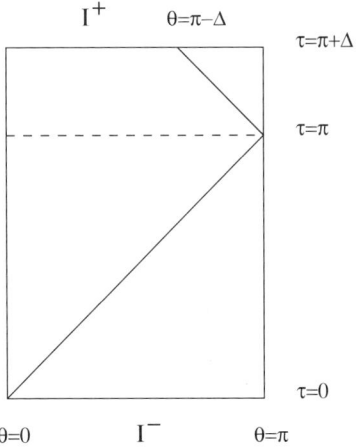

Figure 8. Conformal diagrams of a perturbed de Sitter space. The excess height is represented by Δ.

The discussion of the previous section made clear that an essential ingredient in finding a simple, local mapping of boundary data on I^- to that on I^+ in dS space was the refocusing of the above null cone precisely at the future boundary. Even in that case, we pointed out that the time evolution of the scalar generically introduces nonlocality in the mapping because the retarded Green's function smears a point-like source on I^- out over a finite region on I^+. Here we see that in a tall spacetime, a nonlocal map is inevitable since the causal connection between the past and future boundaries is itself nonlocal. So we should expect that even in the special cases found to have a local map for pure dS space, the mapping should become nonlocal for these same theories in a tall background. That is, for these more general asymptotically dS spacetimes, the relation between the dual field theory operators defined at each of the boundaries becomes nonlocal. Hence we are naturally lead to consider a nonlocal self-duality of the CFT. Further we note that, given the results of Gao and Wald [68], this would be the generic situation. For example, injecting a single scalar field quantum into dS space would actually lead to backreaction effects which would produce a tall spacetime.

7.3   The conformally coupled massless scalar

We now turn to consider conformally coupled massless scalar field theory as an example which illustrates several of the points discussed above. In particular, it is an example where the mapping between the past and future boundaries is local in pure dS space, but becomes nonlocal in a tall background. Another useful feature is that one can perform explicit calculations in a tall spacetime without referring to the detailed evolution of the conformal factor $C(\tau)$. Rather, a knowledge of the boundary conditions (69) is sufficient.

The conformally coupled massless scalar corresponds to the curvature coupling $\xi = \frac{n-1}{4n}$, and $m^2 = 0$ in eq. (38). Hence in pure dS space or in an asymptotically dS region, $M^2 \ell^2 = (n^2 - 1)/4$ and the corresponding scaling exponents (39) become $h_\pm = (n \pm 1)/2$, independent of the value of the cosmological constant. As one might infer from the real exponents, this field lies in the complementary series for any value of the cosmological constant. The remarkable property of this scalar field theory is that the solutions of the wave equation (37) transform in a simple way under local conformal scalings of the background metric [60].

The backgrounds of interest (4) are conformally flat[12] and therefore the Green's function describing the evolution in the tall background is simply the flat space Green's function for a *massless* scalar field, up to some overall time dependent factors. In particular then, for d even (n odd), the Green's function will have support precisely on the light cone. For example, in four-dimensional

---

[12]Note that the coordinate transformation $T = e^\tau$ puts the metric (4) in the form of the flat Milne universe, up to a conformal factor.

dS space, the retarded Green's function can be written as

$$G_R(\tau, \Omega; \tau', \Omega') = -\frac{\sin\tau\sin\tau'}{4\pi\ell^2}\,\delta(w^i w'^i - \cos(\tau' - \tau))\,\theta(\tau' - \tau)\,. \quad (70)$$

Similarly, in higher even-dimensional dS spaces, the Green's function will contain δ-functions (and derivatives of δ-functions) with support only on the light cone [77]. Given this form of the retarded Green's functions, the evolution of the scalar field (62) from I^- to I^+ will produce precisely the antipodal mapping for all of these cases. Note that this result is confirmed by the mode analysis in the first part of this section. The conformally coupled massless scalar has $\mu = 1/2$ and we are considering even dimensions or n odd. This combination matches one of the special cases in which the modes transformed according to the antipodal mapping.

Using the conformal transformation properties of the field [60], the analogous Green's function for any spacetime of the form (4) is easily constructed. For $d = 4$, it may be written as

$$G_R(\tau, \Omega; \tau', \Omega') = -\frac{1}{4\pi}\frac{1}{\sqrt{C(\tau)}\sqrt{C(\tau')}}\,\delta(w^i w'^i - \cos(\tau' - \tau))\,\theta(\tau' - \tau)\,. (71)$$

For other even values of d, the corresponding Green's function has a similar form. For the conformally coupled scalar in such tall spaces, the delocalization of the boundary map does not depend on the detailed evolution, *i.e.,* the details of $C(\tau)$. Rather the nonlocality is completely characterized by Δ, the excess in the range of the conformal time. For example, a source current placed at the north pole ($\theta = 0$) on I^- is smeared over an $(n{-}1)$-sphere centered at the south pole ($\theta = \pi$) and of angular radius $\delta\theta = \Delta$ on I^+.

Using eq. (71), we can make this discussion completely explicit for four dimensions. Consider an arbitrary tall space (4), with $n = 3$, satisfying the boundary conditions given in eq. (69). First, with the conformal time coordinate, the asymptotic boundary conditions (44) for the scalar field at I^- become

$$\lim_{\tau \to 0} \phi(\Omega, \tau) \simeq \phi_{0+}(\Omega)\,\tau^{h_+} + \phi_{0-}(\Omega)\,\tau^{h_-}\,, \quad (72)$$

and similarly at I^+, we have

$$\lim_{\tau \to \pi + \Delta} \phi(\Omega, \tau) \simeq \tilde{\phi}_{0+}(\Omega)\,(\pi + \Delta - \tau)^{h_+} + \tilde{\phi}_{0-}(\Omega)\,(\pi + \Delta - \tau)^{h_-}\,. \quad (73)$$

These boundary conditions apply for a general scalar field theory. In the present case of a conformally coupled massless scalar with $n = 3$, we have $h_+ = 2$ and $h_- = 1$. Hence, inserting (71) and (72) into (62), we may evaluate the result at a point $(\Omega', \tau' = \pi + \Delta - \epsilon)$ near I^+ and compare to eq. (73). The final result for the boundary fields on I^+ is

$$\tilde{\phi}_{0+}(\Omega') = \frac{|\sin\Delta|}{\sin\Delta}\frac{\ell}{\tilde{\ell}}\left\{ \sin\Delta \langle\phi_{0-}\rangle_\Delta(J\Omega') - \cos\Delta \langle\phi_{0+}\rangle_\Delta(J\Omega') \right\}, \tag{74}$$

$$\tilde{\phi}_{0-}(\Omega') = \frac{|\sin\Delta|}{\sin\Delta}\frac{\ell}{\tilde{\ell}}\left\{ \cos\Delta \langle\phi_{0-}\rangle_\Delta(J\Omega') + \sin\Delta \langle\phi_{0+}\rangle_\Delta(J\Omega') + \sin\Delta\, \partial_\theta\langle\phi_{0-}\rangle_\Delta(J\Omega') \right\},$$

where J is the antipodal map on the two-sphere and $\langle\phi_{0\pm}\rangle_\Delta(J\Omega')$ denote the average of $\phi_{0\pm}$ on the two-sphere separated from $J\Omega'$ by an angle Δ. The factors $\frac{|\sin\Delta|}{\sin\Delta}$ are to be understood as being continuous from below; *i.e.*, this factor is -1 at $\Delta = 0$ and $+1$ at $\Delta = \pi$.

This expression simplifies tremendously in the case of dS space with $\Delta = 0$ (as well as $\tilde{\ell} = \ell$) to yield

$$\tilde{\phi}_{0+}(\Omega') = \phi_{0+}(J\Omega'), \qquad\qquad \tilde{\phi}_{0-}(\Omega') = -\phi_{0-}(J\Omega'). \tag{75}$$

Thus, in pure four-dimensional dS space, the map from I^- to I^+ acts on the conformally coupled massless scalar field as simply the antipodal map on ϕ_{0+} and -1 times the antipodal map on ϕ_{0-}. Note that the time reflection symmetry of de Sitter allows solutions for the mode functions to be decomposed into even and odd parts and, furthermore, both even and odd solutions will exist. Thus, with our conventions, and $h_\pm$ real, when evolution from I^- to I^+ leads to the antipodal map it will be associated with a phase $z = +1$ for one set of modes and the opposite phase $z = -1$ for the other.

8.   Discussion of dS/CFT

The dS/CFT correspondence is a striking proposal which carries the potential for extraordinary new insights into cosmology and the cosmological constant problem. Unfortunately, the outstanding problem remains to find a concrete example where the bulk gravity theory and the dual field theory are understood or at least known — see, however, [15]. Lacking the guidance that such a working model would provide, one is left to study various aspects of physics in (asymptotically) dS spacetimes from this new point of view and to determine properties which this correspondence implies for the dual Euclidean CFT.

Such investigations have yielded a number of unusual properties for the dual field theory. It is likely to be nonunitary, *e.g.*, if the bulk theory involves scalars in the principal series [5]. A nonstandard inner product is required to reproduce ordinary quantum field theory in the bulk [11, 47]. One might also observe that this Euclidean field theory should not simply be a standard Wick rotation of a conventional field theory since attempting to 'un-Wick rotate' would produce a bulk theory with two time directions and all of the associated confusions. We may add to this list the observation of section 4.2 that, since bulk correlators are not symmetric in Lorentzian signature quantum field theory, a straightforward duality would require non-symmetric correlation functions in the dual Euclidean theory. But correlators generated by functional differentiation of a partition function are always symmetric, so the Euclidean theory could have no definition through a partition sum. Finally, in the present paper, we have also inferred the existence of unusual nonlocal dualities within the field theory itself.

Our investigation focussed on the mapping of operators between I^+ and I^- provided by time evolution in the bulk spacetime. The essential point is that the time evolution of the scalar generically introduces nonlocality in the mapping because the retarded Green's function smears a point-like source on I^- out over a finite region on I^+. However, despite this nonlocal reorganization of the degrees of freedom within the dual field theory, one reproduces the same boundary correlators. Hence we referred to this relation as a nonlocal duality within the field theory. While this nonlocality already applies for many fields in pure dS space, it seems unavoidable in tall spacetimes because the causal connection between I^+ and I^- is inherently nonlocal. We emphasize that tall spacetimes are quite generic as a result of the theorem in [68]. As soon as one perturbs dS even slightly by, *e.g.,* the introduction of matter fields or gravitational waves, the resulting background solution will have the property that its conformal diagram is taller than it is wide. As the inferred self-duality is nonlocal, *i.e.,* local operators are mapped to nonlocal operators, it seems that the underlying field theory does not have a unique concept of locality. That is, one has a specific dictionary whereby the same short distance singularities can be reproduced by a set of local or nonlocal operators.

Faced with the daunting task of consolidating all of these unusual characteristics in a single Euclidean field theory, one is tempted to revise the interpretation of the dS/CFT correspondence. One suggestion [47] is that the duality should involve two CFT's and that dS spacetime is defined as a correlated state in the Hilbert space of the two field theories. The correlated state is constructed so as to preserve a single $SO(n + 1, 1)$ symmetry group, which is then reflected in the isometries of the dS space. As discussed in section 7, we still feel that our experience with the AdS/CFT is highly suggestive that the two boundaries should not be associated with distinct field theories. Further, it is difficult to see how this framework could incorporate big bang or big crunch backgrounds with a single asymptotic dS region. Note that the latter spacetimes will still give rise to horizons, as well as the associated thermal radiation and entropy.

However, this approach with two CFT's remains an intriguing suggestion. Within this context, the mapping of the boundary data between $I^{\pm}$ would provide information about correlations in the field theory state. Hence our calculations would still find application in this context. The nonlocalities discussed here, while not unnatural, give an indication of the complexity of these correlations.

We should also remark that, in all of our investigations, we treated only the time evolution of a free scalar field theory. The mapping of boundary operators will become even more complex if one was to consider an interacting field theory. Of course, in accord with the discussion here, we would still expect that time evolution of the fields or operators in an interacting theory would still provide the basis for this mapping.

While it is amusing to speculate on such matters, we note that the central thesis of [78] is that one cannot successfully understand the physics of dS space within the context of quantum field theory in curved spacetime. It is interesting to consider how their comments may relate our discussion. Essentially, they suggest that bulk properties of dS space should be analogous to the physics in a thermal system with a finite number of states and deduce from this that the evolution map of linearized quantum field theory should not be trusted in detail near the past and future boundaries. As a result, they suggest that a dual theory may not be as local as one might expect by studying limits of bulk correlation functions in background quantum field theory. The comments of [79] raise further questions about correlation functions between points with a large separation in time. In particular, the problematic correlators would include precisely those between operators on I^- and I^+. Here the smearing observed in the tall spacetimes is likely to play a role since, if backreaction is properly accounted for, even injecting a single scalar field quantum into dS should deform it to a (slightly) tall spacetime. It may be that the nonlocalities discussed here may be a hint that the 'correct physical observables' are themselves nonlocal[13] so that the boundary map would preserve the form of such operators.

Note that there is a certain tension between our strong reliance on time evolution, through which observables near any two Cauchy surfaces can be related, and the idea that the bulk evolution is related to a renormalization group flow in the dual theory [9, 10]. The point is that time evolution naturally produces a scaling of distances on Cauchy surfaces (at least in simple examples) and so these surfaces are naturally associated with different distance scales in the dual theory. However, the time evolution map relating different surfaces is invertible. In contrast, the usual notion of the renormalization group is actually that of a semi-group, in which different scales are related by integrating out modes, *i.e.,* by throwing away short distance details so that the descriptions at two different scales are not fully equivalent.

To gain some perspective on this issue, we would like to return briefly to the AdS/CFT case and the interpretation of renormalization group flows. Recall that the primary assumption is that the relevant asymptotically AdS spacetime is in fact dual to the vacuum of some field theory. The important point is that one

---

[13] Similar implications can be drawn from the finite time resolution discussed in [80].

begins by placing the entire spacetime in correspondence with the vacuum of some single theory. One then uses the IR/UV connection to argue that different regions of the bulk spacetime are naturally related to different energy regimes in the dual theory. The suggestion that this description at differing energy scales is somehow connected to a renormalization group flow seems natural and, in that context, there was no evolution map relating the inner and outer regions to provide such an obvious tension.

In the present dS/CFT context, such a tension does exist. However, the more primitive association of different parts of the spacetime with the behavior of the field theory at differing energy scales still seems plausible. A more concrete version of this idea is suggested by the behavior of the bulk evolution map itself. As we have seen, the evolution map from t to t' 'coarse grains' the observables on t' in the sense that the theory is now presented in terms of variables (those that are local at time t) which are nonlocal averages over the intersection of past light cones from time t with the original hypersurface at t'. However, a sufficient number of overlapping coarse grainings are considered so that no information is lost. Such a procedure can also be performed in a Euclidean field theory and one might speculate that keeping only the simplest terms in the resulting action might bear some similarity to those obtained from more traditional renormalization group methods. This would be in keeping with the identification of a c-function [2, 9, 10] in which a spacetime region is associated with a copy of de Sitter space by considering only the metric and extrinsic curvature on a hypersurface.

Note that this interpretation readily allows us to run our flow both 'forward' (toward the IR) and 'backward' (toward the UV). However, it is far from clear that the coarse graining procedure is unique. This fits well with the interpretation suggested in [2] for 'renormalization group flow spacetimes' with spherical homogeneity surfaces. There, one naturally considers two UV regions (one at I^- and one at I^+) which both 'flow' to the same theory at some minimal sphere where the two parts of the spacetime join. One simply reads the flow as starting in the UV, proceeding toward the IR, but then reversing course. Interestingly, it is possible to arrive at a different UV theory from which one began. Such an odd state of affairs is more natural when one recalls that we have already argued that the theory must possess a nonlocal duality, so that, in fact, it has two distinct local descriptions.

In any event, the c-theorem suggests that the effective number of degrees of freedom in the CFT increases in a generic solution, as it evolves toward an asymptotically dS regime in the future. We would like to point out, however, that this does not necessarily correspond to the number degrees of freedom accessible to observers in experiments. Here we are thinking in terms of holography and Bousso's entropy bounds [34]. Consider a four-dimensional inflationary model with $k=0$ and consider also the causal domain relevant for an experiment beginning at $t = -\infty$ and ending at some arbitrary time $t=t_o$. For a sufficiently small t_o, it is not hard to show that the number of accessible states is given by $3\pi/G\Lambda_{initial}$. Naively, one expects that this number of states will grow to $3\pi/G\Lambda_{final}$ as $t_o \to \infty$. However, this behavior is not universal. It is not hard to construct examples [2] where, in fact, the initial cosmological constant still fixes the number of accessible states for arbitrarily large t_o. This behavior arises because the apparent horizon is spacelike in these geometries. Hence, in such models, the number of degrees of freedom required to describe physical processes throughout a given time slice grows with time, while the number of states that are accessible to experimental probing by a given physicist remains fixed.

This discussion reminds us of the sharp contrast in the 'degrees of freedom' in the dS/CFT duality [5] and in the Λ-N correspondence [7, 38]. In the Λ-N framework, the physics of asymptotically de Sitter universes is to be described by a finite dimensional space of states. This dimension is precisely determined as the number of states accessible to probing by a single observer. (The latter is motivated in part by the conjecture of black hole complementary [37].) In contrast, in the dS/CFT context, one would expect that a conformal field theory with a finite central charge should have an infinite dimensional Hilbert space,[14] and that these states are all involved in describing physical phenomena across entire time slices. Further, as shown above, the central charge, while a measure of number of degrees of freedom on a given time slice, need not be correlated with the number of states experimentally accessible to observers on that slice.

Acknowledgments

We would like to congratulate the organizers of the the School on Quantum Gravity at the Centro de Estudios Científicos, Valdivia for arranging an extremely successful gathering. I would also like to thank them for giving me the opportunity to lecture at their school, as well as the opportunity to enjoy the pleasant surroundings of Valdivia. I must also thank my collaborators, Raphael Bousso, Frédéric Leblond, Don Marolf and Oliver de Wolfe for the opportunity to work with them on the projects discussed here. I would like to thank David Winters for carefully reviewing this manuscript. This work was supported in part by NSERC of Canada and Fonds FCAR du Québec. (Viva CECS!)

Appendix: Scalar field modes in dS space

In this appendix, we present a detailed analysis of the bulk physics of massive scalar fields propagating in a dS space of arbitrary dimension, emphasizing characteristics of their evolution

---

[14]Though this infinity might perhaps be removed if one imposes, as described in [6], that the conformal generators vanish on physical states.

which should be relevant to the proposed dS/CFT correspondence. Our aim is to characterize fully the mode mixing phenomenon inherent to physics in dS space. While the details of this analysis are readily available in the literature for the modes of the principal series (see, for example, ref. [11]), we did not find explicit accounts of the complementary and discrete series.

Field equation and asymptotic behavior

The spherical foliation of $(n+1)$-dimensional dS space is given by the metric

$$ds^2 = -dt^2 + \cosh^2 t \, d\Omega_n^2 \,, \tag{A.1}$$

where in this appendix we set the dS radius to unity ($\ell = 1$). We consider a massive scalar field propagating in this background according to

$$\left[\Box - M^2\right]\phi(x) = 0 \,. \tag{A.2}$$

It is convenient to write the solutions to eq. (A.2) in the form

$$\phi(x) = y_L(t)\, Y_{Lj}(\Omega) \,, \tag{A.3}$$

where the Y_{Lj}'s are spherical harmonics on the n-sphere satisfying

$$\nabla^2 Y_{Lj} = -L(L+n-1)Y_{Lj} \,, \tag{A.4}$$

where ∇^2 is the standard Laplacian on the n-sphere. The differential equation for $y_L(t)$ is then

$$\ddot{y}_L + n \tanh t \, \dot{y}_L + \left[M^2 + \frac{L(L+n-1)}{\cosh^2 t}\right] y_L = 0 \,. \tag{A.5}$$

As discussed in section 2.2, of particular relevance to the dS/CFT correspondence is the behavior of the scalar field near the boundaries I^+ and I^- as $t \to \pm\infty$. In these limits, eq. (A.5) becomes

$$\ddot{y}_L \pm n\dot{y}_L + M^2 y_L = 0 \,, \tag{A.6}$$

which implies that

$$\lim_{t \to -\infty} y_L \sim e^{h_\pm t} \,, \qquad \lim_{t \to +\infty} y_L \sim e^{-h_\pm t} \,, \tag{A.7}$$

where the weights $h_\pm$ are defined by

$$h_\pm = \frac{n}{2} \pm \sqrt{\frac{n^2}{4} - M^2} \equiv \frac{n}{2} \pm \mu \,. \tag{A.8}$$

Formally, one may classify such a scalar field according to the irreducible representations of SO($n + 1, 1$), the isometry group of de Sitter space, which are labelled by the eigenvalues associated with the Casimir operator[15] $Q = \Box$, which simply corresponds to the mass parameter M^2. The *principal series* is defined by the inequality $M^2 > n^2/4$. In this case, the weights

---

[15]In fact, there are two coordinate invariant Casimir operators associated with the de Sitter isometry group but only one is relevant in characterizing massive scalar fields. The other Casimir operator automatically vanishes for all spin-zero fields but may play a role in the classification of higher spin representations [81]. Another interesting formal question is the behavior of these representations in the limit where the cosmological constant is taken to zero. A complete treatment of representation contraction in de Sitter space can be found in ref. [82].

$h_{\pm}$ have an imaginary part, and the corresponding modes, while still being damped near the boundaries, have an oscillatory behavior in the bulk. For the *complementary series*, the effective mass falls in the range $0 < M^2 \leq n^2/4$. As will be made more explicit later, the modes are non-oscillatory asymptotically in this case since both $h_{\pm}$ are real quantities. The remaining *discrete series* corresponds to $M^2 < 0$. This last condition means that $h_- < 0$ (and $h_+ > n$), which implies that the tachyonic fields scaling like $y_L \sim e^{\mp h_- t}$ are growing without bound as one approaches $I^{\pm}$. Still, one is able to find 'normalizable' modes in a certain limited number of cases, as will be discussed below.

The case of $M^2 = 0$, *i.e.*, a massless scalar field, is interesting and deserves further comment. One finds that in this case it is impossible to construct a vacuum state which is invariant under the full de Sitter group $SO(n + 1, 1)$. A great deal of discussion about the peculiar nature of this quantum field theory can be found in the literature [83, 84, 63]. The weights associated with the $M^2 = 0$ field are $h_+ = n$ and $h_- = 0$. Dual to the latter, there should be a marginal operator in the CFT, *i.e.*, a deformation which does not scale under conformal transformations.

To fully solve eq. (A.5), we make the change of variables

$$y_L(t) = \cosh^L t \, e^{(L + \frac{n}{2} + \mu)t} g_L(t) \,. \tag{A.9}$$

Setting $\sigma = -e^{2t}$, this equation for the time-dependent profile takes the form of the hypergeometric equation:

$$\sigma(1 - \sigma)g'' + [c - (1 + a + b)\sigma] \, g' - abg = 0 \,, \tag{A.10}$$

where a 'prime' denotes a derivative with respect to σ and the coefficients are

$$a = L + \frac{n}{2} \,, \quad b = L + \frac{n}{2} + \mu \,, \quad c = 1 + \mu \,. \tag{A.11}$$

The two independent solutions can then be expressed in terms of hypergeometric functions,

$$y_{L+}(t) = N_+ \cosh^L t \, e^{(L + h_+)t} F(L + \frac{n}{2}, L + h_+; 1 + \mu; -e^{2t}) \,, \tag{A.12}$$

$$y_{L-}(t) = N_- \cosh^L t \, e^{(L + h_-)t} F(L + \frac{n}{2}, L + h_-; 1 - \mu; -e^{2t}) \,, \tag{A.13}$$

where $N_{\pm}$ are normalization constants, which will be fixed below. More specifically, we have here chosen the two linearly independent solutions of eq. (A.10) in the neighborhood of the singular point $-e^{2t} = 0$ [85], which corresponds to one of the two limits of interest, *i.e.*, $t \to -\infty$. Following eq. (A.3), we denote the complete mode functions as $\phi_{L\pm} = y_{L\pm}(t)Y_{Lj}(\Omega)$.

One important aspect of the time evolution of the scalar field in the bulk is the mode mixing that occurs between the two boundaries, $I^{\pm}$. For example, this would be related to particle production in the dS space [11, 63, 64, 65]. In the following, we emphasize the differences between the principal, complementary and discrete series.

Principal series

The principal series is frequently discussed in the physics literature, *e.g.*, [11, 63, 64, 65], and would seem to be the most relevant case for the particle spectrum observed in nature. We review some of the salient points here for comparison with the other representations in the following subsection. For the principal series, it is useful to introduce $\omega \equiv -i\mu$. Then the above modes become

$$y_{L-}(t) = \frac{2^{L + (n-1)/2}}{\sqrt{\omega}} \cosh^L t \, e^{(L + \frac{n}{2} - i\omega)t} F(L + \frac{n}{2}, L + \frac{n}{2} - i\omega; 1 - i\omega; -e^{2t}) \,, \tag{A.14}$$

$$y_{L+}(t) = \frac{2^{L+(n-1)/2}}{\sqrt{\omega}} \cosh^L t \, e^{(L+\frac{n}{2}+i\omega)t} F(L+\frac{n}{2}, L+\frac{n}{2}+i\omega; 1+i\omega; -e^{2t}) \, ,(\text{A.15})$$

where $y_{L-}^*(t) = y_{L+}(t)$. Here the normalization constants have been fixed by imposing $(\phi_{L+}, \phi_{L+}) = 1 = (\phi_{L-}, \phi_{L-})$, as usual, with the standard Klein-Gordon inner product [60]. As emphasized above, these solutions have the simple time-dependence of eq. (A.7) in the asymptotic region $t \to -\infty$ near I^-. Because the differential equation (A.5) is invariant under $t \to -t$, one can easily define another pair of linearly independent solutions by applying this transformation to the above modes. We label the resulting modes $y_L^-(t)$ and $y_L^+(t) = y_L^{-*}(t)$ where

$$y^-(t) = y_+^*(-t) \, . \tag{A.16}$$

It readily follows that $y_L^- \sim e^{-h_-t}$ and $y_L^+ \sim e^{-h_+t}$ near I^+. The two sets of modes $y_{L\pm}(t)$ and $y_L^\pm(t)$ can respectively be used to construct the 'in' and 'out' vacua with no incoming and outgoing particles. The Bogolubov coefficients relating these two sets of modes are defined through

$$y_{L-}(t) = C_-^-(\omega) \, e^{-2i\theta_L} \, y_L^-(t) + C_-^+(\omega) \, y_L^+(t) \, , \tag{A.17}$$

with a similar expression for y_{L+} (with $C_+^+(\omega) = C_-^-(\omega)$ and $C_+^-(\omega) = C_-^+(\omega)$). When n is even [11], one finds that $C_-^-(\omega) = 1$ and $C_-^+(\omega) = 0$. This corresponds to the physical statement that there is no particle creation (no mode mixing) in dS space for an odd number of spacetime dimensions. For n odd, there is nontrivial mode mixing with

$$C_-^-(\omega) = \coth \pi\omega \, , \qquad C_-^+(\omega) = (-1)^{\frac{n+1}{2}} \frac{1}{\sinh \pi\omega} \, , \tag{A.18}$$

where $|C_-^-(\omega)|^2 - |C_-^+(\omega)|^2 = 1$ holds since the modes are properly normalized throughout their evolution. The expression for the phase in eq. (A.17) is

$$e^{-2i\theta_L} = (-1)^{L-\frac{n}{2}} \frac{\Gamma(-i\omega)\Gamma(L+\frac{n}{2}+i\omega)}{\Gamma(i\omega)\Gamma(L+\frac{n}{2}-i\omega)} \, . \tag{A.19}$$

It is clear that, for large enough ω, the mixing coefficient $C_-^+(\omega)$ becomes negligible, which is in accord with the intuition that there will be limited particle production in high energy modes. We will find that in the other two series there is no phase comparable to eq. (A.19). This complicates the expressions for mode mixing between the boundaries and will lead to interesting features.

Complementary series and tachyonic fields

For the modes of both the complementary and the tachyonic series, the weights h_+ and h_- are real and so the above mode functions are entirely real,

$$y_{L+}(t) = \bar{N}_+ \cosh^L t \, e^{(L+\frac{n}{2}+\mu)t} F(L+\frac{n}{2}, L+\frac{n}{2}+\mu; 1+\mu; -e^{2t}) \, , \tag{A.20}$$

$$y_{L-}(t) = \bar{N}_- \cosh^L t \, e^{(L+\frac{n}{2}-\mu)t} F(L+\frac{n}{2}, L+\frac{n}{2}-\mu; 1-\mu; -e^{2t}) \, . \tag{A.21}$$

Hence, with respect to the usual Klein-Gordon product, these two solutions have zero norm, *i.e.,* $(\phi_{L+}, \phi_{L+}) = 0 = (\phi_{L-}, \phi_{L-})$.

Of course, this is not unnatural. One gains intuition by considering the usual plane wave decomposition in flat spacetime. There, one may choose between two bases, the one involving complex exponentials and the one involving cosines and sines. The latter basis in fact has the same characteristics as the present modes, in the complementary series, in terms of normalization with respect to the Klein-Gordon inner product. Consequently, to define a reasonable normalization for the mode functions (A.20) and (A.21), we require $(\phi_{L-}, \phi_{L+}) = i$ and

$$\bar{N}_+ = \frac{2^{L+\frac{n-1}{2}}}{\sqrt{\mu}} = \bar{N}_- \, , \tag{A.22}$$

where we have resolved the remaining ambiguity by simply demanding that $\bar{N}_+ = \bar{N}_-$. With this choice of normalization, it is clear that upon quantizing the scalar field in the dS background the corresponding mode coefficients will have commutation relations analogous to those of coordinate and momentum operators, rather than raising and lowering operators.

As in the previous subsection, by substituting $t \to -t$, we define modes $y_L{}^\pm(t) \equiv y_{L\pm}(-t)$ which have the simple time-dependence of eq. (A.7) in the asymptotic region approaching I^+. Using a simple identity of hypergeometric functions [85], one can relate the two sets of modes as

$$
\begin{aligned}
y_{L-}(t) &= \bar{C}_-^-(\mu)\, y_L{}^-(t) + \bar{C}_-^+(\mu)\, y_L{}^+(t) \, , \\
y_{L+}(t) &= \bar{C}_+^-(\mu)\, y_L{}^-(t) + \bar{C}_+^+(\mu)\, y_L{}^+(t) \, ,
\end{aligned}
\tag{A.23}
$$

where the elements of the mixing matrix C (the Bogolubov coefficients) are given by

$$\bar{C}_-^+(\mu) = \frac{\Gamma(1-\mu)\Gamma(-\mu)}{\Gamma(\frac{2-n}{2}-\mu-L)\Gamma(\frac{n}{2}-\mu+L)} \, , \qquad \bar{C}_-^-(\mu) = -(-1)^L \frac{\sin(\frac{\pi n}{2})}{\sin \pi\mu} \, , \tag{A.24}$$

$$\bar{C}_+^-(\mu) = -\frac{\Gamma(1+\mu)\Gamma(\mu)}{\Gamma(\frac{2-n}{2}+\mu-L)\Gamma(\frac{n}{2}+\mu+L)} \, , \qquad \bar{C}_+^+(\mu) = -(-1)^L \frac{\sin(\frac{\pi n}{2})}{\sin \pi\mu} \, . \tag{A.25}$$

We now describe some features of the resulting mode mixing for the complementary series. In this case, recall that $0 < M^2 \leq n^2/4$, which implies that $0 \leq \mu < n/2$. Of course, certain features depend on the spacetime dimension $n+1$ as before:

a) n odd: Generically for the case of an even spacetime dimension, there is nontrivial mode mixing. An exception occurs for $\mu = (2m + 1)/2$, with m a positive integer. For these special cases, there is no mixing, since $\bar{C}_-^+ = 0 = \bar{C}_+^-$ and one finds that $\bar{C}_-^- = \bar{C}_+^+ = (-1)^{\frac{n}{2}+\mu}(-1)^L$.

b) n even: Generically for this case of an odd number of spacetime dimensions, one finds $\bar{C}_+^+ = 0 = \bar{C}_-^-$ and $\bar{C}_-^+ \bar{C}_+^- = -1$ (where $\bar{C}_-^+$ and $\bar{C}_+^-$ both have a nontrivial dependence on L). This means that a mode that is scaling like $e^{h\pm t}$ on I^- will have the 'opposite' scaling $e^{-h\mp t}$ on I^+. We refer to this phenomenon as 'maximal mixing'. This phenomenon is absent when μ is an integer. This case must be treated with some care as the solution for y_{L-} appearing in eq. (A.21) breaks down.[16] The correct solution [85] has an additional logarithmic singularity near I^-, *i.e.,* , subdominant power law behavior in t. In any event, the final result for n even and μ integer is: $\bar{C}_-^- = \bar{C}_+^+ = (-1)^{\frac{n}{2}+\mu+1}(-1)^L$ and $\bar{C}_-^+ = 0 = \bar{C}_+^-$.

Finally we briefly consider the tachyonic or discrete series [62]. Recall that in this case $M^2 < 0$, so that $h_- < 0$ and the modes scaling as $e^{\pm h_- t}$ diverge as one approaches either I^- or I^+, depending on the sign of the exponent. Generically there is nontrivial mode mixing and so even if a mode is convergent at one asymptotic boundary it will be divergent at the opposite

---

[16] Similar remarks apply for n odd and μ integer, but in that case one still finds nontrivial mode mixing.

boundary. However, an interesting exceptional case is when a y_{L+} mode (scaling like e^{h+t} as $t \rightarrow -\infty$) evolves to the corresponding $y_L{}^+$ mode (with e^{-h+t} behavior for $t \rightarrow \infty$). Such a mode would have convergent behavior towards both the future and past boundaries. This behavior would result when $\bar{C}_+^-$ vanishes. A brief examination of eq. (A.25) shows that this requires that $1 + |h_-| - L$ is zero or a negative integer. As above, this constrains μ to be an integer or half-integer, depending on the spacetime dimension. We express this constraint, in terms of the (tachyonic) mass, as[17]

$$-M^2 = \begin{cases} \frac{1}{4}((2m+1)^2 - n^2) & \text{for } n \text{ odd} \quad \text{with } m = (n-1)/2, (n+1)/2, \cdots \\ m^2 - \frac{n^2}{4} & \text{for } n \text{ even} \quad \text{with } m = n/2, n/2+1, \ldots \end{cases} \quad \text{(A.26)}$$

However, the above constraint is not sufficient; rather we must also impose a constraint on the 'angular momentum' quantum number L, namely,

$$L \geq 1 + |h_-| . \quad \text{(A.27)}$$

Hence the completely convergent modes only appear for sufficiently large angular momenta. Note that it is still true that, using the usual Klein-Gordon inner product, these modes have a vanishing norm $(\phi_{L+}, \phi_{L+}) = 0$. However, in the mathematics literature (*e.g.*, [62]), these modes are singled out by having finite norm in the sense given by the spacetime integral: $\int d^{n+1}x \sqrt{-g} |\phi_{L+}|^2 = 1$.

This construction shows that even in the tachyonic mass range, one can find certain normalizable modes (in the above sense) for special choices of parameters. However, we reiterate that while these formal results for the discrete series may be interesting mathematically, they are not useful in understanding the physics of dS space. As emphasized above, in the discussion of the dS/CFT correspondence, one must consider the full space of solutions, and presently even in the exceptional cases, the normalizable modes are accompanied by modes diverging at both asymptotic boundaries. Thus, the normalizable modes do not form a complete set of modes on a Cauchy surface. Such divergences, which occur in the generic case as well, are simply an indication that a linearized analysis of tachyonic fields is inappropriate. Of course, nonlinear field theories with potentials including unstable (or metastable) critical points may play an important role in the paradigm of inflationary cosmology, and such theories can produce interesting asymptotically dS spacetimes [2]. Our point here is simply that one should consider the full nonlinear evolution of such fields, including their backreaction on the spacetime geometry.

References

[1] R. Bousso, O. DeWolfe and R.C. Myers, "Unbounded entropy in spacetimes with positive cosmological constant," arXiv:hep-th/0205080.

[2] F. Leblond, D. Marolf and R.C. Myers, "Tall tales from de Sitter space I: Renormalization group flows," JHEP **0206** (2002) 052 [arXiv:hep-th/0202094].

[3] F. Leblond, D. Marolf and R.C. Myers, "Tall tales from de Sitter space II: Field theory dualities," [arXiv:hep-th/0211025].

[4] S. Perlmutter *et al.* [Supernova Cosmology Project Collaboration], "Measurements of the Cosmological Parameters Ω and Λ from the First Seven Supernovae at z$\geq$0.35," Astrophys. J. **483**, 565 (1997) [arXiv:astro-ph/9608192];
S. Perlmutter *et al.* [Supernova Cosmology Project Collaboration], "Discovery of a Super-nova Explosion at Half the Age of the Universe and its Cosmological Implications," Nature

---

[17]This constraint gives rise to the nomenclature in which these modes are referred to as the discrete series.

391, 51 (1998) [arXiv:astro-ph/9712212];

A.G. Riess *et al.* [Supernova Search Team Collaboration], "Observational Evidence from Supernovae for an Accelerating Universe and a Cosmological Constant," Astron. J. **116**, 1009 (1998) [arXiv:astro-ph/9805201].;

N.A. Bahcall, J.P. Ostriker, S. Perlmutter and P.J. Steinhardt, "The Cosmic Triangle: Revealing the State of the Universe," Science **284**, 1481 (1999) [arXiv:astro-ph/9906463];

P. de Bernardis *et al.*, "Multiple peaks in the angular power spectrum of the cosmic microwave background: Significance and consequences for cosmology," Astrophys. J. **564** (2002) 559 [arXiv:astro-ph/0105296];

R. Stompor *et al.*, "Cosmological implications of the MAXIMA-I high resolution Cosmic Microwave Background anisotropy measurement," Astrophys. J. **561** (2001) L7 [arXiv:astro-ph/0105062];

A.H. Jaffe *et al.* [Boomerang Collaboration], "Cosmology from Maxima-1, Boomerang and COBE/DMR CMB Observations," Phys. Rev. Lett. **86** (2001) 3475 [arXiv:astro-ph/0007333];

M.E. Abroe *et al.*, "Frequentist Estimation of Cosmological Parameters from the MAXIMA-1 Cosmic Microwave Background Anisotropy Data," Mon. Not. Roy. Astron. Soc. **334** (2002) 11 [arXiv:astro-ph/0111010].

[5] A. Strominger, "The dS/CFT correspondence," JHEP **0110** (2001) 034 [arXiv:hep-th/0106113].

[6] E. Witten, "Quantum gravity in de Sitter space," arXiv:hep-th/0106109.

[7] T. Banks, "Cosmological breaking of supersymmetry or little Lambda goes back to the future. II," arXiv:hep-th/0007146.

[8] W. Fischler, "Taking de Sitter seriously," talk given at *Role of Scaling Laws in Physics and Biology (Celebrating the 60th Birthday of Geoffrey West)*, Santa Fe, December 2000.

[9] A. Strominger, "Inflation and the dS/CFT correspondence," JHEP **0111** (2001) 049 [arXiv:hep-th/0110087].

[10] V. Balasubramanian, J. de Boer, and D. Minic, "Mass, Entropy, and Holography in Asymptotically de Sitter spaces" Phys. Rev. D **65** (2002) 123508 [arXiv:hep-th/0110108].

[11] R. Bousso, A. Maloney and A. Strominger, "Conformal vacua and entropy in de Sitter space," Phys. Rev. D **65**(2002) 104039 [arXiv:hep-th/0112218].

[12] M. Spradlin and A. Volovich, "Vacuum states and the S-matrix in dS/CFT," Phys. Rev. D **65** (2002) 104037 [arXiv:hep-th/0112223].

[13] V. Balasubramanian, P. Horava, and D. Minic, "Deconstructing De Sitter," JHEP **0105** (2001) 043 [arXiv:hep-th/0103171].

[14] A.M. Ghezelbash and R.B. Mann, "Action, mass and entropy of Schwarzschild-de Sitter black holes and the de Sitter/CFT correspondence," JHEP **0201** (2002) 005 [arXiv:hep-th/0111217].

[15] D. Klemm and L. Vanzo, "De Sitter gravity and Liouville theory," JHEP **0204** (2002) 030 [arXiv:hep-th/0203268].

[16] D. Klemm, "Some aspects of the de Sitter/CFT correspondence," Nucl. Phys. B **625** (2002) 295 [arXiv:hep-th/0106247];

S. Cacciatori and D. Klemm, "The asymptotic dynamics of de Sitter gravity in three dimensions," Class. Quant. Grav. **19** (2002) 579 [arXiv:hep-th/0110031].

[17] M.K. Parikh, I. Savonije and E. Verlinde, "Elliptic de Sitter space: dS/Z(2)," arXiv:hep-th/0209120.

[18] G.W. Gibbons and S. Hawking, "Cosmological event horizons, thermodynamics and particle creation," Phys. Rev. D **7** (1977) 2738.

[19] A. Strominger and C. Vafa, "Microscopic Origin of the Bekenstein-Hawking Entropy," Phys. Lett. B **379** (1996) 99 [arXiv:hep-th/9601029].

[20] A.W. Peet, "TASI lectures on black holes in string theory," arXiv:hep-th/0008241;
S.R. Das and S.D. Mathur, "The Quantum Physics Of Black Holes: Results From String Theory," Ann. Rev. Nucl. Part. Sci. **50** (2000) 153 [arXiv:gr-qc/0105063];
J.R. David, G. Mandal and S.R. Wadia, "Microscopic formulation of black holes in string theory," Phys. Rept. **369** (2002) 549 [arXiv:hep-th/0203048].

[21] J.M. Maldacena, "Black holes in string theory," arXiv:hep-th/9607235.

[22] R.C. Myers, "Black holes and string theory," arXiv:gr-qc/0107034.

[23] C.V. Johnson, *D-branes* (Cambridge University Press, 2002).

[24] R.C. Myers, "Nonabelian D-branes and noncommutative geometry," Int. J. Mod. Phys. A **16** (2001) 956 [arXiv:hep-th/0106178].

[25] R.C. Myers, "Dielectric-branes," JHEP **9912** (1999) 022 [arXiv:hep-th/9910053].

[26] http://www.hasd.org/ges/talltale/talltale.htm

[27] See, for example: S.W. Hawking and G.F.R. Ellis, *The large scale structure of space-time* (Cambridge: Cambridge University Press, 1980).

[28] J.D. Bekenstein, "Generalized Second Law Of Thermodynamics In Black Hole Physics," Phys. Rev. D **9** (1974) 3292;
J.D. Bekenstein, "Black Holes And Entropy," Phys. Rev. D **7** (1973) 2333;
J.D. Bekenstein, "Black Holes And The Second Law," Lett. Nuovo Cim. **4** (1972) 737.

[29] S. Hawking, J.M. Maldacena and A. Strominger, "DeSitter entropy, quantum entanglement and AdS/CFT," JHEP **0105** (2001) 001 [arXiv:hep-th/0002145].

[30] S. Carlip, *Quantum Gravity In 2+1 Dimensions,* (Cambridge University Press, 1998).

[31] J.M. Maldacena and A. Strominger, "Statistical entropy of de Sitter space," JHEP **9802** (1998) 014 [arXiv:gr-qc/9801096];
F.L. Lin and Y.S. Wu, "Near-horizon Virasoro symmetry and the entropy of de Sitter space in any dimension," Phys. Lett. B **453** (1999) 222 [arXiv:hep-th/9901147].

[32] L. Smolin, "Quantum gravity with a positive cosmological constant," arXiv:hep-th/0209079.

[33] G. 't Hooft, "Dimensional Reduction In Quantum Gravity," arXiv:gr-qc/9310026;
L. Susskind, "The World as a hologram," J. Math. Phys. **36** (1995) 6377 [arXiv:hep-th/9409089].

[34] R. Bousso, "Holography in general space-times," JHEP **9906**, 028 (1999) [arXiv:hep-th/9906022]; "A Covariant Entropy Conjecture," JHEP **9907**, 004 (1999) [arXiv:hep-th/9905177].

[35] R. Bousso, "The holographic principle," Rev. Mod. Phys. **74** (2002) 825 [arXiv:hep-th/0203101].

[36] R. Bousso, "Bekenstein bounds in de Sitter and flat space," JHEP **0104** (2001) 035 [arXiv:hep-th/0012052].

[37] L. Susskind, L. Thorlacius and J. Uglum, "The Stretched horizon and black hole complementarity," Phys. Rev. D **48**, 3743 (1993) [arXiv:hep-th/9306069];
L. Susskind, "String theory and the principles of black hole complementarity," Phys. Rev. Lett. **71**, 2367 (1993) [arXiv:hep-th/9307168];
L. Susskind and L. Thorlacius, "Gedanken experiments involving black holes," Phys. Rev. D **49**, 966 (1994) [arXiv:hep-th/9308100].

[38] R. Bousso, "Positive vacuum energy and the N-bound," JHEP **0011** (2000) 038 [arXiv:hep-th/0010252].

[39] P.G. Freund and M.A. Rubin, "Dynamics Of Dimensional Reduction," Phys. Lett. B **97** (1980) 233.

[40] E. Silverstein, "(A)dS backgrounds from asymmetric orientifolds," arXiv:hep-th/0106209;
      A. Maloney, E. Silverstein and A. Strominger, "De Sitter space in noncritical string theory,"
      arXiv:hep-th/0205316.

[41] C.M. Hull, "De Sitter space in supergravity and M theory," JHEP **0111** (2001) 012
      [arXiv:hep-th/0109213];
      C.M. Hull, "Domain wall and de Sitter solutions of gauged supergravity," JHEP **0111**
      (2001) 061 [arXiv:hep-th/0110048];
      G.W. Gibbons and C.M. Hull, "de Sitter space from warped supergravity solutions,"
      arXiv:hep-th/0111072.

[42] M. Gutperle and A. Strominger, "Spacelike branes," JHEP **0204** (2002) 018 [arXiv:hep-
      th/0202210];
      C.M. Chen, D.V. Gal'tsov and M. Gutperle, "S-brane solutions in supergravity theories,"
      Phys. Rev. D **66** (2002) 024043 [arXiv:hep-th/0204071];
      M. Kruczenski, R.C. Myers and A.W. Peet, "Supergravity S-branes," JHEP **0205** (2002)
      039 [arXiv:hep-th/0204144].

[43] C. Herdeiro, S. Hirano and R. Kallosh, "String theory and hybrid inflation / acceleration,"
      JHEP **0112** (2001) 027 [arXiv:hep-th/0110271];
      K. Dasgupta, C. Herdeiro, S. Hirano and R. Kallosh, "D3/D7 inflationary model and M-
      theory," Phys. Rev. D **65** (2002) 126002 [arXiv:hep-th/0203019].

[44] J.D. Brown and C. Teitelboim, "Neutralization Of The Cosmological Constant By Mem-
      brane Creation," Nucl. Phys. B **297** (1988) 787;
      J.D. Brown and C. Teitelboim, "Dynamical Neutralization Of The Cosmological Constant,"
      Phys. Lett. B **195** (1987) 177.

[45] T. Banks, "Heretics of the false vacuum: Gravitational effects on and of vacuum decay. II,"
      arXiv:hep-th/0211160.

[46] O. Aharony, S.S. Gubser, J. Maldacena, H. Ooguri and Y. Oz, "Large N field theories,
      string theory and gravity," Phys. Rept. **323**,183 (2000) [arXiv:hep-th/9905111].

[47] V. Balasubramanian, J. de Boer and D. Minic, "Exploring de Sitter space and holography,"
      arXiv:hep-th/0207245; to appear in Annals of Physics.

[48] E. Witten, "Anti-de Sitter space and holography," Adv. Theor. Math. Phys. **2** (1998) 253
      [arXiv:hep-th/9802150].

[49] V. Balasubramanian, P. Kraus and A.E. Lawrence, "Bulk vs. boundary dynamics in anti-de
      Sitter spacetime," Phys. Rev. D **59** (1999) 046003 [arXiv:hep-th/9805171].

[50] S.S. Gubser, I.R. Klebanov and A.M. Polyakov, "Gauge theory correlators from non-critical
      string theory," Phys. Lett. B **428** (1998) 105 [arXiv:hep-th/9802109].

[51] P. Breitenlohner and D.Z. Freedman, "Positive Energy In Anti-De Sitter Backgrounds And
      Gauged Extended Supergravity," Phys. Lett. B **115** (1982) 197;
      "Stability In Gauged Extended Supergravity," Annals Phys. **144** (1982) 249.

[52] L. Mezincescu and P.K. Townsend, "Stability At A Local Maximum In Higher Dimensional
      Anti-De Sitter Space And Applications To Supergravity," Annals Phys. **160** (1985) 406.

[53] I.R. Klebanov and E. Witten, "AdS/CFT correspondence and symmetry breaking," Nucl.
      Phys. B **556** (1999) 89 [arXiv:hep-th/9905104].

[54] L. Susskind and E. Witten, "The holographic bound in anti-de Sitter space," arXiv:hep-
      th/9805114;
      A.W. Peet and J. Polchinski, "UV/IR relations in AdS dynamics," Phys. Rev. D **59**, 065011
      (1999) [arXiv:hep-th/9809022].

[55] R. Emparan, C.V. Johnson and R.C. Myers, "Surface terms as counterterms in the AdS/CFT
      correspondence," Phys. Rev. D **60**, 104001 (1999) [arXiv:hep-th/9903238].

[56] V. Balasubramanian and P. Kraus, "Spacetime and the holographic renormalization group," Phys. Rev. Lett. **83** (1999) 3605 [arXiv:hep-th/9903190].

[57] N.R. Constable and R.C. Myers, "Exotic scalar states in the AdS/CFT correspondence," JHEP **9911**, 020 (1999) [arXiv:hep-th/9905081];
K.A. Intriligator, "Maximally supersymmetric RG flows and AdS duality," Nucl. Phys. B **580**, 99 (2000) [arXiv:hep-th/9909082];
M.S. Costa, "Absorption by double-centered D3-branes and the Coulomb branch of N=4 SYM theory," JHEP **0005**, 041 (2000) [arXiv:hep-th/9912073];
M.S. Costa, "A test of the AdS/CFT duality on the Coulomb branch," Phys. Lett. B **482**, 287 (2000) [Erratum-ibid. B **489**, 439 (2000)] [arXiv:hep-th/0003289];
U.H. Danielsson, A. Guijosa, M. Kruczenski and B. Sundborg, "D3-brane holography," JHEP **0005**, 028 (2000) [arXiv:hep-th/0004187];
N. Evans, C.V. Johnson and M. Petrini, "Clearing the throat: Irrelevant operators and finite temperature in large N gauge theory," JHEP **0205**, 002 (2002) [arXiv:hep-th/0112058].

[58] J. Distler and F. Zamora, "Non-supersymmetric conformal field theories from stable anti-de Sitter spaces," Adv. Theor. Math. Phys. **2**, 1405 (1999) [arXiv:hep-th/9810206];
L. Girardello, M. Petrini, M. Porrati and A. Zaffaroni, "Novel local CFT and exact results on perturbations of N=4 super Yang-Mills from AdS dynamics," JHEP **9812**, 022 (1998) [arXiv:hep-th/9810126];
D.Z. Freedman, S.S. Gubser, K. Pilch and N.P. Warner, "Continuous distributions of D3-branes and gauged supergravity," JHEP **0007**, 038 (2000) [arXiv:hep-th/9906194];
N.R. Constable and R.C. Myers, "Exotic scalar states in the AdS/CFT correspondence," JHEP **9911**, 020 (1999) [arXiv:hep-th/9905081];
L. Girardello, M. Petrini, M. Porrati and A. Zaffaroni, "The supergravity dual of N = 1 super Yang-Mills theory," Nucl. Phys. B **569**, 451 (2000) [arXiv:hep-th/9909047].

[59] D.Z. Freedman, S.S. Gubser, K. Pilch and N.P. Warner, "Renormalization group flows from holography supersymmetry and a c-theorem," Adv. Theor. Math. Phys. **3**, 363 (1999) [arXiv:hep-th/9904017];
J. de Boer, E. Verlinde and H. Verlinde, "On the holographic renormalization group," JHEP **0008**, 003 (2000) [arXiv:hep-th/9912012];
V. Balasubramanian, E.G. Gimon and D. Minic, "Consistency conditions for holographic duality," JHEP **0005**, 014 (2000) [arXiv:hep-th/0003147];
V. Sahakian, "Holography, a covariant c-function and the geometry of the renormalization group," Phys. Rev. D **62**, 126011 (2000) [arXiv:hep-th/9910099];
E. Alvarez and C. Gomez, "Geometric holography, the renormalization group and the c-theorem," Nucl. Phys. B **541**, 441 (1999) [arXiv:hep-th/9807226].

[60] See, for example: N.D. Birrell and P.C.W. Davies, *Quantum fields in curved space* (Cambridge, 1982).

[61] E.A. Tagirov, "Consequences Of Field Quantization In De Sitter Type Cosmological Models," Annals Phys. **76** (1973) 561.

[62] R. Raczka, N. Limic and J. Niederle, "Discrete Degenerate Representations of Noncompact Rotation Groups. I," J. Math. Phys. **7** (1966) 1861;
N. Limic, J. Niederle and R. Raczka, J. Math. Phys. **8** (1966) 2026;
N. Limic, J. Niederle and R. Raczka, "Eigenfunction Expansions Associated with the Second-Order Invariant Operator on Hyperboloids and Cones. III," J. Math. Phys. **8** (1967) 1079.

[63] B. Allen and A. Folacci, "The Massless Minimally Coupled Scalar Field In De Sitter Space," Phys. Rev. D **35** (1987) 3771.

[64] E. Mottola, "Particle Creation In De Sitter Space," Phys. Rev. D **31** (1985) 754.

[65] B. Allen, "Vacuum States In De Sitter Space," Phys. Rev. D **32** (1985) 3136.

[66] M. Spradlin and A. Volovich, "Vacuum states and the S-matrix in dS/CFT," Phys. Rev. D **65** (2002) 104037 [arXiv:hep-th/0112223].

[rgflows] K.R. Kristjansson and L. Thorlacius, "Cosmological models and renormalization group flow," JHEP **0205** (2002) 011 [arXiv:hep-th/0204058];
E. Halyo, "Holographic inflation," arXiv:hep-th/0203235;
A.J. Medved, "How not to construct an asymptotically de Sitter universe," Class. Quant. Grav. **19** (2002) 4511 [arXiv:hep-th/0203191];
A.J. Medved, "dS-holographic C-functions with a topological, dilatonic twist," Phys. Rev. D **66** (2002) 064001 [arXiv:hep-th/0202193];
R. Argurio, "Comments on cosmological RG flows," arXiv:hep-th/0202183;
F. Larsen, J.P. van der Schaar and R.G. Leigh, "de Sitter holography and the cosmic microwave background," JHEP **0204** (2002) 047 [arXiv:hep-th/0202127].

[67] A.B. Zamolodchikov, "Irreversibility of the Flux of the Renormalization Group in a 2D Field Theory," JETP Lett. **43**, 730 (1986) [Pisma Zh. Eksp. Teor. Fiz. **43**, 565 (1986)].

[68] S. Gao and R.M. Wald, "The physical process version of the first law and the generalized second law for charged and rotating black holes," Phys. Rev. D **64** (2001) 084020 [arXiv:gr-qc/0106071].

[69] A. Borde and A. Vilenkin, "Eternal Inflation And The Initial Singularity," Phys. Rev. Lett. **72** (1994) 3305 [arXiv:gr-qc/9312022];
A. Borde and A. Vilenkin, "The Impossibility Of Steady State Inflation," arXiv:gr-qc/9403004;
A. Borde and A. Vilenkin, "Singularities in inflationary cosmology: A review," Int. J. Mod. Phys. D **5** (1996) 813 [arXiv:gr-qc/9612036];
A. Borde, A.H. Guth and A. Vilenkin, "Inflation is not past-eternal," arXiv:gr-qc/0110012.

[70] T. Vachaspati and M. Trodden, "Causality and cosmic inflation," Phys. Rev. D **61** (2000) 023502 [arXiv:gr-qc/9811037].

[71] A. Strominger, private communication.

[72] J.M. Maldacena, "Eternal black holes in Anti-de-Sitter," [arXiv:hep-th/0106112].

[73] V. Balasubramanian, P. Kraus and A.E. Lawrence, "Bulk vs. boundary dynamics in anti-de Sitter spacetime," Phys. Rev. D **59** (1999) 046003 [arXiv:hep-th/9805171].

[74] J. Louko and D. Marolf, "Single-exterior black holes and the AdS-CFT conjecture," Phys. Rev. D **59** (1999) 066002 [arXiv:hep-th/9808081].

[75] G.J. Galloway, K. Schleich, D. Witt and E. Woolgar, "The AdS/CFT correspondence conjecture and topological censorship," Phys. Lett. B **505** (2001) 255 [arXiv:hep-th/9912119]; "Topological censorship and higher genus black holes," Phys. Rev. D **60** (1999) 104039 [arXiv:gr-qc/9902061].

[76] M. Spradlin, A. Strominger and A. Volovich, "Les Houches lectures on de Sitter space," arXiv:hep-th/0110007.

[77] For example, the required flat space Green's functions may be found in section 3 of: D.V. Galtsov, "Radiation reaction in various dimensions," Phys. Rev. D **66** (2002) 025016 [arXiv:hep-th/0112110].

[78] L. Dyson, J. Lindesay and L. Susskind, "Is there really a de Sitter/CFT duality," JHEP **0208** (2002) 045 [arXiv:hep-th/0202163].

[79] L. Dyson, M. Kleban and L. Susskind, "Disturbing implications of a cosmological constant," JHEP **0210** (2002) 011 [arXiv:hep-th/0208013].

[80] T. Banks, W. Fischler and S. Paban, "Recurrent nightmares?: Measurement theory in de Sitter space," arXiv:hep-th/0210160.

[81] J.P. Gazeau and M.V. Takook, "Massive vector field in de Sitter space," J. Math. Phys. **41** (2000) 5920 [arXiv:gr-qc/9912080].

[82] J. Mickelsson and J. Niederle, "Contractions of representations of de Sitter groups," Commun. Math. Phys. **27** (1972) 167.

[83] J.P. Gazeau, J. Renaud and M.V. Takook, "Gupta-Bleuler quantization for minimally coupled scalar fields in de Sitter space," Class. Quant. Grav. **17** (2000) 1415 [arXiv:gr-qc/9904023].

[84] A.J. Tolley and N. Turok, "Quantization of the massless minimally coupled scalar field and the dS/CFT correspondence," arXiv:hep-th/0108119.

[85] M. Milton and I.A. Abramowitz, *Handbook of mathematical functions* (Dover, 1965).

CAUSAL SETS: DISCRETE GRAVITY

Rafael D. Sorkin

Department of Physics, Syracuse University, Syracuse, NY 13244-1130, U.S.A.
sorkin@physics.syr.edu

Abstract   These are some notes in lieu of the lectures I was scheduled to give, but had to cancel at the last moment. In some places, they are more complete, in others much less so, regrettably. I hope they at least give a feel for the subject and convey some of the excitement felt at the moment by those of us working on it.

An extensive set of references and a glossary of terms can be found at the end of the notes. For a philosophically oriented discussion of some of the background to the causal set idea, see reference [1]. For general background see [2] [3] [4] [5] [6] [7].

1.   Introduction

It seems fair to say that causal set theory has reached a stage in which questions of phenomenology are beginning to be addressed meaningfully. This welcome development is due on one hand to improved astronomical observations which shed light on the magnitude of the cosmological constant (in apparent confirmation of a long-standing prediction of the theory) and on the other hand to theoretical advances which for the first time have placed on the agenda the development of a quantum dynamical law for causal sets (and also for a scalar field residing on a background causal set). What we have so far are: (i) an apparently confirmed order of magnitude prediction for the cosmological constant; (ii) a method of counting black hole horizon "states" at the kinematical level; (iii) the beginnings of a framework in which two-dimensional Hawking radiation can be addressed; (iv) a *classical* causal set dynamics which arguably is the most general consistent with the discrete analogs of general covariance and relativistic causality; and in consequence of this, both (v) the formulation of a "cosmic renormalization group" which indicates how one might in principle solve some of the large number puzzles of cosmology without recourse to a post-quantum era of "inflation"; and (vi) a hint of how non-gravitational matter might arise at the fundamental level from causal sets rather than having to be added in by hand or derived at a higher level à la Kaluza-Klein from

305

an effective spacetime topology arising from the fundamental structures via coarse-graining. In addition, a good deal of computer code has been written for use in causal set simulations, including a library of over 5000 lines of Lisp code that can be used by anyone with access to the Emacs editor. At present, the principal need, in addition to fleshing out the developments already outlined, is for a quantum analog of the classical dynamics alluded to above. It looks as if a suitable quantum version of Bell causality (see below) could lead directly to such a dynamics, that is to say, to a theory of quantum spacetime, and in particular to a theory of quantum gravity.

The remainder of these notes rapidly reviews the subject in its current state, progressing broadly from kinematics to dynamics to phenomenology. Although this sequence does not always reflect exactly the chronological development of the theory, it is not far off, and it also fits in well with Taketani's "3-stages" schema of scientific discovery [8].

2.   Origins of the causet idea

The tradition of seeing the causal order of spacetime as its most fundamental structure is almost as old as the idea of spacetime itself (in its Relativistic form). In [9], Robb presented a set of axioms for Minkowski space analogous to Euclid's axioms for plane geometry. In so doing, he effectively demonstrated that, up to an overall conformal factor, the geometry of 4-dimensional flat spacetime (which I'll denote by $\mathbb{M}^4$, taken always with a definite time-orientation) can be recovered from nothing more than the underlying point set and the order relation $\prec$ among points (where $x \prec y \iff$ the vector from x to y is timelike or lightlike and future-pointing). Later, Reichenbach [10] from the side of philosophy and Zeeman [11] from the side of mathematics emphasized the same fact, the latter in particular by proving the theorem, implicit in [9], that any order-isomorphism of $\mathbb{M}^4$ onto itself must — up to an overall scaling — belong to the (isochronous) Poincaré group.

In a certain sense, however, these results appear to say more than they really do. Informally, they seem to tell us that $\mathbb{M}^4$ can be reconstructed from the relation $\prec$, but in actually carrying out the reconstruction (see below), one needs to know that what one is trying to recover is a flat spacetime and not just a conformally flat one. Clearly, there's nothing in the relation $\prec$ *per se* which can tell us that. This difficulty shows itself, in a sense, in the failure of Zeeman's theorem for $\mathbb{M}^2$ and $\mathbb{M}^1$ (i.e. 1+1 and 0+1 dimensional Minkowski space). But it shows up still more clearly with the curved spacetimes of General Relativity, where the natural generalization of the flat space theorems is that a Lorentzian geometry M can be recovered from its causal order only up to a *local* conformal factor.

Notice that when one says that a Lorentzian manifold M is recovered, one is talking about *all* the mathematical structures that go into the definition of a spacetime geometry: its topology, its differential structure and its metric. Various special results show how to recover, say, the topology (see e.g. [12]) but the most complete theorems are those of [13] and [14], the latter delineating very precisely how close M can come to violating causality without the theorem breaking down.

The upshot of all these reconstruction theorems is that, in the continuum, something is lacking if we possess only the causal order, namely the *conformal factor* or equivalently the "volume element" $\sqrt{-g}\,d^4x$. On the other hand, if we do give the volume element (say in the form of a measure μ on M) then it is clear that metric g_{ab} will be determined in full. The causal order alone, however, is — in the continuum — incapable of furnishing such a measure.

This failing is perhaps one reason to question the reality of the continuum, but of course it is not the only one. In modern times, doubts show up clearly in Riemann's inaugural lecture (Habilitationsschrift) [15], where he contrasts the idea of what he calls a *discrete manifold* with that of a continuous manifold, which latter he takes to be a relatively unfamiliar and unintuitive idea in comparison with the former! The most evocative quotes from this lecture are perhaps the following:

> Grössenbegriffe sind nur da möglich, wo sich ein allgemeiner Begriff vorfindet, der verschiedene Bestimmungsweisen zulässt. Je nachdem unter diesen Bestimmungsweisen von einer zu einer andern ein stetiger Übergang stattfindet oder nicht, bilden sie eine stetige oder discrete Mannigfaltigkeit; die enzelnen Bestimmungsweisen heissen im ersten Falle Punkte, im letzten Elemente dieser Mannigfaltigkeit. (p.273)

or in translation,[1]

> Concepts of magnitude are only possible where a general concept is met with that admits of different individual instances [Bestimmungsweisen]. According as, among these individual instances, a continuous passage from one to another takes place or not, they form a continuous or discrete manifold; the individual instances are called in the first case points, in the second elements of the manifold;

> Die Frage über die Gültigkeit der Voraussetzungen der Geometrie im Unendlichkleinen hängt zusammen mit der Frage nach dem innern Grunde der Massverhältnisse des Raumes. Bei dieser Frage, welche wohl noch zur Lehre vom Raume gerechnet werden darf, kommt die obige Bemerkung zur Anwendung, dass bei einer discreten Mannigfaltigkeit das Princip der Massverhältnisse schon in dem Begriffe dieser Mannigfaltigkeit enthalten ist, bei einer stetigen aber anders woher hinzukommen muss. Es muss also entweder das dem Raume zu Grunde liegende Wirkliche eine discrete Mannigfaltigkeit bilden, oder der Grund der Massverhältnisse ausserhalb, in darauf wirkenden bindenden Kräften, gesucht werden.

---

[1] These translations are not guaranteed, but they're a lot better than what "Google" did (try it for fun!).

or in translation,

> The question of the validity of the presuppositions of geometry in the infinitely
> small hangs together with the question of the inner ground of the metric rela-
> tionships of space. [I almost wrote "spacetime"!] In connection with the latter
> question, which probably [?] can still be reckoned to be part of the science
> of space, the above remark applies, that for a discrete manifold, the principle
> of its metric relationships is already contained in the concept of the manifold
> itself, whereas for a continuous manifold, it must come from somewhere else.
> Therefore, either the reality which underlies physical space must form a discrete
> manifold or else the basis of its metric relationships must be sought for outside
> it, in binding forces [bindenden Kräfte] that act on it;

and finally,

> Bestimmte, durch ein Merkmal oder eine Grenze unterschiedene Theile einer
> Mannigfaltigkeit heissen Quanta. Ihre Vergleichung der Quantität nach geschieht
> bei den discreten Grössen durch Zählung, bei den stetigen durch Messung. (p.274)

or in translation,

> Definite portions of a manifold, distinguished by a criterion [Merkmal] or a
> boundary, are called quanta. Their quantitative comparison happens for discrete
> magnitudes through counting, for continuous ones through measurement.

With the subsequent development of physics, more compelling reasons emerged
for questioning the continuum, including the singularities and infinities of General Relativity, of Quantum Field Theory (including the standard model), and
of black hole thermodynamics. Einstein, for example, voiced doubts of this
sort very early [16]:

> But you have correctly grasped the drawback that the continuum brings. If
> the molecular view of matter is the correct (appropriate) one, i.e., if a part of
> the universe is to be represented by a finite number of moving points, then the
> continuum of the present theory contains too great a manifold of possibilities. I
> also believe that this too great is responsible for the fact that our present means
> of description miscarry with the quantum theory. The problem seems to me
> how one can formulate statements about a discontinuum without calling upon a
> continuum (space-time) as an aid; the latter should be banned from the theory
> as a supplementary construction not justified by the essence of the problem,
> which corresponds to nothing "real". But we still lack the mathematical structure
> unfortunately. How much have I already plagued myself in this way!

and at a later stage stresseindentd the importance of the causal order in this connection, writing [17] that it would be "especially difficult to derive something
like a spatio-temporal quasi-order" from a purely algebraic or combinatorial
scheme.

The causal set idea is, in essence, nothing more than an attempt to combine
the twin ideas of discreteness and order to produce a structure on which a theory
of quantum gravity can be based. That such a step was almost inevitable is indicated by the fact that very similar formulations were put forward independently
in [4], [5] and [2], after having been adumbrated in [18]. The insight underlying these proposals is that, in passing from the continuous to the discrete,

one actually *gains* certain information, because "volume" can now be assessed (as Riemann said) *by counting*; and with both order *and* volume information present, we have enough to recover geometry.

In this way the topology, the differential structure and, the metric of continuum physics all become unified with the causal order (much as mass is unified with energy in Special Relativity). Moreover the Lorentzian signature (namely $(-+++)$ in 4 dimensions) is singled out as the only one compatible with a consistent distinction between past and future, hence the only one that can make contact with the idea of causal order. indent

   indent

To see how these basic ideas work themselves out, we need first a more precise statement of what a causal set is.

3.   What is a causal set?

As a mathematical structure, a causal set (or *causet* for short) is simply a *locally finite ordered set*. In other words, it is a set C endowed with a binary relation $\prec$ possessing the following three properties:

(i) transitivity: $(\forall x, y, z \in C)(x \prec y \prec z \Rightarrow x \prec z)$

(ii) irreflexivity: $(\forall x \in C)(x \not\prec x)$

(iii) local finiteness: $(\forall x, z \in C)\,(\mathrm{card}\,\{y \in C \mid x \prec y \prec z\} < \infty)$

where 'card' stands for "cardinality".In the presence of transitivity, irreflexivity automatically implies acyclicity, i.e. the absence of cycles $x_0 \prec x_1 \prec x_2 \prec \cdots \prec x_n = x_0$, and this is often taken as an axiom in place of (ii). The condition (iii) of local finiteness is a formal way of saying that a causet is *discrete*. Thus the real number line, for example does not qualify as a causet, although it is a partial order.[2]

A structure satisfying the above axioms can be thought of as a graph, and in this sense is conveniently represented as a so-called Hasse diagram in which the elements of C appear as vertices and the relations appear as edges. (The sense of the relation is usually shown, just as in the spacetime diagrams of Relativity theory, by making the line between x and y be a rising one when $x \prec y$.) Actually, it is not necessary to draw in all the relations, but only those not implied by transitivity (the "links"), and this convention is almost always adopted to simplify the appearance of the diagram. A causet can also be thought of as a matrix M (the "causal matrix") with the rows and columns labeled by

---

[2]The above definition utilizes indentthe so called "irreflexive convention" that no element precedes itself. Axioms (i) and (ii) define what is variously called an "order", a "partial order", a "poset", an "ordered set" or an "acyclic transitive digraph". Axiom (iii), which expresses the condition of local finiteness, can also be stated in the form "every order-interval has finite cardinality".

the elements of C and with the matrix element M_{jk} being 1 if $j \prec k$ and 0 otherwise. Perhaps, however, the most suggestive way to think of a causet for the purposes of quantum gravity is as a relation of "descent", effectively a family tree that indicates which of the elements C are "ancestors" of which others.

The multiplicity of imagery associated with causets (or partial orders more generally) is part of the richness of the subject and makes it natural to use a variety of language in discussing the structural relationships induced by the basic order relation $\prec$. Thus, the relationship $x \prec y$ itself, is variously described by saying that x *precedes* y, that x is an *ancestor* of y, that y is a *descendant* of x, or that x lies to the *past of* y (or y to the *future* of x). Similarly, if x is an *immediate* ancestor of y (meaning that there exists no intervening z such that $x \prec z \prec y$) then one says that x is a *parent* of y, or y a *child* of x, or that y *covers* x, or that $x \prec y$ is a *link*. (See the glossary.)

Still other interpretations of the relation $\prec$ are possible and can also be useful. For example a causet of finite cardinality is equivalent to a T_0 topological space of finite cardinality, allowing one to use the language of topology in talking about causets (which indeed may turn out to have more than just a metaphorical significance). A causet can also be treated as a *function* by identifying C with the function 'past' that associates to each $x \in C$ the set $\mathrm{past}(x)$ of all its ancestors, and this is in fact the representation on which the Lisp code of [19] is based.

For the purposes of quantum gravity, a causal set is, of course, meant to be the deep structure of spacetime. Or to say this another way, the basic hypothesis is that spacetime ceases to exist on sufficiently small scales and is superseded by an ordered discrete structure to which the continuum is only a coarse-grained, macroscopic approximation.

Now, at first sight, a structure based purely on the concept of order might seem to be too impoverished to reproduce the geometrical and topological attributes in terms of which general relativistic spacetime is normally conceived. However, if one reflects that light cones can be defined in causal terms and that (in the continuum) the light cones determine the metric up to a conformal rescaling, then it becomes understandable that (given minimal regularity conditions like the absence of closed causal curves) the causal order of a Lorentzian manifold (say J^+ in the usual notation) captures fully the conformal metric, as well as the topology and the differential structure. The volume element $\sqrt{-g}d^n x$ cannot be recovered from J^+, but in the context of a *discrete* order, it can be obtained in another way — by equating the *number* of causet elements to the *volume* of the corresponding region of the spacetime continuum that approximates C. As discussed above, these observations provide the kinematical starting point for a theory of discrete quantum gravity based on causal sets. The dynamics must then be obtained in the form of a "quantum law of motion" for the causet. Let us consider the kinematics further.

4. Causal set kinematics in general

Both the study of the mathematics of causets for its own sake and its study for the sake of clarifying how the geometrical and topological properties of a continuous spacetime translate into order properties of the underlying causet can be regarded as aspects of causal set kinematics: the study of causets without reference to any particular dynamical law.

A large amount is known about causet kinematics as a result of extensive work by both physicists and mathematicians. (See for example [20] [21] [22] [23] [24] [25] [26] [27].) (Some of the mathematicians were directly influenced by the causal set idea, others were studying ordered sets for their own reasons.) We know for example, that the length of the longest chain[3] provides a good measure of the proper time (geodesic length) between any two causally related elements of a causet that can be approximated by a region of Minkowski space [28]. And for such a causet, we also possess at least two or three good dimension estimators, one of which is well understood analytically [29].

The next few sections are devoted to some of these topics.

5. "How big" is a causet element?

Of course the question is badly worded, because a causet element has no size as such. What it's really asking for is the conversion factor v_0 for which $N = V/v_0$. Only if we measure length in units such that $v_0 \equiv 1$ can we express the hypothesis that number=volume in the form $N = V$. On dimensional grounds, one naturally expects $v_0 \sim (G\hbar)^2$ [where I've taken $c \equiv 1$]. But we can do better than just relying on dimensional analysis *per se*. Consider first the entropy of a black hole horizon, which is given by $S = A/4G\hbar = 2\pi A/\kappa$, with $\kappa = 8\pi G$, the rationalized gravitational constant. This formula suggests forcefully that about one bit of entropy belongs to each horizon "plaquette" of size $\kappa\hbar$, and that the effectively finite size of these "plaquettes" reflects directly an underlying spacetime discreteness. Consideration of the so called entanglement entropy leads to the same conclusion, namely that there exists an effective "ultraviolet cutoff" at around $l = \sqrt{\kappa\hbar}$. [30]

A related but less direct train of thought starts by considering the gravitational action-integral $\frac{1}{2\kappa} \int R dV$. Here the "coupling constant" $1/2\kappa$ is an inverse length^2 and conventional Renormalization Group wisdom suggests that, barring any "fine tuning", the order of magnitude of such a coupling constant will be set by the underlying "lattice spacing", or in this case, the fundamental discreteness

---

[3]The term 'chain' is defined in the glossary.

scale, leading to the same conclusion as before that $l \equiv (v_0)^{1/4}$ is around $l = \sqrt{\kappa\hbar} \sim 10^{-32} cm$.

A noteworthy implication of the formula $l \sim \sqrt{\kappa\hbar}$ is that $l \to 0$ if $\hbar \to 0$ (κ being fixed). That is, the classical limit is necessarily a continuum limit: spacetime discreteness is inherently quantal.

6.   The reconstruction of $\mathbb{M}^4$

In order to get a better feel for how it is that "geometry = order + number", it is useful to work through the reconstruction — in the continuum — of $\mathbb{M}^4$ from its causal order and volume-element. The proof can be given in a quite constructive form which I'll only sketch here.

We start with a copy M of $\mathbb{M}^4$ and let $\prec$ be its causal order. We construct in turn: light rays l, null 3-planes, spacelike 2-planes, spacelike lines, arbitrary 2-planes, arbitrary lines, parallel lines, parallelograms, vectors. Once we have vector addition (affine structure) it is easy to get quadratic forms and in particular the flat metric η_{ab}. (The normalization of η_{ab} uses the volume information.)

You may enjoy working these constructions out for yourself, so I won't give them all here. At the risk of spoiling your fun however, let me give just the first two, which perhaps are less straightforward than the rest. We define a light ray l to be a maximal chain such that $\forall x, y \in l$, interval(x, y) is *also* a chain; and from any such l we get then the null hyperplane $N(l) = l \cup l^{\natural}$, where $l^{\natural}$ is the set of all points of M spacelike to l.

7.   Sprinkling, coarse-graining, and the "Hauptvermutung"

A basic tenet of causet theory is that spacetime does not exist at the most fundamental level, that it is an "emergent" concept which is relevant only to the extent that some manifold-with-Lorentzian-metric M furnishes a good approximation to the physical causet C. Under what circumstances would we want to say that this occurred? So far the most promising answer to this question is based on the concepts of *sprinkling* and *coarse-graining*.

Given a manifold M with Lorentzian metric g_{ab} (which is, say, globally hyperbolic) we can obtain a causal set $C(M)$ by selecting points of M and endowing them with the order induced from that of M (where in M, $x \prec y$ iff there is a future causal curve from x to y). In order to realize the equality $N = V$, the selected points must be distributed *with unit density* in M. One way to accomplish this (and conjecturally the *only* way!) is to generate the points of $C(M)$ by a *Poisson process*. (To realize the latter, imagine dividing M up into small regions of volume ϵ and independently putting a point into each region with probability ϵ. In the limit $\epsilon \to 0$ this is the Poisson process of unit intensity in M.) Let us write $M \approx C$ for the assertion that M is a good approximation to C. The idea then is that $M \approx C$ if C "might have been produced by a sprinkling of M" (in a sense to be specified more fully below).

It's important here that the elements of $C(M)$ are selected at random from M. In particular, this fact is an ingredient in the heuristic reasoning leading to the prediction of a fluctuating cosmological constant (see below). But such a *kinematic* randomness might seem gratuitous. Wouldn't a suitable *regular* embedding of points into M yield a subset that was equally uniformly distributed, if not more so? To see what goes wrong, consider the "diamond lattice" in $\mathbb{M}^2$ consisting of all points with integer values of the null coordinates $u = t - x$ and $v = t + x$. This would *seem* to be a uniform lattice, but under a boost $u \to \lambda u$, $v \to v / \lambda$ it goes into a distribution that looks entirely different, with a very high density of points along the u=constant lines (say) and large empty spaces in between. In particular, our diamond lattice is far from Lorentz invariant, which a truly uniform distribution *should be* — and which $C(M)$ produced by a Poisson process *actually is*. Examples like this suggest strongly that, in contrast to the situation for Euclidean signature, only a random sprinkling can be uniform for Lorentzian signature.

I have just argued that the idea of random sprinkling must play a role in making the correspondence between the causet and the continuum, and for purely kinematic reasons. A second concept which might be needed as well, depending on how the dynamics works out, is that of *coarse-graining*. Indeed, one might expect that, on very small scales, the causet representing our universe will no more look like a manifold than the trajectory of a point particle looks microscopically like a smooth curve in nonrelativistic quantum mechanics. Rather, we might recover a manifold only after some degree of "averaging" or "coarse-graining" (assuming also that we keep away from the big bang and from black hole interiors, etc., where we don't expect a manifold at all). That is, we might expect not $C \approx M$ but $C' \approx M'$, where C' is some coarse-graining of C and M' is M with a correspondingly rescaled metric. The relevant notion of coarse-graining here seems to be an analog of sprinkling applied to C itself: let C' be obtained from C by selecting a subset at random, keeping each element $x \in C$ with some fixed probability, say $1/2$ if we want a 2:1 coarse-graining.

Implicit in the idea of a manifold approximating a causet is that the former is relatively unique; for if two very different manifolds could approximate the same C, we'd have no objective way to understand why we observe one particular spacetime and not some very different one. (On the other hand, considering things like AdS/CFT duality, who knows...!) The conjecture that such ambiguities don't occur has been called the "Hauptvermutung". In the $G\hbar \to 0$ limit, it has already been proven in [31]. Moreover, the fact that we know how to obtain dimensional and proper time information in many situations (see below) is strong circumstantial evidence for its truth at finite $G\hbar$. Nevertheless it would be good to prove it in full, in something like the following form.

CONJECTURE If $M_1 \approx C$ and $M_2 \approx C$ then $M_1 \approx M_2$

Here $M_1 \approx M_2$ means that the manifolds M_1 and M_2 are "approximately isometric". As the quotation marks indicate, it is surprisingly difficult to give this conjecture rigorous meaning, due ultimately to the Lorentzian signature of g_{ab} (cf. [32]). Here's a sketch of how it might be done: interpret $M \approx C$ to mean that $\mathrm{Prob}(C(M) = C)$ is relatively large (in comparison with $\mathrm{Prob}(C(M) = C')$) for most of the other C'; and interpret $M_1 \approx M_2$ to mean that the random variables $C(M_1)$ and $C(M_2)$ share similar probability distributions (on the space of causets). These definitions almost make the conjecture look tautological, but it isn't! (cf. [33])

8.   Dimension and length

Assuming that — as seems very likely — causal sets do possess a structure rich enough to give us back a macroscopically smooth Lorentzian geometry, it is important to figure out how in practice one can extract geometrical information from an order relation. But before we can speak of a geometry we must have a manifold, and the most basic aspect of a manifold's topology is its dimension. So an obvious first question is whether there is a good way to recognize the effective continuum dimension of a causal set (or more precisely of a causal set that is sufficiently "manifold like" for the notion of its dimension to be meaningful). In fact several workable approaches exist. Here are three of them. All three estimators will assign a dimension to an interval I in a causet C and are designed for the case where $I \approx A$ for some interval ("double light cone") A in Minkowski space $\mathbb{M}^d$.

Myrheim-Meyer dimension [5] [29]. Let $N = |I|$ be the number of elements in I and let R be the number of relations in I (i.e. pairs x, y such that $x \prec y$). Let $f(d) = \frac{3}{2} \binom{3d/2}{d}^{-1}$. Then $f^{-1}(R/\binom{N}{2})$ is a good estimate of d when $N \gg (27/16)^d$.

REMARK The Myrheim-Meyer estimator is coarse-graining invariant on average (as is the next)

Midpoint scaling dimension. Let $I = \mathrm{interval}(a,b)$ and let $m \in I$ be the (or a) "midpoint" defined to maximize $N' = \min\{|\mathrm{interval}(a,m)|, |\mathrm{interval}(m,b)|\}$. Then $\log_2(N/N')$ estimates d.

A third dimension estimator. Let K be the total number of chains in I. Then $\ln N / \ln \ln K$ estimates d. However, the logarithms mean that good accuracy sets in only for exponentially large N.

9.   A length estimator

Again this is for $C \approx \mathbb{M}^d$ (or some convex subspace of $\mathbb{M}^d$). Let $x \prec y$. The most obvious way to define a distance (or better a time-lapse) from x to y is just to count the number of elements L in the longest chain joining them, where a joining chain is by definition a succession of elements z_i such that $x \prec z_1 \prec z_2 \prec z_3 \cdots \prec y$. Clearly a maximal path in this sense is analogous to a timelike geodesic, which maximizes the proper-time between its endpoints. It is known that this estimator L converges rapidly to a multiple of the true proper time T as the latter becomes large. However the coefficient of proportionality depends on the dimension d and is known exactly only for $d = 1$ and $d = 2$. For $d > 2$ only bounds are known, but they are rather tight.

Thus, it seems that we have workable tools for recovering information on both dimensionality and length (in the sense of timelike geodesic distance). However, these tools have been proven so far primarily in a flat context, and it remains to be shown that they continue to work well in the presence of generic curvature.

10. Dynamics

A priori, one can imagine at least two routes to a "quantum causet dynamics". On one hand, one could try to mimic the formulation of other theories by seeking a causet-invariant analogous to the scalar curvature action, and then attempting to build from it some discrete version of a gravitational "sum-over-histories". On the other hand, one could try to identify certain general principles or rules powerful enough to lead, more or less uniquely, to a family of dynamical laws sufficiently constrained that one could then pick out those members of the family that reproduced the Einstein equations in an appropriate limit or approximation. (By way of analogy, one could imagine arriving at general relativity either by seeking a spin-2 analog of Poisson's equation or by seeking the most general field equations compatible with general covariance and locality.)

The recent progress in dynamics has come from the second type of approach, and with the causet's "time-evolution" conceived as a process of what may be termed *sequential growth*. That is, the causet is conceived of as "developing in time",[4] rather than as "existing timelessly" in the manner of a film strip. At the same time the growth process is taken to be random rather than deterministic — classically random to start with, but ultimately random in the quantum sense familiar from atomic physics and quantum field theory. (Thus the quantum dynamical law is being viewed as more analogous to a classical stochastic process like Brownian motion than to a classical deterministic dynamics like that of the harmonic oscillator. [34] [35]) Expressed more technically, the idea is to seek a quantum causet dynamics by first formulating the causet's growth as a classical stochastic process and then generalizing the formulation to the case of a "quantum measure" [36] or "decoherence functional" [37].

---

[4]It might be more accurate to say that the growth of the causet *is time*.

The growth process in question can be viewed as a sequence of "births" of new causet elements and each such birth is a transition from one partial causet to another. A dynamics or "law of growth" is then simply an assignment of probabilities to each possible sequence of transitions. Without further restriction, however, there would be a virtually limitless set of possibilities for these probabilities. The two principles that have allowed us to narrow this field of possibilities down to a (hopefully) manageable number are *discrete general co-variance* and *Bell causality*. To understand the first of these, notice that taking the births to be sequential implicitly introduces a labeling of the causet elements (the first-born element being labeled 0, the second-born labeled 1, etc). Discrete general covariance is simply the requirement that this labeling be "pure gauge", that it drop out of the final probabilities in the same way that the choice of coordinates drops out of the equations of general relativity. This requirement has the important side effect of rendering the growth process Markovian, so that it is fully definable in terms of *transition probabilities* obeying the Markov sum rule. The requirement of Bell causality is slightly harder to explain, but it is meant to capture the intuition that a birth taking place in one region of the causet cannot be influenced by other births that occur in regions spacelike to the first region.

Taken together, these assumptions lead to a set of equations and inequalities that — remarkably — can be solved explicitly and in general [38] [39]. The resulting probability for a transition $C \to C'$ in which the new element is born with ϖ ancestors and m parents (immediate ancestors) is given by the ratio

$$\frac{\lambda(\varpi, m)}{\lambda(n, 0)} , \tag{1}$$

where $n = \mathrm{card}(C)$ is the number of elements before the birth in question and where the function λ is defined by the formula

$$\lambda(\varpi, m) = \sum_{k=m}^{\varpi} \binom{\varpi - m}{k - m} t_k \tag{2}$$

with $t_n \geq 0$ and $t_0 > 0$. A particular dynamical law, then, is determined by the sequence of "coupling constants" t_n (or more precisely, by their ratios).

(It turns out that the probabilities resulting from these rules can be re-expressed in terms of an "Ising model" whose spins reside on the relations $x \prec y$ of the causet, and whose "vertex weights" are governed directly by the parameters t_n [38]. In this way, a certain form of "Ising matter" emerges indirectly from the dynamical law, albeit its dynamics is rather trivial if one confines oneself to a fixed background causet. This illustrates how one might hope to recover in an appropriate limit, not only spacetime and gravity, but also certain forms of non-gravitational matter (here unified with gravity in a way reminiscent of earlier proposals for "induced gravity" [40]).)

With some progress in hand concerning both the kinematics and dynamics of causets, it is possible to start to think about applications, or if you will "phenomenology". Several projects of this nature are under way, with some interesting results already obtained. In the remaining sections I will mention some of these results and projects.

11. Fluctuations in the cosmological constant

>From the most basic notions of causal set theory, there follows already an order of magnitude prediction for the value of the cosmological constant Λ. More precisely, one can argue that Λ should *fluctuate* about its "target value", with the magnitude of the fluctuations decreasing with time in proportion to $N^{-1/2}$, where N is the relevant number of ancestors (causet elements) at a given cosmological epoch. If one assumes that (for reasons yet to be understood) the target value for Λ is zero, and if one takes for the N of today the spacetime volume of the currently visible universe from the big bang until the present, then one deduces for Λ a magnitude consistent with the most recent observations, as predicted already in [41] (and with refined arguments in [35] and [42]).

Four basic features of causet theory enter as ingredients into the refined version of the argument: the fundamental discreteness, the relation $n = V$, the Poisson fluctuations in n associated with sprinkling, and the fact that n serves as a parameter time in the dynamics of sequential growth.

>From the first of these we derive a finite value of n (at any given cosmic time). From the fourth we deduce that, since time is not summed over in the path-integral of non-relativistic quantum mechanics, neither should one expect to sum over n in the gravitational path integral that one expects to result as an approximation to the still to be formulated quantum dynamics of causets. But holding n fixed means holding spacetime volume V fixed, a procedure that leads in the continuum to what is called "unimodular gravity".

In the classical limit, this unimodular procedure leads to the action principle

$$\delta \left(\int (\frac{1}{2\kappa}R - \Lambda_0)dV - \lambda V \right) = 0$$

where Λ_0 is the "bare" cosmological constant, $V = \int dV$, $\kappa = 8\pi G$, and λ is a Lagrange multiplier implementing the fixation of V. Plainly, the last two terms combine into $-\int \Lambda dV$ where $\Lambda = \Lambda_0 + \lambda$, turning the effective cosmological constant Λ into a free constant of integration rather than a fixed microscopic parameter of the theory. Moreover, the fact that Λ and V enter into the action-integral in the combination $-\Lambda V$ means that they are conjugate in the quantum mechanical sense, leading to the indeterminacy relation

$$\delta\Lambda\, \delta V \sim \hbar \, .$$

Finally the Poisson fluctuations in n of size $\delta n \sim \sqrt{n}$ at fixed V imply that, at fixed n, there will be fluctuations in V of the same magnitude: $\delta V \sim \sqrt{n} \sim \sqrt{V}$, which correspond to fluctuations in Λ of magnitude $\delta\Lambda \sim \hbar/\delta V \sim 1/\sqrt{V}$ (taking $\hbar = 1$). The observed Λ would thus be a sort of residual quantum gravity effect, even though one normally associates the quantum with the very small, rather than the very big!

Of course, this prediction of a fluctuating Λ remains at a heuristic level until it can be grounded in a complete "quantum causet dynamics". Nevertheless, given its initial success, it seems worthwhile to try to extend it by constructing a model in which not only the instantaneous magnitude of the fluctuations could be predicted, but also their correlations between one time and another. In this way, one could assess whether the original prediction was consistent with important cosmological constraints such as the extent of structure formation and the abundances of the light nuclei. In this respect, it is worth noting that current fits of nucleosynthesis models to the observed abundances favor a non-integer number of light neutrinos falling between two and three. If this indication holds up, it will require some form of effective negative energy density at nucleosynthesis time, and a negative fluctuation in the contemporaneous Λ is perhaps the simplest way to realize such an effective density.

Added note: A concrete model of the sort suggested above has been developed in [43].

12.   Links across the horizon

An important question on which one can hope to shed light while still remaining at the level of kinematics is that of identifying the "horizon states" that underpin the entropy of a black hole. Indeed, just as the entropy of a box of gas is, to a first approximation, merely counting the molecules in the box, one might anticipate that the entropy of a black hole is effectively counting suitably defined "molecules" of its horizon. With this possibility in mind, one can ask whether any simply definable sub-structures of the causets associated with a given geometry could serve as candidates for such "horizon molecules" in the sense that counting them would approximately measure the "information content" of the black hole.

Perhaps the most obvious candidates of this sort are the causal links crossing the horizon in the neighborhood of the hypersurface Σ for which the entropy is sought. (Recall that a link is an irreducible relation of the causet.) Of course, the counting of any small scale substructures of the causet is prone to produce a result proportional to the area of the horizon, but there is no reason *a priori* why the coefficient of proportionality could not be divergent or vanishing, or why, if it is finite, it could not depend on the details of the horizon geometry.

Djamel Dou [44] has investigated this question for two very different black hole geometries, one in equilibrium (the 4 dimensional Schwarzschild metric) and one very far from equilibrium (the conical horizon that represents the earliest portion of a black hole formed from the collapse of a spherical shell of matter). For the Schwarzschild case, he made an ad hoc approximation that reduced the problem to 2 dimensional Schwarzschild and found, for a certain definition of "near horizon link", that the number N of such links has an expectation value which reduces in the $\hbar \to 0$ limit to $c(\pi^2/6)A$, where A is the horizon area and c is a constant arising in the dimensional reduction. (By $\hbar \to 0$ I mean equivalently $l^2/R^2 \to 0$ where l is the fundamental causet scale and R the horizon radius.) For the expanding horizon case (again dimensionally reduced from 4 to 2) he obtained exactly the same answer, $c(\pi^2/6)A$, despite the very different geometries. Not only is this a nontrivial coincidence, it represents the first time, to my knowledge, that something like a number of horizon states has been evaluated for any black hole far from equilibrium. The first step in solidifying and extending these results would be to control the dimensional reduction from 4 to 2, evaluating in particular the presently unknown coefficient c. Second, one should check that both null and spacelike hypersurfaces Σ yield the same results. (The null case is the best studied to date. Conceptually, it is important for possible proofs of the generalized second law [45].) Also one should assess the sensitivity of the answer to changes in the definition of "near horizon link", since there exist examples showing that the wrong definition can lead to an answer of either zero or infinity. And of course one should extend the results to other black hole geometries beyond the two studied so far.

Here is one of the definitions of "near horizon link" investigated by Djamel: Let H be the horizon of the black hole and Σ, as above, the hypersurface for which we seek the entropy S. The counting is meant to yield the black hole contribution to S, corresponding to the section $H \cap \Sigma$ of the horizon. We count pairs of sprinkled points (x, y) such that

(i) $x \prec \Sigma, H$ and $y \succ \Sigma, H$.

(ii) $x \prec y$ is a link[5]

(iii) x is maximal‡ in (past Σ) and y is minimal‡ in (future Σ) $\cap$ (future H).

13. What are the "observables" of quantum gravity?

Just as in the continuum the demand of diffeomorphism-invariance makes it harder to formulate meaningful statements,[6]

---

[5] see glossary

[6] Think, for example, of the statement that light slows down when passing near the sun.

so also for causets the demand of discrete general covariance has the same consequence, bringing with it the risk that, even if we succeeded in characterizing the covariant questions in abstract formal terms, we might never know what they meant in a physically useful way. I believe that a similar issue will arise in every approach to quantum gravity, discrete or continuous (unless of course general covariance is renounced).[7]

Within the context of the classical growth models described above, this problem has been largely solved [47], the "observables" being generated by "stem-predicates". ('stem' is defined in the glossary).

Added note: The conjecture in [47] has been settled in the affirmative by [48].

14.   How the large numbers of cosmology might be understood: a "Tolman-Boltzmann" cosmology

Typical large number is ratio r of diameter of universe to wavelength of CMB radiation. Idea is cycling of universe renormalizes [49] coupling constants such that r automatically gets big after many bounces (no fine tuning). Large numbers thus reflect large age of universe. See [50] and [51].

15.   Fields on a background causet

See [52], [53], [54].

16.   Topology change

See [55].

Acknowledgments

This research was partly supported by NSF grant PHY-0098488, by a grant from the Office of Research and Computing of Syracuse University, and by an EPSRC Senior Fellowship at Queen Mary College, University of London. I would like to express my warm gratitude to Goodenough College, London for providing a splendid living and working environment which greatly facilitated the preparation of these notes.

References

[1] R.D. Sorkin, "A Specimen of Theory Construction from Quantum Gravity", in J. Leplin (ed.), *The Creation of Ideas in Physics: Studies for a Methodology of Theory Construction* (Proceedings of the Thirteenth Annual Symposium in Philosophy, held Greensboro, North

---

[7]In the context of canonical quantum gravity, this issue is called "the problem of time". There, covariance means commuting with the constraints, and the problem is how to interpret quantities which do so in any recognizable spacetime language. For an attempt in string theory to grapple with similar issues see [46].

Carolina, March, 1989) pp. 167-179 (Number 55 in the University of Western Ontario Series in Philosophy of Science) (Kluwer Academic Publishers, Dordrecht, 1995) ⟨gr-qc/9511063⟩

[2] L. Bombelli, J. Lee, D. Meyer and R.D. Sorkin, "Spacetime as a causal set", *Phys. Rev. Lett.* **59:** 521-524 (1987);
C. Moore, "Comment on 'Space-time as a causal set'," *Phys. Rev. Lett.* **60:** 655 (1988);
L. Bombelli, J. Lee, D. Meyer and R.D. Sorkin, "Bombelli et al. Reply", *Phys. Rev. Lett.* **60:** 656 (1988).

[3] L. Bombelli, *Space-time as a Causal Set*, Ph.D. thesis, Syracuse University (1987)

[4] G. 't Hooft, "Quantum gravity: a fundamental problem and some radical ideas", in *Recent Developments in Gravitation* (Proceedings of the 1978 Cargese Summer Institute) edited by M. Levy and S. Deser (Plenum, 1979)

[5] J. Myrheim, "Statistical geometry," CERN preprint TH-2538 (1978)

[6] G. Brightwell and R. Gregory, "The Structure of Random Discrete Spacetime", *Phys. Rev. Lett.* **66:** 260-263 (1991)

[7] David D. Reid, "Discrete Quantum Gravity and Causal Sets", *Canadian Journal of Physics* **79:** 1-16 (2001) ⟨gr-qc/9909075⟩

[8] Mituo Taketani, "On formation of the Newton Mechanics", *Suppl. Prog. Theor. Phys.* **50:** 53-64 (1971)

[9] Alfred Arthur Robb, *Geometry of Time and Space* (Cambridge University Press, 1936) (a revised version of *A theory of Time and Space* (C.U.P. 1914))

[10] Hans Reichenbach:
Physikal. Zeitschr. **22:** 683 (1921); and
Axiomatik der relativistische Raum-Zeit-Lehre, translated into English as *Axiomatization of the theory of relativity* (Berkeley, University of California Press, 1969).

[11] E.C. Zeeman, "Causality Implies the Lorentz Group", *J. Math. Phys.* **5:** 490-493 (1964)

[12] Roger Penrose, *Techniques of Differential Topology in Relativity*, AMS Colloquium Publications (SIAM, Philadelphia, 1972)

[13] S.W. Hawking, A.R. King and P.J. McCarthy, "A New Topology for Curved Space-Time which Incorporates the Causal, Differential and Conformal Structures", *J. Math. Phys.* **17:** 174 (1976)

[14] David Malament, "The class of continuous timelike curves determines the toplogy of space-time", *J. Math. Phys* **18:** 1399 (1977)

[15] Bernhard Riemann, "Über die Hypothesen, welche der Geometrie zu Grunde liegen", in *The Collected Works of B. Riemann* (Dover NY 1953)

[16] A. Einstein, Letter to Walter Dällenbach, Nov. 1916, Item 9-072 translated and cited by Stachel in the following article, page 379.

[17] A. Einstein, Letter to H.S. Joachim, August 14, 1954, Item 13-453 cited in J. Stachel, "Einstein and the Quantum: Fifty Years of Struggle", in *From Quarks to Quasars, Philosophical Problems of Modern Physics*, edited by R.G. Colodny (U. Pittsburgh Press, 1986), pages 380-381.

[18] David Finkelstein, "The space-time code", *Phys. Rev.* **184:** 1261 (1969)

[19] R.D. Sorkin, "Lisp Functions for Posets", http://www.physics.syr.edu/~sorkin (version 1.0 1997, version 2.0 2002).
Another poset package (using Maple) is [56].

[20] Alan Daughton, "An investigation of the symmetric case of when causal sets can embed into manifolds" *Class. Quant. Grav.* **15**: 3427-3434 (1998)

[21] G. Brightwell and P. Winkler, "Sphere Orders", *Order* **6**: 235-240 (1989)

[22] Graham Brightwell, "Models of Random Partial Orders", in *Surveys in Combinatorics, 1993*, London Math. Soc. Lecture Notes Series **187**: 53-83, ed. Keith Walker (Cambridge Univ. Press 1993)

[23] B. Pittel and R. Tungol, "A Phase Transition Phenomenon in a Random Directed Acyclic Graph", (Ohio State preprint, 1998) (submitted to *Combinatorics, Probability and Computing*)

[24] Klaus Simon, Davide Crippa and Fabian Collenberg, "On the Distribution of the Transitive Closure in a Random Acyclic Digraph", *Lecture Notes in Computer Science* **726**: 345-356 (1993)

[25] Charles M. Newman, "Chain Lengths in Certain Random Directed Graphs", *Random Structures and Algorithms* **3**: 243-253 (1992)

[26] Jeong Han Kim and Boris Pittel, "On tail distribution of interpost distance" (preprint, Ohio State University and Microsoft Corp., 1999)

[27] Ioannis Raptis, "Algebraic Quantization of Causal Sets" *Int. J. Theor. Phys.* **39**: 1233-1240 (2000) ⟨gr-qc/9906103⟩

[28] B. Bollobas and G. Brightwell, "The Height of a Random Partial Order: Concentration of Measure", *Annals of Applied Probability* **2**: 1009-1018 (1992)

[29] D.A. Meyer:
"Spherical containment and the Minkowski dimension of partial orders", *Order* **10**: 227-237 (1993); and
The Dimension of Causal Sets, Ph.D. thesis, M.I.T. (1988).

[30] R.D. Sorkin, "On the Entropy of the Vacuum Outside a Horizon", in B. Bertotti, F. de Felice and A. Pascolini (eds.), *Tenth International Conference on General Relativity and Gravitation (held Padova, 4-9 July, 1983), Contributed Papers*, vol. II, pp. 734-736 (Roma, Consiglio Nazionale Delle Ricerche, 1983);
G. 't Hooft, "On the quantum structure of a black hole", *Nuclear Phys. B* **256**: 727-745 (1985)

[31] L. Bombelli and D.A. Meyer, "The origin of Lorentzian geometry," *Physics Lett. A* **141**: 226-228 (1989)

[32] Johan Noldus, "A new topology on the space of Lorentzian metrics on a fixed manifold" *Class. Quant. Grav* **19**: 6075-6107 (2002)

[33] Luca Bombelli, "Statistical Lorentzian geometry and the closeness of Lorentzian manifolds", *J. Math. Phys.* **41**: 6944-6958 (2000) ⟨gr-qc/0002053⟩

[34] R.D. Sorkin:
"On the Role of Time in the Sum-over-histories Framework for Gravity", paper presented to the conference on The History of Modern Gauge Theories, held Logan, Utah, July 1987, published in *Int. J. Theor. Phys.* **33**: 523-534 (1994); and
"A Modified Sum-Over-Histories for Gravity", reported in *Proceedings of the International Conference on Gravitation and Cosmology, Goa, India, 14-19 December, 1987*, edited by B. R. Iyer, Ajit Kembhavi, Jayant V. Narlikar, and C. V. Vishveshwara, see pages 184-186 in the article by D. Brill and L. Smolin: "Workshop on quantum gravity and new directions", pp 183-191 (Cambridge University Press, Cambridge, 1988)

[35] R.D. Sorkin, "Forks in the Road, on the Way to Quantum Gravity", talk given at the conference entitled "Directions in General Relativity", held at College Park, Maryland, May, 1993, published in *Int. J. Th. Phys.* **36:** 2759–2781 (1997) ⟨gr-qc/9706002⟩

[36] R.D. Sorkin, "Quantum Mechanics as Quantum Measure Theory", *Mod. Phys. Lett. A* **9** (No. 33): 3119-3127 (1994) ⟨gr-qc/9401003⟩

[37] J.B. Hartle, "Spacetime Quantum Mechanics and the Quantum Mechanics of Spacetime", in B. Julia and J. Zinn-Justin (eds.), *Les Houches, session LVII, 1992, Gravitation and Quantizations* (Elsevier Science B.V. 1995)

[38] David P. Rideout and Rafael D. Sorkin, "A Classical Sequential Growth Dynamics for Causal Sets", *Phys. Rev. D* **61:** 024002 (2000) ⟨gr-qc/9904062⟩

[39] David P. Rideout, *Dynamics of Causal Sets*, Ph.D. thesis (Syracuse University 2001)

[40] T. Jacobson, "Black Hole Entropy and Induced Gravity", ⟨gr-qc/9404039⟩;
S.L. Adler, "Einstein gravity as a symmetry-breaking effect in quantum field theory", *Reviews of Modern Physics* **54:** 729-766 (1982);
V.A. Kazakov, "The appearance of matter fields from quantum fluctuations of 2D-gravity", *Mod. Phys. Lett. A* **4:** 2125-2139 (1989) .

[41] R.D. Sorkin:
"First Steps with Causal Sets", in R. Cianci, R. de Ritis, M. Francaviglia, G. Marmo, C. Rubano, P. Scudellaro (eds.), *General Relativity and Gravitational Physics* (Proceedings of the Ninth Italian Conference of the same name, held Capri, Italy, September, 1990), pp. 68-90 (World Scientific, Singapore, 1991); and
"Spacetime and Causal Sets", in J.C. D'Olivo, E. Nahmad-Achar, M. Rosenbaum, M.P. Ryan, L.F. Urrutia and F. Zertuche (eds.), *Relativity and Gravitation: Classical and Quantum* (Proceedings of the *SILARG VII Conference*, held Cocoyoc, Mexico, December, 1990), pages 150-173 (World Scientific, Singapore, 1991).

[42] R.D. Sorkin, Two Talks given at the 1997 Santa Fe workshop: "A Review of the Causal Set Approach to Quantum Gravity" and "A Growth Dynamics for Causal Sets", presented at: "New Directions in Simplicial Quantum Gravity" July 28 - August 8, 1997. The scanned in transparencies may be viewed at http://t8web.lanl.gov/people/emil/Slides/sf97talks.html ;
Y. Jack Ng and H. van Dam, "A small but nonzero cosmological constant", (Int. J. Mod. Phys D., to appear) ⟨hep-th/9911102⟩

[43] Maqbool Ahmed, Scott Dodelson, Patrick Greene and Rafael D. Sorkin, "Everpresent Λ", ⟨astro-ph/0209274⟩

[44] Djamel Dou, "Causal Sets, a Possible Interpretation for the Black Hole Entropy, and Related Topics", Ph. D. thesis (SISSA, Trieste, 1999) ⟨gr-qc/0106024⟩.
See also Dj. Dou and R.D. Sorkin, "Black Hole Entropy as Causal Links", *Foundations of Physics* (2003, to appear).

[45] R.D. Sorkin, "Toward an Explanation of Entropy Increase in the Presence of Quantum Black Holes", *Phys. Rev. Lett.* **56:** 1885-1888 (1986)

[46] Edward Witten, "Quantum Gravity in De Sitter Space" ⟨hep-th/0106109⟩

[47] Graham Brightwell, H. Fay Dowker, Raquel S. García, Joe Henson and Rafael D. Sorkin, "General Covariance and the 'Problem of Time' in a Discrete Cosmology", in K.G. Bowden, Ed., *Correlations*, Proceedings of the ANPA 23 conference, held August 16-21, 2001, Cambridge, England (Alternative Natural Philosophy Association, London, 2002), pp 1-17 ⟨gr-qc/0202097⟩

[48] Graham Brightwell, Fay Dowker, Raquel Garcia, Joe Henson and Rafael D. Sorkin, ""Observables" in Causal Set Cosmology", ⟨gr-qc/0210061⟩

[49] Xavier Martin, Denjoe O'Connor, David Rideout and Rafael D. Sorkin, "On the "renormalization" transformations induced by cycles of expansion and contraction in causal set cosmology", *Phys. Rev. D* **63:** 084026 (2001) ⟨gr-qc/0009063⟩

[50] Rafael D. Sorkin, "Indications of causal set cosmology", *Int. J. Theor. Ph.* **39 (7):** 1731-1736 (2000) (an issue devoted to the proceedings of the Peyresq IV conference, held June-July 1999, Peyresq France) ⟨gr-qc/0003043⟩

[51] Avner Ash and Patrick McDonald, "Moment problems and the causal set approach to quantum gravity", ⟨gr-qc/0209020⟩

[52] A.R. Daughton, *The Recovery of Locality for Causal Sets and Related Topics*, Ph.D. thesis, Syracuse University (1993)

[53] Roberto Salgado, Ph.D. thesis, Syracuse University (in preparation)

[54] Richard F. Blute, Ivan T. Ivanov, Prakash Panangaden, "Discrete Quantum Causal Dynamics", ⟨gr-qc/0109053⟩

[55] Fay Dowker, "Topology Change in Quantum Gravity", ⟨gr-qc/0206020⟩

[56] John R. Stembridge (jrs@math.lsa.umich.edu), *A Maple Package for Posets: Version 2.1*, (Ann Arbor 1998-may-10) available at www.math.lsa.umich.edu/˜jrs

[57] David P. Rideout and Rafael D. Sorkin, "Evidence for a continuum limit in causal set dynamics", *Phys. Rev. D* **63:** 104011 (2001) ⟨gr-qc/0003117⟩

[58] F. Markopoulou and L. Smolin, "Causal evolution of spin networks", *Nuc. Phys.* **B508:** 409-430 (1997) ⟨gr-qc/9702025⟩

[59] R.D. Sorkin, "Percolation, Causal Sets and Renormalizability", unpublished talk delivered at the RG2000 conference, held in Taxco, Mexico, January 1999

[60] David P. Rideout and Rafael D. Sorkin, "Evidence for a Scaling Behavior in Causal Set Dynamics", (in preparation)

[61] L. Smolin, "The fate of black hole singularities and the parameters of the standard models of particle physics and cosmology", ⟨gr-qc/9404011⟩

[62] Roberto B. Salgado, "Some Identities for the Quantum Measure and its Generalizations", *Mod. Phys. Lett.* **A17:** 711-728 (2002) ⟨gr-qc/9903015⟩

[63] Fay Dowker and Rafael D. Sorkin, "Toward an Intrinsic Definition of Relativistic Causality" (in preparation)

[64] A. Criscuolo and H. Waelbroeck, "Causal Set Dynamics: A Toy Model", *Class. Quant. Grav.* **16:** 1817-1832 (1999) ⟨gr-qc/9811088⟩

[65] Joohan Lee and Rafael D. Sorkin, "Limits of Sequential Growth Dynamics" (in preparation)

[66] T. Jacobson, "Black-hole evaporation and ultrashort distances", *Phys. Rev. D* **44:** 1731-1739 (1991))

[67] Curtis Greene et al., *A Mathematica Package for Studying Posets* (Haverford College, 1990, 1994) available at www.haverford.edu/math/cgreene.html

[68] C.D. Broad, *Scientific Thought* (Harcourt Brace and Company, 1923)

GLOSSARY

Major deviations from these definitions are rare in the literature but minor ones are common. (In this glossary, we use the symbol $<$ rather than $\prec$.)

ancestor/descendant   If $x < y$ then x is an ancestor of y and y is a descendant of x.

antichain   A trivial order in which no element is related to any other (cf. 'chain').

causet   = causal set = locally finite order

chain   = linear order.
An order, any two of whose elements are related. In particular, any linearly ordered subset of an order is a chain. n-chain = chain of n elements.

child/parent   If $x < y$ is a link we can say that x is a parent of y and y a child of x. One also says that "y covers x".

comparable   See 'related'.

covering relation, covers   See 'child', 'link'.

descendant   See 'ancestor'.

down-set   = downward-set = past-set = order-ideal = ancestral set.
A subset of an order that contains all the ancestors of its members

full stem   A partial stem whose complement is the (exclusive) future of its top layer.

future   See 'past'.

inf   = greatest lower bound (cf. 'sup'.)

interval   See 'order-interval'.

level   In a past-finite causet the level of an element x is the number of links in the longest chain $a < b < ... < c < x$. Thus, level 0 comprises the minimal elements, level 1 is level 0 of the remainder, etc.

linear extension   Let S be a set and $<$ an order-relation on S. A linear extension of $<$ is a second order-relation $\prec$ which extends $<$ and makes S into a chain.

link=covering relation   An irreducible relation of an order, that is, one not implied by the other relations via transitivity. Of course we exclude pairs (xx) from being links, in the case where such pairs are admitted into the order relation at all.
(There's no inconsistency here with the notion that a "chain" ought to be made up of "links": the links in a chain are indeed links relative to the chain itself, even if they aren't links relative to the enveloping order.)

locally finite   An order is locally finite iff all its order-intervals are finite. (cf. 'past-finite'.)

maximal/minimal   A maximal/minimal element of an order is one without descendants/ancestors.

natural labeling   A natural labeling of a past-finite order is an assignment to its elements of labels 0 1 2 ... such that $x < y \Rightarrow label(x) < label(y)$. Thus it is essentially the same thing as a locally finite linear extension.

order-interval   (or just plain interval)
The interval determined by two elements a and b is the set, interval$(a, b) = \{x | a < x < b\}$.

order   = ordered set = poset = partially ordered set = partial order.
An order is a set of elements carrying a notion of "ancestry" or "precedence". Perhaps the simplest way to express this concept axiomatically is to define an order as a transitive, irreflexive relation $<$. Many other, equivalent definitions are possible. In particular, many authors use the reflexive convention, in effect taking $\leq$ as the defining relation.
It is convenient to admit the empty set as a poset.

order-isomorphic   Isomorphic as posets

origin   = minimum element.
A single element which is the ancestor of all others.

originary   A poset possessing an origin is originary.

parent   See 'child'.

p   artial stem (or just plain stem) A past set of finite cardinality.

partial post   An element x of which no descendant has an ancestor spacelike to x. The idea is x is the progenitor of a "child universe".

partially ordered set   See 'order'.

past/future   past$(x) = \{y | y < x\}$, future$(x) = \{y | x < y\}$.

past-finite   An order is past-finite iff all its down-sets are finite. (cf. 'locally finite'.)

path = saturated chain   A chain all of whose links are also links of the enveloping poset (saturated means it might be "extended" but it can't be "filled in")

poset   See 'order'.

post   An element such that every other element is either its ancestor or its descendant: a one-element slice.

preorder   = preposet = acyclic relation = acyclic digraph.
A preorder is a relation whose transitive closure is an order.

pseudo-order   = transitive relation (possibly with cycles).

related   = comparable.
Two elements x and y are 'related' (or 'comparable') if $x < y$ or $y < x$.

slice = maximal antichain. (where maximal means it can't be enlarged and remain an antichain.)

Equivalently, every x in the causet is either in the slice or comparable to one of its elements.

Equivalently, its inclusive past is a full stem .

spacelike = incomparable.

Two elements x and y are spacelike to each other ($x \natural y$) iff they are unrelated (ie neither $x < y$ nor $y < x$).

stem See 'partial stem'.

sup = least upper bound (cf. 'inf'.)

transitively reduced The "transitive reduction" of an order is its Hasse digraph, an acyclic relation containing only links.

THERMAL DECAY OF THE COSMOLOGICAL CONSTANT INTO BLACK HOLES

Andrés Gomberoff
Centro de Estudios Científicos (CECS), Valdivia, Chile

Marc Henneaux
Physique Théorique et Mathématique, Université Libre de Bruxelles, Campus Plaine C.P. 231,
B–1050 Bruxelles, Belgium, and
Centro de Estudios Científicos (CECS), Valdivia, Chile

Claudio Teitelboim
Centro de Estudios Científicos (CECS), Valdivia, Chile

Frank Wilczek
Center for Theoretical Physics, Massachusetts Institute of Technology, Cambridge, MA 02139-
4307, USA
and Centro de Estudios Científicos (CECS), Valdivia, Chile

Abstract   We show that the cosmological constant may be reduced by thermal production of membranes by the cosmological horizon, analogous to a particle "going over the top of the potential barrier," rather than tunneling through it. The membranes are endowed with charge associated with the gauge invariance of an antisymmetric gauge potential. In this new process, the membrane collapses into a black hole; thus, the net effect is to produce black holes out of the vacuum energy associated with the cosmological constant. We study here the corresponding Euclidean configurations ("thermalons") and calculate the probability for the process in the leading semiclassical approximation.

1. Introduction

One of the outstanding open problems of theoretical physics is to reconcile the very small observational bound on the cosmological constant Λ with the very large values that standard high energy physics theory predicts for it [1]. This challenge has led us to consider the cosmological constant as a dynamical variable, whose evolution is governed by equations of motion. In that context mechanisms have been sought that would enable Λ to relax from a large initial value to a small one during the course of the evolution of the Universe. The simplest context in which this idea may be analyzed is through the introduction of an antisymmetric gauge potential, a three-form [2] in four spacetime dimensions. The three-form couples to the gravitational field with a term proportional to the square of the field strength. In the absence of sources, the field strength is constant in space and time and provides a contribution to the cosmological term, which then becomes a constant of motion rather than a universal constant.

Changes in the cosmological constant occur when one brings in sources for the three-form potential. These sources are two-dimensional membranes which sweep a three-dimensional history during their evolution ("domain walls"). The membranes carry charge associated with the gauge invariance of the three-form field, and they divide spacetime into two regions with different values of the cosmological term.

The membranes may be produced spontaneously in two physically different ways. One way is by tunneling through a potential barrier as happens in pair production in two-dimensional spacetime. The other is by a thermal excitation of the vacuum analogous to going "over the top" of the potential barrier rather than tunneling through it.

The tunneling process was originally studied in [3] and was further explored in [4]. The purpose of this article is to study the spontaneous decay of the cosmological constant through the other process, namely, the production of membranes due to the thermal effects of the cosmological horizon.

It is useful to visualize the decay process in terms of its simplest context, which is a particle in a one-dimensional potential barrier, as recalled in Fig. 1. When the barrier is in a thermal environment, the particle can go from one side of the barrier to the other by "climbing over the top" rather than tunneling. There is a probability given by the Boltzmann factor $e^{-\beta E}$ for the particle to be in a state of energy E. If E is greater than the height of the barrier the particle will move from one side to the other even classically. The effect is optimized when the energy is just enough for the particle to be at the top of the barrier and roll down to the other side. In this case, the Boltzmann factor is as large as possible while still allowing for the process without quantum mechanical tunneling. It turns out that, in the leading approximation, the probability is given by the exponential of the Euclidean action evaluated on an appropriate classical solution, just as for tunneling [5]. In the case of tunneling, the classical solution is called an instanton and it is time dependent [6]. In the present case, the classical solution corresponds to the configuration in which the particle sits at the top of the barrier, and thus it is time independent. Since the solution is unstable, when slightly perturbed the particle will fall half of the time to the left side and half of the time to the right side.

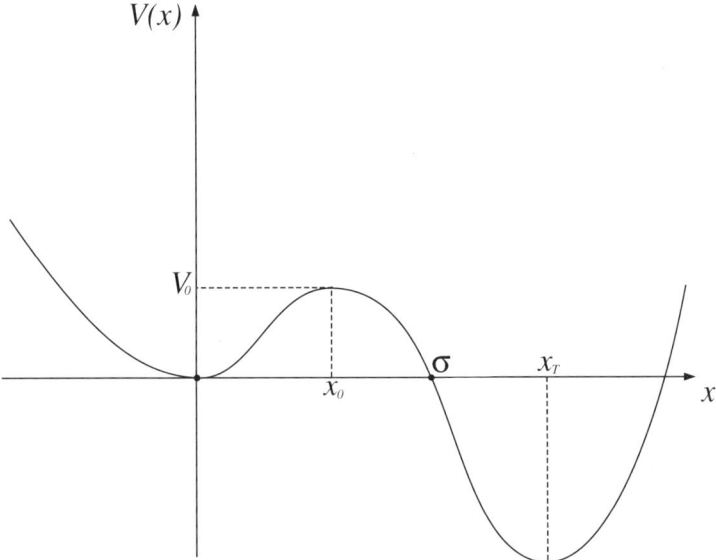

Figure 1.   The figure shows a one-dimensional potential barrier. If the particle is initially at the minimum $x = 0$ it may end up on the other side of the barrier, reaching the lower minimum $x = x_T$ by two different mechanisms. It can either (i) tunnel through the potential quantum mechanically or (ii) jump over it by a thermal fluctuation when the barrier is in a thermal environment.

In more complex situations, "sitting at the top of the barrier" is replaced by a "time independent classical solution with one instability mode" [5]. In the context of gauge theories such solutions appear in the analysis of violation of baryon-number conservation and have been called "sphalerons" [7]. In the present case we will use the name "thermalon," to emphasize that the static solutions will be intimately connected with the intrinsic thermal properties of event horizons in gravitational theory. We show below that there exists a thermalon which reduces the cosmological constant through production of membranes in de Sitter space due to the thermal properties of the cosmological horizon. In this thermalon, a membrane is emitted by the cosmological horizon and collapses, forming a black hole.

2.   The Thermalon

2.1   Basic geometry and matching equations

Once produced, the membrane divides space into two regions with the interior having a lower value of Λ. Subsequently, the membrane evolves with the region of lower Λ filling more and more of the space, thus lowering the value of the cosmological constant.

In Ref. [8] we gave the equations of motions of a charged membrane of tension μ and charge q coupled to the gravitational field and employed them to analyze instantons associated with tunneling. The same equations will be used here for the study of thermalons. We reproduce them verbatim here, together with the corresponding explanations, to make the discussion self-contained.

We consider a Euclidean spacetime element of the form

$$ds^2 = f^2(r)dt^2 + f^{-2}dr^2 + r^2\left(d\theta^2 + \sin^2\theta d\phi^2\right). \tag{1}$$

The antisymmetric field strength tensor takes the form

$$F_{\mu\nu\lambda\rho} = (dA)_{\mu\nu\lambda\rho} = E\sqrt{g}\epsilon_{\mu\nu\lambda\rho}. \tag{2}$$

The history of the membrane will divide the spacetime into two regions, one that will be called the interior, labeled by the subscript "$-$" and the other the exterior, labeled by the subscript "$+$." The exterior is the initial region and defines the "background," while the interior is the final region. The boundary may be described by the parametric equations

$$r = R(\tau), \qquad t_\pm = T_\pm(\tau), \tag{3}$$

where τ is the proper length in the r-t sector, so that its line element reads

$$ds^2 = d\tau^2 + R^2(\tau)\left(d\theta^2 + \sin^2\theta d\phi^2\right), \tag{4}$$

with

$$1 = f_\pm^2\left(R(\tau)\right)\dot{T}_\pm^2 + f_\pm^{-2}\left(R(\tau)\right)\dot{R}^2. \tag{5}$$

In the "$+$" and "$-$" regions the solution of the field equations read

$$f_\pm^2 = 1 - \frac{2M_\pm}{r} - \frac{r^2}{l_\pm^2}, \tag{6}$$

$$E_\pm^2 = \frac{1}{4\pi}\left(\frac{3}{l_\pm^2} - \lambda\right). \tag{7}$$

The actual cosmological constant $\Lambda = 3/l^2$ is thus obtained by adding λ, normally taken to be negative, coming from "the rest of physics" and not subject to change, and the contribution $4\pi E^2$, which is subject to dynamical equations. The discontinuities in the functions f^2 and E across the membrane are given by

$$f_-^2 \dot{T}_- - f_+^2 \dot{T}_+ = \mu R \,. \tag{8}$$

$$E_+ - E_- = q \,. \tag{9}$$

Here μ and q are the tension and charge on the membrane, respectively. Equation (9) follows from integrating the Gauss law for the antisymmetric tensor across the membrane, whereas Eq. (8) represents the discontinuity in the extrinsic curvature of the membrane when it is regarded as embedded in either the "$-$" or the "$+$" spaces [9]. In writing these equations, the following orientation conventions have been adopted, and will be maintained from here on: (i) The coordinate t increases anticlockwise around the cosmological horizon, and (ii) the variable τ increases when the curve is traveled along leaving the interior on its right side.

Equation (8) may be thought of as the first integral of the equation of motion for the membrane, which is thus obtained by differentiating it with respect to τ ("equations of motion from field equations"). Hence, satisfying Eqs. (6)–(9) amounts to solving all the equations of motion and, therefore, finding an extremum of the action. More explicitly, the first integral of the equation of motion for the membrane may be written as

$$\Delta M = \frac{1}{2}(\alpha^2 - \mu^2)R^3 - \mu f_+^2 \dot{T}_+ R^2 = \frac{1}{2}(\alpha^2 + \mu^2)R^3 - \mu f_-^2 \dot{T}_- R^2 \,, \tag{10}$$

where $\Delta M \equiv M_- - M_+$ is the mass difference between the initial and final geometries, and

$$\alpha^2 = \frac{1}{l_+^2} - \frac{1}{l_-^2} \,. \tag{11}$$

The Euclidean evolution of the membrane lies between two turning points. Once the initial mass M_+ and the initial cosmological constant $\Lambda_+ = 3/l_+^2$ are given, the turning points R are determined through

$$\Delta M = \frac{1}{2}(\alpha^2 - \mu^2)R^3 - \varepsilon_+ \mu f_+ R^2 = \frac{1}{2}(\alpha^2 + \mu^2)R^3 - \varepsilon_- \mu f_- R^2, \tag{12}$$

by setting $\dot{R} = 0$ in Eq. (10). Here we have defined $\varepsilon_\pm = \mathrm{sgn}\dot{T}_\pm$.

The graph of the function (12) for fixed M_+, shown in Fig. 2, will be referred to as the "mass diagram" for the decay of Schwarzschild–de Sitter space. There are two branches which merge smoothly, corresponding to taking both signs in Eq. (12), much as $x = \pm\sqrt{1 - y^2}$ gives the smooth circle $x^2 + y^2 = 1$. The lower and upper branches merge at the intersections with the curve $\Delta M = (1/2)(\alpha^2 - \mu^2)R^3$ that determines the sign of $\dot{T}_+$, which is positive below the curve and negative above it. This curve, the "$\dot{T}_+ = 0$ curve," plays an important role because it determines which side of the "+" geometry must be retained: According to the conventions established above, if $\dot{T}_+$ is positive, one must retain the side of the membrane history with (locally) greater values of the radial coordinate; if $\dot{T}_+$ is negative, one must keep the other side. The curve $\dot{T}_+ = 0$ is also shown on the mass diagram, as is the curve $\dot{T}_- = 0$, which, as can be seen from Eq. (10), has for equation $\Delta M = (1/2)(\alpha^2 + \mu^2)R^3$. A similar rule applies for determining which side of the minus geometry is retained: it is the side of increasing r if $\dot{T}_-$ is negative and the other side if it is positive.

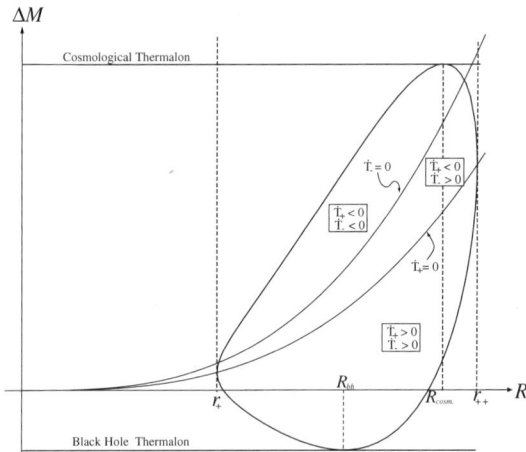

Figure 2.   The closed curve shows the points R where $\dot{R} = 0$ for a given mass gap ΔM. The thermalons are located at the maximum (cosmological thermalon) and minimum (black hole thermalon) of the curve. The curves where R is such that, for a given ΔM, $\dot{T}_\pm = 0$ are also shown. All the "+" parameters are held fixed.

2.2   Black hole and cosmological thermalons

There are four distinguished points in the mass diagram, two for which the tangent is vertical and two for which it is horizontal. The former, located at r_+ and r_{++}, correspond to membrane creation through tunneling and they give rise to the instantons discussed in [8]. The latter are thermalons. The thermalon at the top of the diagram will be called a "cosmological thermalon" because it is associated with the cosmological event horizon. The thermalon at the bottom of the mass diagram will be called a "black hole thermalon." It needs an initial black hole to provide the thermal environment.

For the particular values of ΔM corresponding to the thermalons, the two turning points coalesce and the membrane trajectory is a circle. The geometry of the thermalons is depicted in Fig. 3.

For the cosmological thermalon, the sign of $\dot{T}_-$ may be either positive or negative, depending on the values of the parameters. If $\dot{T}_-$ is negative, one must glue the region of the minus geometry that contains the cosmological horizon r_{--} to the original background geometry. The thermalon then has one black hole horizon (r_+) and one cosmological horizon (r_{--}). If, on the other hand, $\dot{T}_-$ is positive, one must glue the other side of the membrane history to the original background geometry. This produces a solution with two black hole horizons, one at r_+ and one at r_-. The inversion of the sign of $\dot{T}_-$ happens when $r_- = r_{--}$, that is, when the "$-$" geometry becomes the Nariai geometry. Therefore we will call this particular case the "Nariai threshold" and will discuss it in Sec. 4 below. Note that $\dot{T}_+$ is always negative; hence it is always the r_+ side of the plus geometry that must be kept.

For the black hole thermalon $\dot{T}_-$ is always positive. This is because $\dot{T}_-$ is greater than $\dot{T}_+$, which is positive in this case.

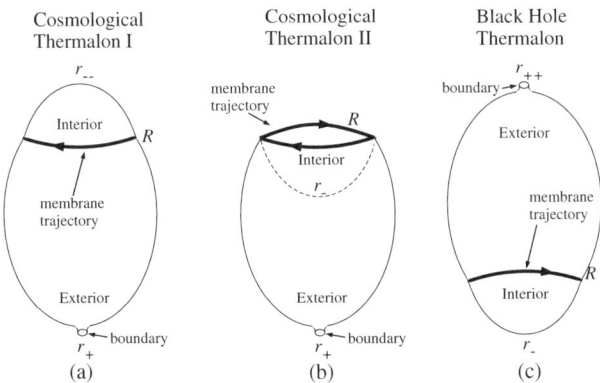

Cosmological Thermalon I   Cosmological Thermalon II   Black Hole Thermalon

(a)   (b)   (c)

Figure 3.   Thermalon geometry. (a) and (b) represent the geometry of the cosmological thermalon below and above the Nariai threshold, respectively, while (c) depicts the black hole thermalon. Only the r-t section is shown. In each figure, the radial coordinate increases if one moves upward.

For both thermalons, the Minkowskian solution is unstable and the Euclidean solution is stable. When slightly perturbed, the Euclidean solution oscillates around the thermalon. For the Minkowskian case, the membrane can evolve in two ways: it can start accelerating (rolling down) toward either the exterior or the interior. If the acceleration is directed toward the exterior, spacetime will become filled with the interior ("$-$") geometry, and the cosmological term will decrease. If, on the other hand, it accelerates toward the interior, then, and as time increases, the exterior will become the whole geometry and the cosmological constant will return to its original value. In this case the process will not have changed anything.

The minimum and the maximum of the mass curve are Euclidean stable points, because both correspond to maxima of the potential in the sense of Fig. 1. The cosmological thermalon corresponds to a maximum of the potential ultimately because, as established in [10, 13], the internal energy of the cosmological horizon is $-M$ rather than M. Thus, in the analogy with the potential problem one should plot $-\Delta M$ on the vertical axis, so that the potential in Fig. 2 is upside down and the analogy holds. The instability of both thermalons is proven explicitly in Appendix A.

Lastly, there is another important distinction between black hole and cosmological thermalons which resides in how they behave when gravity is decoupled, that is, when Newton's constant G is taken to vanish. In that case the black hole thermalon becomes the standard nucleation of a bubble of a stable phase within a metastable medium, whereas the cosmological thermalon no longer exists. The decoupling of gravity is discussed in Appendix B.

3.   Lorentzian continuation

The prescription for obtaining the Lorentzian signature solution, which describes the decay process in actual spacetime, is to find a surface of time symmetry in the Euclidean signature solution and evolve the Cauchy data on that surface in Lorentzian time. The fact that the surface chosen is one of time symmetry ensures that the Lorentzian signature solution will be real. Another way of describing the same statement is to say that one matches the Euclidean and Lorentzian signature solutions on a surface of time symmetry.

For the cosmological thermalon that induces decay of de Sitter space we will take the surface of time symmetry as the line in r-t space which, described from the Euclidean side, starts from $t = t_0$, $r = 0$, proceeds by increasing r, keeping $t = t_0$, until it reaches the cosmological horizon, and then descends back to $r = 0$ along the line $t = t_0 + \beta/2$. Thus, the surface of time symmetry crosses the membrane formation radius R twice, which implies that actually *two* membranes, of opposite polarities, are formed. The cosmological constant is decreased in the finite-volume region between them. The Penrose diagram for the Lorentzian section, below the Nariai threshold, is given in Fig. 4.

To obtain the diagram above the Nariai threshold one should replace, in Fig. 4, r_{--} by r_- and $r = \infty$ by the black hole singularity $r = 0$ of the "$-$" geometry.

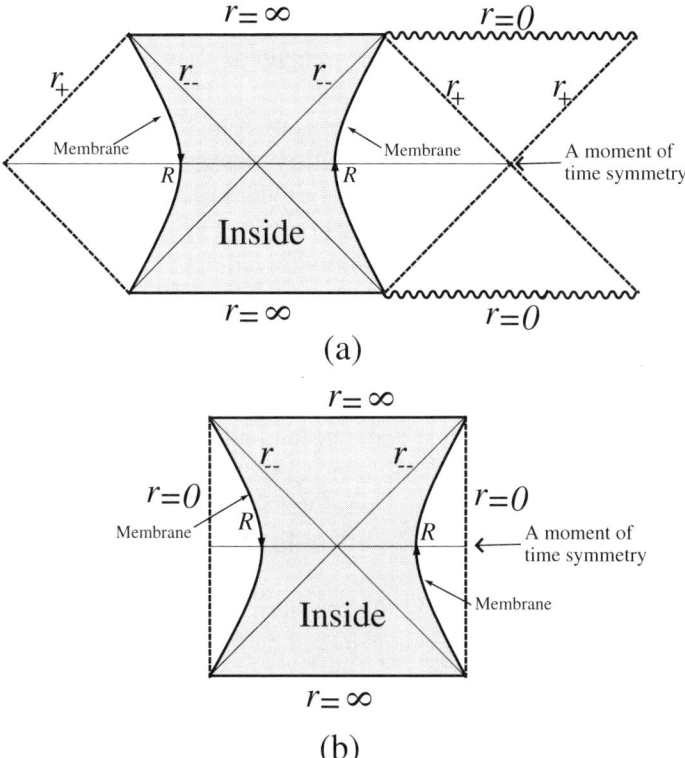

Figure 4. Cosmological thermalon Penrose diagrams. (b) is the Penrose diagram for the Lorentzian geometry of the cosmological thermalon when the initial mass is zero. The generic case $M_+ \neq 0$ below the Nariai threshold is shown for completeness in (a). The dotted lines are the points $r = r_+$, where the boundary is located in the Euclidean version of this geometry. Each diagram has two static membranes with opposite polarities at $r = R$.

4.   Action and Probability

We will be interested in the probability Γ per unit of time and unit of spatial volume for the production of a thermalon. In the leading semiclassical approximation, that probability is given by

$$\Gamma = Ae^{-B/\hbar}[1 + \mathcal{O}(\hbar)] \,, \qquad (13)$$

where $-B = I_{\text{thermalon}}$ is the value of the Euclidean action on the thermalon solution (in our conventions, the sign of the Euclidean action is such that, in the semiclassical limit, it corresponds to $-\beta F$, the inverse temperature times the free energy). The prefactor A is a slowly varying function with dimensions of a length to the negative fourth power built out of l_+, α^2, and μ. The choice of the action depends on the boundary conditions used, and also involves "background subtractions," which ensure that when the coupling with the membrane is turned off ($\mu = q = 0$), the probability (13) is equal to unity, since, in that case, P becomes the probability for things to remain as they are when nothing is available to provoke a change.

To address the issue of boundary conditions we refer the readers to Figs. 3(a), 3(b), and 3(c). There we have drawn small circles around the exterior black hole and cosmological horizons, respectively, to indicate that the corresponding point is treated as a boundary, as discussed in [10, 13]. The point in question takes the place in each case of spatial infinity for a black hole in asymptotically flat space. It represents the "platform" on which the "external observer" sits. On the boundary there is no demand that the equations of motion should hold, and thus it is permissible for a conical singularity to appear there, as indeed happens.

Once a boundary is chosen one sums in the path integral over all possible configurations elsewhere. As a consequence, no conical singularity is allowed anywhere else but at the boundary. Therefore, the gravitational action should include a contribution equal to one-fourth of the area of that horizon which is not at the boundary.

To be definite, consider the cosmological thermalon. In that case the boundary is placed at r_+, and therefore one is including the thermodynamical effects of the cosmological horizon (which is the reason for the term "cosmological thermalon"). We then fix at the boundary the value of r_+ itself or, equivalently, the mass M_+, in addition to fixing the cosmological constant Λ_+.

With these boundary conditions the total action for the problem is just the standard "bulk" Hamiltonian action of the coupled system formed by the gravitational field, the antisymmetric tensor field and the membrane, with one-fourth of the corresponding horizon area added [14]. This action includes the minimal coupling term of the three-form field to the membrane, by which the canonical momentum of the membrane differs from the purely kinetic ("mass times the velocity") term, and whose evaluation needs a definition of the potential on the membrane. Such a definition requires a mild form of regularization which is dictated by the problem itself, and which we proceed to analyze now.

The minimal coupling term is

$$q \int_{V_3} A \,, \tag{14}$$

where the integral extends over the membrane history V_3 and A is evaluated on V_3. The potential A is such that the magnitude E of its exterior derivative jumps by q when passing from the interior ("$-$" region) of V_3 to the exterior ("$+$" region) as stated in Eq. (9). We will impose the following conditions on A. (i) A must be regular at that origin (horizon) which is in the interior of V_3. More precisely, A_- should be equal to zero up to a regular gauge transformation, in order for the integral over a very small loop enclosing the horizon to vanish as the loop shrinks to a point. This is quite straightforward. The subtlety comes in with A_+, the value of A in the exterior of V_3. The fields in the exterior should be those corresponding to the solution of the equations of motion that would hold everywhere if the transition where never to occur. Therefore, we will demand that (ii) the function A_+ will be the same, for a given value of E_+, as when the membrane is absent. This means that A_+ should be regular at the horizon in the absence of the membrane, which implies—as one may show—that A is discontinuous across the membrane. With the above definition of A one may now rewrite the minimal coupling term (14) as an integral over the interior V_4 of the membrane, by means of the Stokes formula. This gives

$$
\int_{V_3} A = q E_{\text{av}} = (E_+ - E_-) \frac{1}{2} (E_+ + E_-) = \frac{1}{2} \left(E_+^2 - E_-^2 \right) = \frac{3}{8\pi} \alpha^2 ,
$$

(15)

with α^2 given by Eq. (11). The appearance of the average field E_{av}, may be thought of as coming from defining the integral over V_4 as the average of the integrals obtained when the boundary of the V_4 is displaced infinitesimally toward the interior and exterior of the membrane worldsheet. This is equivalent to "thickening" the membrane and taking the boundary half way inside.

It is interesting to point out that, in the case of membrane production in flat space, the definition of the potential A on the membrane through the "thickening" employed in Eq. (15) is equivalent to subtracting the background field action $\int F_+^2$. Thus, we will assume that when Eq. (15) is used it remains only to subtract the gravitational background action, which is equal to one-fourth of the horizon area in the absence of the membrane (the background Hamiltonian is zero). The difference in the horizon areas with and without the membrane may be thought of as the change in the available phase space of horizon states induced by the creation of the membrane.

To be able to write explicitly the form of I, one further notices that, since on shell the Hamiltonian constraints hold, the bulk Hamiltonian action reduces to the "$p\dot{q}$" term. Furthermore, for both the gravitational and antisymmetric fields, which are time independent in the exterior and interior of the membrane, $\dot{q}$ vanishes, and, therefore, only the membrane contribution which contains both the membrane kinetic term and the minimal coupling terms, remains.

With all these observations taken into account, the action that appears in the probability of the cosmological thermalon becomes

$$
I_{\text{thermalon I}} = \frac{1}{4} [A(r_{--}) - A(r_{++})] - \frac{\mu}{4\pi} V_3 + \alpha^2 \frac{3}{8\pi} V_4
$$

(16)

below the Nariai threshold, and

$$
I_{\text{thermalon I}} = \frac{1}{4} [A(r_-) - A(r_{++})] - \frac{\mu}{4\pi} V_3 + \alpha^2 \frac{3}{8\pi} V_4
$$

(17)

above it.

For the central case of interest in this paper, namely, the decay of de Sitter space through the cosmological thermalon, we set the initial mass M_+ equal to zero. Note that the cosmological thermalon lies in the upper branch of the mass curve, while the tunneling decay investigated in [3] lies on the lower branch. Therefore, the thermalon process discussed here and the tunneling of [3] are not to be thought of as happening in the same potential barrier in the sense of the analogy illustrated by Fig. 1. Yet the two probabilities may be compared, and the comparison is of interest. Since there is no black hole in the initial state, the initial geometry has the full $O(5)$ symmetry and the nucleation process can occur anywhere in spacetime. Hence, the computed probability is a probability per unit of spacetime volume. Note that the thermalon solution breaks the symmetry down to $O(3) \times O(2)$, while the tunneling solution of [3] breaks it to the larger symmetry group $O(4)$. Hence it is expected to have higher probability [11], which is indeed confirmed by the analysis given below.

After de Sitter decays through the thermalon once, the process may happen again and again. If the black hole formed after the decay is small, one may, to a first approximation, ignore its presence and use the same formula for the probability, taking the final cosmological radius l_- of the previous step as the l_+ of the new step. It is quite all right to ignore the presence of the black hole when r_- is small because the probability of the second black hole being near the first one will be very small, since it is a probability per unit of volume. The approximation will break down when the Nariai bound is in sight.

5.   Nariai threshold

For given membrane parameters μ, q, bare cosmological constant λ, and initial mass M_+, there is a value l_N of l_+ for which the final geometry becomes the Nariai solution. In that case, $R = r_- = r_{--}$, since there is nowhere else for R to be, and thus the curve $\dot{T}_- = 0$, which always crosses the mass curve at a root of f_-^2, does it now precisely at the cosmological thermalon nucleation radius R. If one starts from a small l_+ one finds the situation illustrated in Fig. 3a. As l_+ increases, the size of the black hole present in the "$-$" region also increases, and at $l_+ = l_N$, the "$-$" geometry becomes the Nariai solution. If l_+ increases further, and we will refer to this further increase as "crossing the Nariai threshold," $\dot{T}_-$ becomes positive, which means that, as seen from the "$-$" side, the orientation of the membrane is reversed. This implies that one must glue the part of the "$-$" region which one was discarding below the Nariai threshold to the "$+$" region, thus giving rise to the situation described in Fig. 3b.

It should be emphasized that crossing the Nariai threshold is not a violent operation. The bound[12]

$$M_- \leq \frac{l_-}{3\sqrt{3}} \qquad (18)$$

for the existence of a de Sitter black hole is maintained throughout [$f_+(R)$ is real since $R < l_+$, which implies, using Eq. (12), that $f_-(R)$ is also real and hence the function f_-^2 has two positive real roots]. The action, and hence the probability, remains continuous. However, the nature of the thermalon geometry is now somewhat different, since the "−" geometry has a black hole horizon instead of a cosmological horizon. As one moves from the boundary r_+ toward the membrane, the radius r of the attached S^2 increases. But, after crossing the membrane, it starts decreasing—in contradistinction to what happens below the threshold—until it reaches its minimun value at the black hole radius of the "−" geometry.

6. Decay of de Sitter space through the cosmological thermalon

6.1 Nucleation radius and mass of final state black hole

We now focus on the central case of interest in this paper, namely, the decay of de Sitter space through the cosmological thermalon. The cosmological thermalon radius of nucleation, R, may be obtained by differentiating Eq. (12) in its plus sign version,

$$\frac{3}{2}(\alpha^2 - \mu^2)R + 2\mu f_+ + \mu f_+' R = 0. \qquad (19)$$

When $M_+ = 0$, Eq. (19) can be rewritten in terms of the dimensionless auxiliary variable $x = f_+ l_+/R$, or, equivalently,

$$R^2 = \frac{l_+^2}{1 + x^2} . \qquad (20)$$

Then, Eq. (19) gives

$$x = \frac{3}{4}\left[-\gamma + \left(\gamma^2 + \frac{8}{9}\right)^{1/2} \right], \qquad (21)$$

where

$$\gamma = \frac{l_+(\alpha^2 - \mu^2)}{2\mu} . \qquad (22)$$

From Eq. (12) we obtain for the mass of the black hole appearing in the final state,

$$M_- = \frac{\mu l_+^2}{3x}(1 + x^2)^{-1/2} . \qquad (23)$$

The Nariai threshold radius $l_+ = l_N$ may be also evaluated explicitly. It is given by

$$\frac{1}{l_N^2} = \frac{1}{2}\left(\mu^2 + 8\pi q^2 + \sqrt{8\pi q^2(6\pi q^2 - 2\lambda + 3\mu^2)}\right). \tag{24}$$

6.2   Response of the final geometry to changes in the initial cosmological constant

For $l_+ < l_N$ one easily verifies that, at R, $t_- < \Delta M$ and hence $\dot{T}_- < 0$. Therefore the thermalon geometry is the one pictured in Fig. 3a. As l_+ increases the final state approaches the Nariai solution. We have plotted in Fig. 5 the quantity $l_- - 3\sqrt{3}M_-$ as a function of l_+. As l_+ crosses the Nariai value l_N, the situation becomes the one depicted in Fig. 3b. If l_+ increases further, the function l_- and thus also $l_- - 3\sqrt{3}M_-$ blow up for some value $l_\infty > l_N$ of l_+; for that value, the final cosmological constant l_- is infinite and Λ_- vanishes. Above l_∞ the final cosmological constant is negative and thus a transition from de Sitter space to Schwarzschild–anti–de Sitter space takes place. The transition probability is well defined because, since $\dot{T}_-$ is still positive, a finite-volume piece of Schwarzschild–anti–de Sitter space enters into the action, namely, the one between the membrane and the black hole horizon r_-.

The radius $l_+ = l_\infty$ for which the final cosmological constant vanishes is given by

$$\frac{1}{l_\infty^2} = \frac{4}{3}q\left(\pi q + \sqrt{-\pi\lambda}\right). \tag{25}$$

Because the horizons r_-, r_{--} and inverse temperature β_- of the final state are determined by algebraic equations of a high degree, we have found it necessary to compute the action $I(l_+)$ by direct numerical attack. The result gives a curve of the form shown in Fig. 6. One sees that the probability decreases very quickly as l_+ increases.

Since, as the graph shows, the action is very small for large cosmological constant, the process is not exponentially suppressed at the beginning. As the process goes on, the action becomes monotonically more negative and the probability becomes exponentially suppressed. The action is continuous both at the Nariai value l_N and at the critical point l_∞ where the final cosmological constant becomes negative.

6.3   Small charge and tension limit

In order to avoid fine-tuning it is necessary to assume that the jumps in the cosmological constant are of the order of the currently observed value Λ_{obs}, which is, in Planck units adopted from now on, $\Lambda_{obs} \sim 10^{-120}$. Thus we

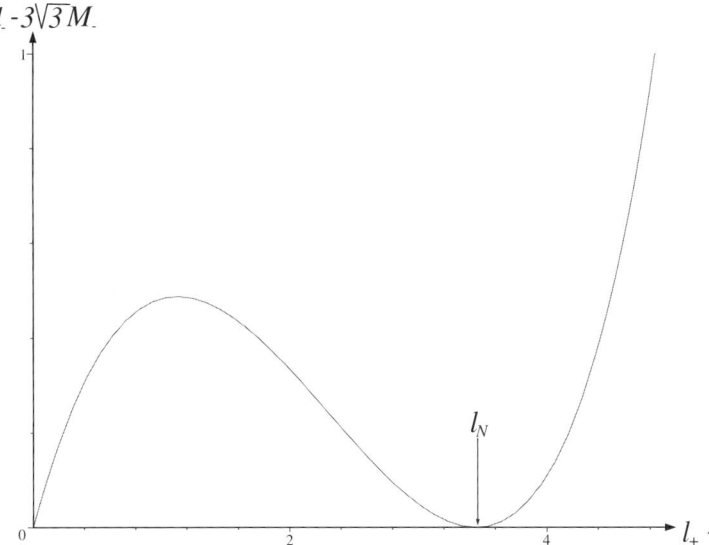

Figure 5.   The quantity $l_- - 3\sqrt{3}M_-$ is shown in the graph as a function of the initial cosmological radius l_+. It vanishes when the final geometry is that of Nariai, at $l_+ = l_N$, and it goes to infinity when the final geometry has vanishing cosmological constant, at $l_+ = l_\infty$. Here we have set, in Planck units, $q = 0.01$, $\mu = 0.3$, and $\lambda = -1$.

take $\alpha^2 \sim 10^{-120}$ ("small jump condition"). The bare cosmological constant itself could be as big as the Planck scale ($|\lambda| \sim 1$) unless protected by some broken symmetry, e.g., $|\lambda| \sim 10^{-60}$ [supersymmetry (SUSY)]. From Eq. (7) we find that the E field is of the order of $\sqrt{-\lambda}$, which gives, using (15), $q \sim 10^{-120}/\sqrt{-\lambda}$: the small jump condition requires a small charge q. As the decay of the cosmological constant proceeds, the radius of the universe l_+ goes from "small" values $\sim (\sqrt{|\lambda|})^{-1}$ to large values $\sim 10^{60}$ and the dimensionless product $\alpha^2 l_+^2$ varies from $\sim 10^{-120}$ (Planck) or $\sim 10^{-60}$ (SUSY) to a quantity of order unity. We will assume that

$$l_+^2 \alpha^2 \ll 1 \,, \tag{26}$$

which holds during the whole decay of the cosmological constant, beginning from early stages $l_+ \sim 1$ (Planck), or $l_+ \sim 10^{30}$ (SUSY), all the way up to "almost" the late stages $l_+ \sim 10^{58}$, say. We shall call Eq. (26) the "small charge limit."

We will also assume

$$l_+\mu \ll 1, \tag{27}$$

which we shall call the "small tension limit."

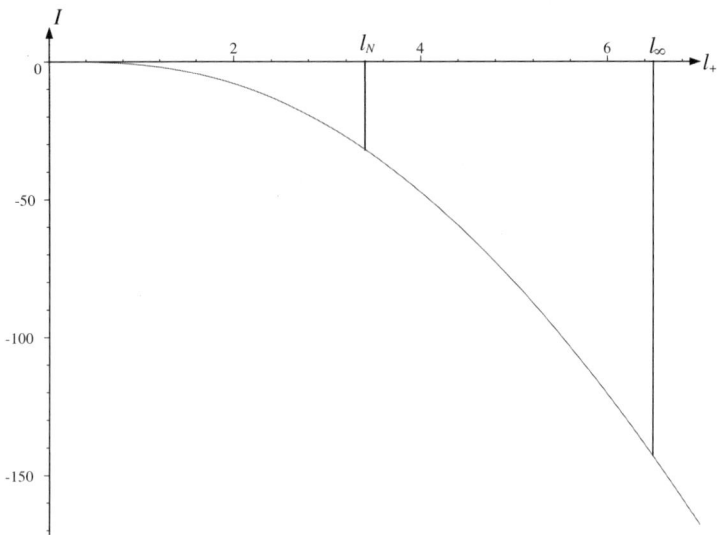

Figure 6.   The graph depicts the value of the action for the cosmological thermalon as a function of the initial cosmological radius. We have used the same values of the parameters as those used in Fig. 5.

Within the small charge and tension limits there are two interesting subcases which are amenable to analytical treatment.

(a) First, we may assume that

$$\frac{l_+\alpha^2}{\mu} \ll 1, \tag{28}$$

This limit may be achieved, for instance, by taking μ to be of the order of α, that is, $\mu \sim 10^{-60}$, again, up to late stages of the evolution of l_+. In this limit,

$$\gamma \ll 1. \tag{29}$$

Direct computations yield

$$R^2 \approx \frac{2}{3}l_+^2\left(1 + \frac{\gamma}{\sqrt{2}}\right), \quad \frac{M_-}{l_+} \approx \frac{2\sqrt{3}}{9}\mu l_+\left(1 + \sqrt{2}\gamma\right). \tag{30}$$

The nucleation radius is roughly $\sqrt{2/3}l_+$ and the initial and final universe radii l_+ and l_- are equal to leading order. Since both $\alpha^2 l_+^2$ and $\mu^2 l_+^2$ are small, Eq. (24) implies $l_+ \ll l_N$ so that the final geometry contains the cosmological horizon r_{--}.

We then get for the action $I_{\text{thermalon}}$

$$-\frac{8\pi\sqrt{3}}{9}l_+^2(\mu l_+), \tag{31}$$

which can be compared to the tunneling-process action with the same parameters

$$-\frac{\pi}{2}l_+^2\left(\mu l_+\right).$$ (32)

We conclude that in this regime the tunneling process is more probable than the cosmological thermalon.

(b) Conversely, we may assume

$$\frac{\mu}{l_+\alpha^2} \ll 1,$$ (33)

which is equivalent to

$$\gamma \gg 1.$$ (34)

This limit may be achieved, for instance, by setting $q \sim \mu$ ("BPS condition") for late stages of the evolution of the cosmological constant $[l_+ > 10$ (Planck), $l_+ > 10^{31}$ (SUSY)]. Here we get for the nucleation radius and final mass,

$$R^2 \approx l_+^2\left(1 - \frac{1}{9\gamma^2}\right), \qquad \frac{M_-}{l_+} \approx \mu l_+ \gamma \approx \alpha^2 l_+^2,$$ (35)

which shows that, in this limit, the nucleation radius gets close to the cosmological horizon. The value of the action for this process is, to leading order,

$$-\frac{4\pi l_+^2}{9}\left(\frac{\mu}{\alpha^2 l_+}\right)^2,$$ (36)

which may be compared with the action for the tunneling process in the same limit,

$$-4\pi l_+^2\left(\frac{\mu}{\alpha^2 l_+}\right)^4\left(\alpha^2 l_+^2\right).$$ (37)

We again see in this regime that the tunneling process is more probable than the cosmological thermalon.

7. Can the thermalon account for the small present value of the cosmological term?

The cosmological thermalon has a distinct advantage over the instanton proposed in [3] as a mechanism for relaxing the cosmological constant in that it does not have the so called "horizon problem." Indeed, since for the thermalon the cosmological constant is reduced in the region with bigger values of the radial coordinate, and then the membrane proceeds to collapse, one is sure that, even though the universe is expanding (and even more because it is so), the whole of the universe will relax its cosmological constant.

The other problem that the instanton has is that the rate of membrane nucleation was too small to account for the present small value of the cosmological term. It is not clear at the moment of this writing whether the thermalon will also be an improvement in this regard. Indeed, to make a mild assessment of whether the rate is sufficiently strong, we first recall that, in Planck units, the observed value of the cosmological constant now is 10^{-120} and the age of the universe 10^{60}. Even though the process of relaxation of the cosmological constant may not have occurred through the entire life span of the post big-bang universe, we may use for very rough estimates 10^{60} as the time available for the process to keep occurring. If we assume that the cosmological radius started at the Planck scale $l_+ = 1$ and at present has the value 10^{60}, and recall that each bubble nucleation reduces l_+^{-2} by α^2, one needs 10^{120} events to occur during 10^{60} Planck time units. Furthermore, since at any given time the volume of space available for the location of the center of the bubble is of the order of l_+^3 (spatial volume in the comoving frame of de Sitter space), we conclude that the rate Γ is given by

$$\Gamma \sim \frac{10^{120}}{10^{60} l_+^3} = \frac{10^{60}}{l_+^3} \quad \text{(Planck units)}. \tag{38}$$

In order for Eq. (38) to be realizable for l_+ in the range $1 < l_+ < 10^{60}$ we get, using Eq. (13),

$$10^{-120} < Ae^{-B} < 10^{60}. \tag{39}$$

One must keep in mind that, as stated above, the parameter α^2 is fixed to be 10^{-120} and that the parameter B should not be too small in order for the semiclassical approximation to be valid, which imposes a constraint relating the other two parameters μ and l_+. It is therefore necessary to perform a careful analysis of the prefactor A to see whether the inequality (39) can be satisfied with an l_+ within the range available in the history of the universe. If this is not so, one could not argue that the thermalon can account for all of the relaxation of the cosmological constant and, *a fortiori*, one could not argue that all of the vacuum energy was condensed into black holes.

8.   Conclusions

In this paper, we have exhibited a new process through which the cosmological constant can decay. At the same time, a black hole is created. The classical solution describing the process is an unstable ("ready-to-fall") static solution which we have called a "cosmological thermalon;" it is an analogue of the sphaleron of gauge theories. Gravity and, in particular, the thermal effects of the cosmological horizon are essential for the existence of the solution, which disappears in the flat space limit.

The net effect of the process is thus to transform nonlocalized dark energy into localized dark matter, thus providing a possible link between the small present value of the cosmological constant and the observed lack of matter in the universe. Of course, the emergence in this way of (nearly) flat space as a natural end point of dynamical evolution is most intriguing in view of cosmological observations.

Note added

After this paper was finished we became aware of the interesting work of Garriga and Megevand [15], where "a "static" instanton, representing pair creation of critical bubbles—a process somewhat analogous to thermal activation in flat space" is discussed.

Acknowledgments

This work was funded by an institutional grant to CECS of the Millennium Science Initiative, Chile, and Fundación Andes, and also benefits from the generous support to CECS by Empresas CMPC. AG gratefully acknowledges support from FONDECYT grant 1010449 and from Fundación Andes. AG and CT acknowledge partial support under FONDECYT grants 1010446 and 7010446. The work of MH is partially supported by the "Actions de Recherche Concertées" of the "Direction de la Recherche Scientifique - Communauté Française de Belgique", by IISN - Belgium (convention 4.4505.86), by a "Pôle d'Attraction Universitaire" and by the European Commission RTN programme HPRN-CT-00131, in which he is associated to K. U. Leuven.

Appendix: A. Instability of thermalons

It is illuminating to see explicitly that both the minimum and the maximum of the mass curve are Euclidean stable points, since "unthoughtful" analogy with a standard potential problem might have suggested that only the minimum of the mass curve should be a stable Euclidean equilibrium.

Consider first the black hole thermalon and perturb it, $R_{\text{BHT}} \rightarrow R_{\text{BHT}} + \eta(\tau)$. The mass equation yields

$$\Delta M(R_{\text{BHT}}) + \delta m = \frac{1}{2}(\alpha^2 - \mu^2)R^3 - \mu\sqrt{f_+^2 - \dot{R}^2}\,R^2 \qquad (A.1)$$

where δm is the mass perturbation. Using that ΔM is an extremum at the thermalon, one finds

$$\delta m = \frac{1}{2}\left.\frac{\partial^2 \Delta M}{\partial R^2}\right|_{\text{BHT}} \eta^2 + \left.\frac{\mu R^2}{2f_+}\right|_{\text{BHT}} \dot{\eta}^2. \qquad (A.2)$$

Because the second derivative of ΔM is positive at the minimum R_{BHT}, the right-hand side of Eq. (A.2) is non-negative, so that the perturbation $\eta(\tau)$ is forced to remain in the bounded range $|\eta| \leq \sqrt{2\delta m(\partial^2\Delta M/\partial R^2)^{-1}}$, which implies stability.

For the cosmological thermalon, it is now the upper branch of the mass curve that is relevant, and Eq. (A.1) is replaced by

$$\Delta M(R_{\text{CT}}) + \delta m = \frac{1}{2}(\alpha^2 - \mu^2)R^3 + \mu\sqrt{f_+^2 - \dot{R}^2}\,R^2 \qquad (A.3)$$

so that, instead of Eq. (A.2), one has

$$\delta m = \frac{1}{2}\left.\frac{\partial^2 \Delta M}{\partial R^2}\right|_{\text{CT}} \eta^2 - \left.\frac{\mu R^2}{2f_+}\right|_{\text{CT}} \dot{\eta}^2. \qquad (A.4)$$

At the cosmological thermalon, ΔM is maximum and so its second derivative is negative. Again the perturbation η is bounded and the Euclidean solution is stable.

The signs in Eq. (A.4) are in agreement with the proof given in [10, 13] that the internal energy of the cosmological horizon is $-M$; thus the "thoughtful" analogy with the potential problem simply amounts to realizing that for the cosmological thermalon one should plot $-\Delta M$ on the vertical axis, so that the "potential" in Fig. 2 is upside down and the analogy holds.

We end this appendix by computing the frequency of oscillations around the cosmological thermalon. From Eq. (A.4) we find that these are given by

$$\omega^2 = \left(\frac{\partial^2 \Delta M}{\partial R^2}\right)\frac{f_+}{\mu R^2}\bigg|_{CT},$$

which may be evaluated in terms of the nucleations radius R of the cosmological thermalon,

$$\omega^2 = \frac{2l_+^2 - R^2}{R^2(l_+^2 - R^2)}. \tag{A.5}$$

Note that there is a resonance in the limit when R approaches the cosmological radius l_+. This may be achieved when the BPS condition of Eq. (33) is satisfied ($\mu/l_+\alpha^2 \ll 1$, which holds, for instance, on setting $q \sim \mu$). In that case we get

$$\omega = \frac{3}{2}\frac{\alpha^2}{\mu}, \tag{A.6}$$

and, from Eq. (33), we see that, in fact, $\omega \gg l_+^{-1}$.

Appendix: B. Gravitation Essential for Existence of Cosmological Thermalons.

An interesting feature of the cosmological thermalon is that it does not exist in the limit of no gravity, where Newton's constant G is taken to vanish. Indeed, when G is explicitly written, the mass equation becomes

$$G\Delta M = \frac{1}{2}(\alpha^2 - G^2\mu^2)R^3 \pm G\mu f_+ R^2 \tag{B.1}$$

with

$$\alpha^2 = \frac{4\pi G}{3}(E_+^2 - E_-^2) \tag{B.2}$$

and

$$f_+^2 = 1 - \frac{2GM_+}{R} - \frac{R^2}{l_+^2}, \quad \frac{3}{l_+^2} = \lambda + 4\pi GE_+^2. \tag{B.3}$$

Note that the bare cosmological constant λ depends on G through

$$\lambda = 8\pi G\rho_{vac}, \tag{B.4}$$

where ρ_{vac} is the vacuum energy density coming from "the rest of physics." Therefore, in the limit $G \to 0$, the lower branch of the mass curve becomes

$$\Delta M = -V(R) = -(\mu R^2 - \nu R^3), \quad \nu = \frac{2\pi}{3}(E_+^2 - E_-^2) > 0. \tag{B.5}$$

The (Minkowskian) potential $V(R)$ appearing in Eq. (B.5) exhibits the competition between surface and volume effects which gives rise to the nucleation of a bubble of a stable phase within a metastable medium. Thus, in the no-gravity limit, the distinguished points in the lower branch of the mass curve have a clear interpretation: the black hole thermalon is the unstable static solution sitting at the maximum of the potential, while the instanton becomes the standard bounce solution for metastable vacuum decay.

On the other hand, the upper branch of the mass curve yields

$$V(R) = \mu R^2 + \nu R^3, \tag{B.6}$$

and has no maximum. More precisely, the maximum, R, has gone to infinity ($R \sim 1/\sqrt{G}$). There is therefore no zero-gravity limit of the cosmological thermalon. One may understand the behavior of the potential by enclosing the bubble in a sphere of radius L and recalling that in this case the change from the metastable to the stable phase occurs in the *exterior* of the bubble (i.e., on the side with bigger values of the radial coordinates). This yields the potential $V(R) = \mu R^2 - \nu(L^3 - R^3)$, which has the correct volume dependence for bubble nucleation. The extra term $-\nu L^3$ is constant, and infinite in the limit $L \to \infty$. Gravity changes the shape of the potential and makes it finite, so that there is an unstable static solution, the cosmological thermalon.

References

[1] S. Weinberg, Rev. Mod. Phys. **61**, 1 (1989); S. M. Carroll, Living Rev. Relativ. **4**, 1 (2001); P. J. Peebles and B. Ratra, Rev. Mod. Phys. **75**, 559 (2003).

[2] A. Aurilia, H. Nicolai and P. K. Townsend, Nucl. Phys. **B176**, 509 (1980); M. J. Duff and P. van Nieuwenhuizen, Phys. Lett. **94B**, 179 (1980); P. G. Freund and M. A. Rubin, *ibid.* **97B**, 233 (1980); S. W. Hawking, *ibid.* **134B**, 403 (1984); M. Henneaux and C. Teitelboim, *ibid.* **143B**, 415 (1984).

[3] J. D. Brown and C. Teitelboim, Phys. Lett. B**195**, 177 (1987). Nucl. Phys. **B297**, 787 (1988).

[4] J. L. Feng, J. March-Russell, S. Sethi, and F. Wilczek, Nucl. Phys. **B602**, 307 (2001); R. Bousso and J. Polchinski, J. High Energy Phys. **06**, 006 (2000).

[5] J. S. Langer, Ann. Phys. (N.Y.) **54**, 258 (1969); A. D. Linde, Phys. Lett. **70B**, 306 (1977); **100B**, 37 (1981); Nucl. Phys. **B216**, 421 (1983); **B223**, 544(E) (1983); I. Affleck, Phys. Rev. Lett. **46**, 388 (1981).

[6] S. R. Coleman, Phys. Rev. D **15**, 2929 (1977); **16**, 1248(E) (1977).

[7] F. R. Klinkhamer and N. S. Manton, Phys. Rev. D **30**, 2212 (1984).

[8] A. Gomberoff, M. Henneaux, and C. Teitelboim, "Turning the Cosmological Constant into Black Holes," hep-th/0111032.

[9] W. Israel, Nuovo Cimento B **44S10**, 1 (1966) ; **48**, 463(E) (1967).

[10] C. Teitelboim, "Gravitational Thermodynamics of Schwarzschild–de Sitter Space," in *Strings and gravity: Tying the forces together*, proceedings of the Fifth Francqui Colloquium, Brussels, 2001, edited by Marc Henneaux and Alexander Sevrin (De Boeck Larcier, Bruxelles, 2003).

[11] S. R. Coleman, V. Glaser, and A. Martin, Commun. Math. Phys. **58**, 211 (1978).

[12] G. W. Gibbons and S. W. Hawking, Phys. Rev. D **15**, 2738 (1977).

[13] A. Gomberoff and C. Teitelboim, Phys. Rev. D **67**, 104024 (2003).

[14] M. Banados, C. Teitelboim, and J. Zanelli, Phys. Rev. Lett. **72**, 957 (1994) .

[15] J. Garriga and A. Megevand, Phys. Rev. D (to be published), hep-th/0310211.